Methods and Protocols in Food Science

Series Editor
Anderson S. Sant'Ana
University of Campinas
Campinas, Brazil

For further volumes:
http://www.springer.com/series/16556

Methods and Protocols in Food Science series is devoted to the publication of research protocols and methodologies in all fields of food science.

Volumes and chapters will be organized by field and presented in such way that the readers will be able to reproduce the experiments in a step-by-step style. Each protocol will be characterized by a brief introductory section, followed by a short aims section, in which the precise purpose of the protocol will be clarified.

Food Chemistry

Basic Protocols

Edited by

Lilian Mariutti

School of Food Engineering, University of Campinas, São Paulo, Brazil

Frank Vriesekoop

Food Tech. and Innovation Dept, Harper Adams University, Edgmond, Shropshire, UK

Humana Press

Editors
Lilian Mariutti
School of Food Engineering
University of Campinas
São Paulo, Brazil

Frank Vriesekoop
Food Tech. and Innovation Dept
Harper Adams University
Edgmond, Shropshire, UK

ISSN 2662-950X ISSN 2662-9518 (electronic)
Methods and Protocols in Food Science
ISBN 978-1-0716-4677-9 ISBN 978-1-0716-4678-6 (eBook)
https://doi.org/10.1007/978-1-0716-4678-6

This Humana imprint is published by the registered company Springer Science+Business Media, LLC, part of Springer Nature.
The registered company address is: 1 New York Plaza, New York, NY 10004, U.S.A.

Preface

Food chemistry is at the core of understanding complex biochemical composition and properties of food and beverages. By investigating the molecular components responsible for food quality, functionality, stability, and nutritional value, this book fills the gap between food science, nutrition, and analytical chemistry. "Basic Protocols in Food Chemistry" offers a comprehensive guide to essential analytical techniques applied in food chemistry laboratories, aiming at supporting students, researchers, and professionals in performing reliable, reproducible, and accurate food analyses.

This book comprises a carefully curated selection of chapters that detail practical protocols and theoretical foundations for analyzing major and minor food components. Each chapter provides step-by-step instructions, troubleshooting tips, and insights into sample preparation, extraction, and analysis. The protocols span across a diverse range of food constituents, including macronutrients, micronutrients, vitamins, pigments, and volatile compounds.

The chapters on macronutrients cover key methodologies such as amino acid profiling by gas chromatography (GC) and biogenic amine analysis, enzymatic assays for starch and dietary fiber, and lipid extraction and characterization. These protocols are designed to address challenges posed by complex food matrices and provide accurate quantitative and qualitative data essential for food product development and nutritional evaluation.

Micronutrient analysis is equally crucial in food chemistry, given the growing global interest in fortification and functional foods. This book includes methods for quantifying water-soluble and fat-soluble vitamins, such as vitamin C, B vitamins (thiamine, riboflavin, pyridoxine, pantothenic acid), and vitamins D and E (tocopherols). Techniques like high-performance liquid chromatography (HPLC) coupled with fluorescence detection or mass spectrometry (LC-MS) are detailed, offering insights into the stability, extraction, and quantification of these vitamins in various food matrices.

Food pigments, which contribute to the color, are another focus of this book. Chapters discuss the analysis of carotenoids, anthocyanins, betalains, and chlorophylls using advanced chromatographic and spectrophotometric techniques.

Additionally, volatile compounds that contribute to food aroma and flavor are addressed in protocols utilizing gas chromatography coupled with flame ionization (GC-FID) and mass spectrometry (GC-MS) detectors. These methods explore the extraction, separation, and quantification of volatile compounds and aroma profiling to enhance sensory quality and detect spoilage.

Beyond compositional analysis, this book includes chapters on food stability, focusing on moisture content and water activity, key parameters in determining shelf life and microbial safety.

Our objective is to offer readers not only technical procedures but also a deeper understanding of the scientific principles underlying these methodologies. This knowledge empowers users to modify protocols for different food applications, interpret results critically, and contribute to advancements in food research and industry practices.

We thank the authors and contributors for their invaluable expertise and dedication in preparing these chapters. We hope this book becomes a trusted resource for students, food chemists, and professionals striving to enhance the safety, quality, and nutritional value of food products worldwide.

Campinas, São Paulo, Brazil *Lilian Regina Barros Mariutti*
Newport, Shropshire, UK *Frank Vriesekoop*

Preface to the Series

The *Methods and Protocols in Food Science* series is devoted to the publication of research protocols and methodologies in all fields of food science. The series is unique as it includes protocols developed, validated, and used by food and related scientists, and theoretical basis is also provided for each protocol. Aspects related to improvements in the protocols and adaptations and further developments in the protocols may also be approached.

The *Methods and Protocols in Food Science* series aims to bring the most recent developments in research protocols in the field as well as very well-established methods. As such the series targets undergraduates, graduates, and researchers in the field of food science and correlated areas. The protocols documented in the series will be highly useful for scientific inquiries in the field of food sciences and presented in such a way that the readers will be able to reproduce the experiments in a step-by-step style.

Each protocol will be characterized by a brief introductory section, followed by a short aims section, in which the precise purpose of the protocol is clarified. Then, an in-depth list of materials and reagents required for employing the protocol is presented, followed by comprehensive and step-by-step procedures on how to perform that experiment. The next section brings the dos and don'ts when carrying out the protocol, followed by the main pitfalls faced and how to troubleshoot them. Finally, template results will be presented and their meaning/conclusions addressed.

The *Methods and Protocols in Food Science* series will fill an important gap, addressing a common complain of food scientists, regarding the difficulties in repeating experiments detailed in scientific papers. With this, the series has a potential to become a reference material in food science laboratories of research centers and universities throughout the world.

Campinas, Brazil ***Anderson S. Sant'Ana***

Contents

Part IV Carbohydrates

Part V Liposoluble Vitamins

Part VI Hydrosoluble Vitamins

Contributors

HENRIQUE SILVANO ARRUDA • *Department of Food Science and Nutrition, School of Food Engineering, University of Campinas, Campinas, São Paulo, Brazil*

MICHEL ROCHA BAQUETA • *Department of Food Science and Nutrition, School of Food Engineering, University of Campinas, Campinas, São Paulo, Brazil*

NEURA BRAGAGNOLO • *Department of Food Science and Nutrition, School of Food Engineering, University of Campinas, Campinas, São Paulo, Brazil*

MARIA ELISA CAETANO-SILVA • *Department of Health and Kinesiology, University of Illinois, EUA, Urbana-Champaign, IL, USA*

CELSO ANDRADE CAMARGO • *Department of Food Science and Nutrition, School of Food Engineering, University of Campinas, Campinas, São Paulo, Brazil*

CINTHIA BAÚ BETIM CAZARIN • *Department of Food Science and Nutrition, School of Food Engineering, Univesity of Campinas, Campinas, São Paulo, Brazil*

LIANG CHEN • *College of Bioengineering, Beijing Polytechnic University, Beijing, China*

MARIA TERESA PEDROSA SILVA CLERICI • *Department of Food Science and Nutrition, School of Food Engineering, University of Campinas, Campinas, São Paulo, Brazil*

LUCIMARA MACH CÔRTES CORDEIRO • *Department of Biochemistry and Molecular Biology, Federal University of Paraná, Curitiba, Brazil*

ROSINELSON DA SILVA PENA • *Graduate Program in Food Science and Technology (PPGCTA), Institute of Technology (ITEC), Federal University of Pará (UFPA), Belém, Pará, Brazil; Faculty of Food Engineering (FEA), Institute of Technology (ITEC), Federal University of Pará (UFPA), Belém, Pará, Brazil*

ADRIANO LUCENA DE ARAÚJO • *Graduate Program in Food Science and Technology (PPGCTA), Institute of Technology (ITEC), Federal University of Pará (UFPA), Belém, Pará, Brazil*

ELEM TAMIRYS DOS SANTOS CARAMÊS • *Department of Microbiology, Institute of Biomedical Sciences, University of São Paulo, São Paulo, Brazil*

PENG DONG • *School of Bioengineering, Beijing Polytechnic University, Beijing, China*

PATRICIA ESQUIVEL • *Escuela de Tecnología de Alimentos, Universidad de Costa Rica, San Pedro, Costa Rica*

MÁRIA HERMINIA FERRARI FELISBERTO • *Department of Food Technology (DTA), Federal University of Viçosa (UFV), Viçosa, MG, Brazil*

LARYSA GURAL • *Department of Food Chemistry and Expertise, Odesa National University of Technology, Odesa, Ukraine*

ADRIELE HACKE • *Department of Food Science and Nutrition, School of Food Engineering, University of Campinas, Campinas, São Paulo, Brazil*

BARRY HUCKER • *Harper Food Innovation, Harper Adams University, Newport, Shropshire, UK; Department of Food Land and Agribusiness Management, Harper Adams University, Newport, Shropshire, UK*

ANDREA IRÍAS-MATA • *CIGRAS/IIA, Universidad de Costa Rica, San Pedro, Costa Rica*

BIN JIANG • *School of Bioengineering, Beijing Polytechnic University, Beijing, China*

VÍCTOR M. JIMÉNEZ • *CIGRAS/IIA, Universidad de Costa Rica, San Pedro, Costa Rica*

ANTONINA KAPUSTIAN • *Department of Food Chemistry and Expertise, Odesa National University of Technology, Odesa, Ukraine*

AYŞE BINNUR KARATAŞ • *Republic of Türkiye Ministry of Agriculture and Forestry, Central Research Institute of Food and Feed Control, Bursa, Türkiye*
SEBNEM KURHAN • *Harper Food Innovation, Harper Adams University, Newport, UK*
IRYNA LEVCHUK • *Scientific and Methodological Laboratory of Chromatographic Studies, State Enterprise "Ukrmetrteststandart", Kyiv, Ukraine*
LINYING LIU • *College of Bioengineering, Beijing Polytechnic University, Beijing, China*
MARA MANDRIOLI • *Department of Agricultural and Food Sciences, Alma Mater Studiorum-University of Bologna, Bologna, Italy*
LILIAN REGINA BARROS MARIUTTI • *Department of Food Science and Nutrition, School of Food Engineering, University of Campinas, Campinas, São Paulo, Brazil*
MARCELLA CAMARGO MARQUES • *Department of Food Science and Nutrition, School of Food Engineering, University of Campinas, Campinas, São Paulo, Brazil*
LIGIA MANOEL MARTINS • *Department of Food Science and Nutrition, School of Food Engineering, University of Campinas, Campinas, São Paulo, Brazil*
MAYARA GALVÃO MARTINS • *Graduate Program in Food Science and Technology (PPGCTA), Institute of Technology (ITEC), Federal University of Pará (UFPA), Belém, Pará, Brazil*
ADRIANA Z. MERCADANTE • *Food Research Center (FoRC-CEPID) and Department of Food Science and Nutrition, School of Food Engineering, University of Campinas, Campinas, São Paulo, Brazil*
DARIO MERCATANTE • *Department of Agricultural and Food Sciences, Alma Mater Studiorum-University of Bologna, Bologna, Italy*
FLAVIA MARIA NETTO • *Department of Food Science and Nutrition, School of Food Engineering, University of Campinas, Campinas, São Paulo, Brazil*
JULIANA AZEVEDO LIMA PALLONE • *Department of Food Science and Nutrition, School of Food Engineering, University of Campinas, Campinas, São Paulo, Brazil*
GUSTAVO ARAUJO PEREIRA • *School of Food Engineering, Institute of Technology, Federal University of Pará, Belém, Para, Brazil*
ANTONIO PÉREZ-GÁLVEZ • *Group of Chemistry and Biochemistry of Pigments. Food Phytochemistry Department, Instituto de la Grasa (CSIC), Sevilla, Spain*
ANA PAULA REBELLATO • *Department of Food Science and Nutrition, School of Food Engineering, University of Campinas, Campinas, São Paulo, Brazil*
MARÍA ROCA • *Group of Chemistry and Biochemistry of Pigments. Food Phytochemistry Department, Instituto de la Grasa (CSIC), Sevilla, Spain*
MARIA TERESA RODRIGUEZ-ESTRADA • *Department of Agricultural and Food Sciences, Alma Mater Studiorum-University of Bologna, Bologna, Italy; CIRI-Agrifood (Interdepartmental Centre of Industrial Agrifood Research), Alma Mater Studiorum-University of Bologna, Cesena, Italy*
ANNE A. SALES • *Department of Food Science and Nutrition, School of Food Engineering, University of Campinas, Campinas, São Paulo, Brazil*
ULLIANA MARQUES SAMPAIO • *Department of Food Science and Nutrition, School of Food Engineering, Universidade Estadual de Campinas, Campinas, São Paulo, Brazil*
LIVIA A. SANTIAGO • *Department of Food Science and Nutrition, School of Food Engineering, University of Campinas, Campinas, São Paulo, Brazil*
ESRA YILDIRIM SERVI • *Department of Molecular Biology and Genetics, Faculty of Engineering and Natural Sciences, Istanbul Sabahattin Zaim University, Istanbul, Türkiye*

JOYCE GRAZIELLE SIQUEIRA STÜRMLINGER • *Department of Food Technology and Bioprocess Engineering, Federal Research Institute of Nutrition and Food, Max Rubner-Institute, Karlsruhe, Germany*

XIAOHONG SUN • *Dalhousie University, Truro, NS, Canada*

GUILHERME M. TAVARES • *Department of Food Science and Nutrition, School of Food Engineering, University of Campinas, Campinas, São Paulo, Brazil*

APOLLINAIRE TSOPMO • *Carleton University, Ottawa, ON, Canada*

CHIBUIKE UDENIGWE • *University of Ottawa, Ottawa, ON, Canada*

HALIME UĞUR • *Department of Nutrition and Dietetics, Faculty of Health Sciences, Istanbul Medeniyet University, Istanbul, Türkiye*

FRANK VRIESEKOOP • *Harper Food Innovation, Harper Adams University, Newport, Shropshire, UK; Department of Food Land and Agribusiness Management, Harper Adams University, Newport, Shropshire, UK*

CUICUI WANG • *College of Biology Engineering, Beijing Polytechnic, Beijing, China*

MUSTAFA YAMAN • *Department of Molecular Biology and Genetics, Faculty of Engineering and Natural Sciences, Istanbul Sabahattin Zaim University, Istanbul, Türkiye*

RAN YU • *School of Bioengineering, Beijing Polytechnic University, Beijing, China*

HUCHENG ZHANG • *College of Bioengineering, Beijing Polytechnic University, Beijing, China*

XINYING ZHAO • *College of Bioengineering, Beijing Polytechnic University, Beijing, China*

Part I

Water

Chapter 1

Analytical Methods to Assess Moisture in Foods

Joyce Grazielle Siqueira Stürmlinger, Michel Rocha Baqueta, Ana Paula Rebellato, Elem Tamirys dos Santos Caramês, and Juliana Azevedo Lima Pallone

Abstract

Water is one of the main compounds in food; the moisture content is important to evaluate the food composition once it impacts food quality. Moisture evaluation is part of the routine of laboratories and could be assessed by direct and indirect methods. This chapter presents an overview of analytical methods to evaluate moisture in several food types and guidelines on how to perform the experiments by direct and indirect methods.

Key words Direct methods, Indirect methods, Drying, Food, Microwave, Near infrared spectroscopy

1 Introduction

The analytical methods to determine the moisture content in food matrices are classified by the analytical principles, such as drying, distillation, chemical, and physical methods, or by direct and indirect procedures. Direct methods typically involve removing moisture from food by drying, distilling, and extracting, and the amount of water removed is measured by weighing, volumetry, or titration to determine its moisture content [1, 2]. For indirect methods, moisture is not removed from the sample and directly quantified, but the properties of the food that depend on the amount of water or the number of hydrogen atoms are measured [3]. Therefore, indirect methods must be calibrated against the standard moisture content that has been accurately determined using one or more direct methods.

The choice of the appropriate method to determine the moisture is important once it is possible to perform several combinations of time and temperature. High temperatures associated with a long time could result in the decomposition of some food components, such as carbohydrates, and the degradation of volatile

Lilian Mariutti and Frank Vriesekoop (eds.), *Food Chemistry: Basic Protocols*, Methods and Protocols in Food Science, https://doi.org/10.1007/978-1-0716-4678-6_1,

compounds. Because of that, the use of high temperatures should be avoided [4]. Moreover, some food components decompose, which could result in water release from this reaction, overestimating the moisture content, such as during carbohydrate degradation at high temperatures, where dehydrated hydrocarbon compounds are formed and water is released from the reaction. According to the food type, an appropriate method could be used, as presented in Table 1.

2 Direct Methods for Moisture Evaluation in Food

2.1 Drying in an Air Oven

Air oven drying is a direct method and the most common and widely used in laboratories for routine moisture analysis. Drying can be carried out in convection or forced draft ovens. These ovens must be thermally regulated and have minimal temperature variations inside [5]. The oven drying principle is based on the weight loss of the food. Thus, the food matrix must be thermally stable and must not contain volatile compounds in significant quantities [3] that can be lost during drying. Thus, the oven drying method for determining moisture involves sample preparation, weighing, drying, cooling, and reweighing, or repeating the sequence until constant weight. The main steps to determine humidity by drying in an oven are described below, and the information adopted follows officially accepted procedures [6].

2.1.1 Material

Air circulation oven; analytical balance (sensitivity of 0.1 mg); desiccator containing dry desiccant; porcelain, nickel, stainless steel, or aluminum crucible with and without covers depending on the sample to be analyzed (metal crucible should not be used in corrosive samples); grinder or blender; glove or forceps for crucibles; spatula and thermometer from 0 to 130 °C.

2.1.2 Procedure

1. The crucible and covers must be washed, rinsed, dried in an oven at 100 °C for 24 h, cooled in a desiccator, weighed, and identified.
2. The food sample must be prepared (crushed, ground) and homogenized. A 2 to 5 g wet sample must be weighed in a previously weighed crucible using an analytical balance. The wet sample must be spread evenly on the bottom of each crucible.
3. Place the crucible containing the wet sample in the oven and avoid contact with the walls.
4. Refer to Table 1 to find out the recommended oven temperature and the time required for drying the selected food matrices. In general, the drying time can be estimated until constant weight (in preliminary tests) for different food matrices.

Table 1
Moisture methods and conditions according to food matrices

Method	Food type	Temperature (°C) Pressure (mmHg) Reagents	Sample weight	Time (h)	AOAC (official number method)
Air oven method	Beverages and beverage materials/malt	103–104	5–10	3	935.29
	Barley/malting barley	130 ± 3	2	1	981.05
	Beverages and beverage materials/cereal adjuncts	103–104	5–10	3	945.15
	Flour	130 ± 3	2	1	925.10
	Meat and meat products	100–102	2	16–18	950.46
	Frozen vegetables/frozen french fried potatoes	103 ± 2	10	16	984.25
Vacuum oven method	Flour	98–100/≤25	2	5	925.09
	Sugars and sugar products	60–70/≤50	2–5	2	925.45
	Gelatin, dessert preparations, and mixes/gelatin	98–100/≤25	2	5	935.46
	Nuts and nut products	95–100/≤100	2	5	925.40
	Oils and fats	20–25/≤100	5 ± 0.2	Until constant weight	926.12
	Dried fruits	60 ± 1/≤100	5–10	6	934.06
	Milk/dried milk products	100/≤100	1–1.5	5	927.05
	Dairy products/cheese	100/≤100	2–3	4	926.08
	Coffee/roasted coffee	98–100/≤25	5	5.5	968.11
	Tea	95–100/≤100	2 (dry material)	5	925.19
Distillation	Foods/spices and condiments	Distillation with toluene	40	1–2	941.11
Karl Fischer	Cocoa bean, derivatives, confectionery, milk chocolate	–		Few minutes	977.10
	Oils and fats		5–25	15 min	984.20
Microwave	Meat and poultry products (rapid microwave drying method)	80–100% power	4 g	3–5 min	985.14
	Cheese				977.11

Data adapted from AOAC [6] and Park et al. (2004) [12]
NE not specified

5. After the selected time, remove the crucible from the oven with tweezers (if using a crucible with a cover, be sure to cover it before removing it from the oven) and place it in the desiccator for at least 30 min or until it reaches room temperature.
6. Weigh the crucible with a dry sample on an analytical balance.
7. Calculate the moisture percentage (wt/wt) as indicated below. The analysis must be carried out in triplicate.

Calculations

$$\%\text{moisture} = \frac{\text{wt of water in sample (g)}}{\text{wt of the wet sample (g)}} \times 100$$

2.2 Vacuum Oven Drying

Using a vacuum oven is usually the standard drying method and is most accurate for determining moisture in food. Some disadvantages observed in drying in an air oven can be overcome in vacuum drying [3, 4]. Generally, foods are heated in a vacuum oven at temperatures ranging from 98 to 102 °C and pressure from 25 to 100 mmHg [1, 5, 7]. As this technique allows reduced temperatures (60–70 °C), it is well accepted for analyzing moisture in sugary products, especially in foods containing fructose, to prevent decomposition. When using reduced pressure, the water contained in the food evaporates more quickly and, therefore, the drying time is also reduced [5].

Official procedures, such as those of AOAC, normally recommend that the moisture determination be carried out in a vacuum oven at 100 °C for 2 to 6 h under a pressure of 25 to 100 mmHg [6]. The materials and the main steps for determining moisture in a vacuum oven are described below (*% moisture*). Table 1 details the drying conditions for some food products.

2.2.1 Material

Vacuum oven; metal crucible with lid and flat bottom; the other materials are the same as drying in an air oven.

2.2.2 Methods

1. The crucible and covers must be washed, rinsed, dried in an oven, cooled in a desiccator, weighed, and identified.
2. The prepared wet sample food (crushed, ground) must be homogenized, and a sample of 3 to 5 g must be weighed in a previously weighed crucible using an analytical balance. The sample must be spread evenly on the bottom of the crucible, and the crucible must be capped.
3. Take the crucibles to the vacuum oven and partially uncover them with tweezers. Select the time and temperature according to the characteristics of the sample to be analyzed. Refer to Table 1 for the drying conditions for different food matrices.

4. After the selected period, turn off the vacuum pump and slowly return the air to the oven. Cover the crucibles with tweezers, transfer them to a desiccator to cool, and weigh them again.
5. Place the crucibles in the oven again for another hour or more to ensure that they reach a constant weight.
6. Weigh the crucibles again on an analytical balance and calculate the moisture percentage according to the calculations presented in the previous section.

2.2.3 Notes

1. The preparation and handling of the crucibles before use need attention. All crucibles must be properly heated in the oven, cooled, and weighed before use. Use only tweezers or gloves to handle the crucibles; the marks of the fingerprints can influence the weight.
2. When disposable aluminum crucibles are used, they must be pre-dried in a vacuum oven at 100 °C for 3 h. It is recommended that the crucibles be pre-dried for 3 to 15 h at 100 °C in a vacuum oven or a forced-draft oven.
3. Dry crucibles must be stored in a desiccator containing dry desiccant before being used for moisture analysis.
4. Some food matrices, such as sticky fruits, tend to form crusts on the surface or clump during drying, which can result in analytical errors. To control this problem, sand or diatomaceous earth is recommended.
5. It is recommended that previous dry liquid samples, such as milk, be dried in a steam bath at 100 °C before being dried in the oven to avoid splashing the sample. In this case, the drying time is reduced to 75–180 min.
6. Transformations of the sample during drying must be considered. Usually, the darkening of the sample during drying suggests erroneous moisture loss. Lipid oxidation and weight gain of the sample may occur due to the high temperature used in a forced draft oven. Samples with a high carbohydrate content should be dried in a vacuum oven at a temperature below 70 °C to avoid sample decomposition.

2.3 Infrared Drying

Another efficient moisture method based on the drying process is the analytical balance coupled to an infrared lamp or infrared moisture balance. Infrared drying involves an infrared lamp that integrates the moisture analyzer (250–500 W) carried out as a thermal source with operation temperatures around 2000 K–2500 K. It generates a high heat penetration, causing an effective drying process into the sample, shorting the drying time required to other dryer processes, such as the oven dryer, finishing the analysis in around 10–25 min [2].

On the other hand, infrared dryers evaporate the moisture content, heating the sample from the outside to the inside by the absorption of the infrared radiation. This approach takes a longer time than other fast dryers, such as microwaves, but a constant drying temperature can be maintained. Consequently, infrared drying can be applied over a large moisture range [8].

Some types of infrared moisture balances are also commercialized with forced ventilation to avoid moist air. During the drying process, the samples lose weight, which is detected and registered by the balance. The weight difference determines the moisture content in the sample. This method is currently approved by AOAC International [2].

2.3.1 Material

Rapid moisture analyzer, where an infrared lamp is coupled to a balance that registers the weight lost; aluminum crucible.

2.3.2 Procedure

1. Weigh approximately 3 g of sample into the aluminum crucible [9].
2. Place the sample on the plate balance.
3. Use the auto-dry mode if available. If this mode is not available on the equipment, it is necessary to optimize the time required to complete drying [10].
4. Turn on the equipment and set it to 105 °C.

2.3.3 Notes

1. The analyst should conduct the analysis carefully, considering the time necessary to achieve a completely dry sample. Overestimating the time can lead to burning the sample.

2.4 Karl Fischer

Karl Fischer is a direct method to determine moisture. It is commonly applied in fats and oils foods once it prevents oxidation and uses automated equipment. It is also used in other low-moisture products such as candies, dried fruit, and vegetables rich in sugars or proteins. This method is considered fast, selective, highly precision, and sensitive. It is based on the reaction of iodine and sulfur dioxide in the presence of water [1]. Table 1 presents applications for food matrices.

2.4.1 Materials

Karl Fischer titration assembly, automated or not. Titration vessel with magnetic agitation device, burette, electrodes, and circuitry for dead stop endpoint determination. Karl Fischer (KF) reagent, 2-methoxyethanol, pyridine, sulfur dioxide, pyridine, iodine, methanol (anhydrous), sodium tartrate dihydrate. The reagents are toxic and harmful; the analyst should use individual equipment to protect the eyes and skin [7].

2.4.2 Karl Fischer Reagent Standardization

Before starting the experiment, the KF reagent should be standardized to calculate its water equivalence. This procedure should be performed each time a new Karl Fischer reagent is prepared or on the day of the experiment.

1. Add 50 mL of anhydrous methanol to the reaction vessel. And start the agitation using the magnetic agitation device.
2. Fill the burette with the KF reagent.
3. Titrate the water in methanol and add Karl Fischer reagent to change the solution's color from clear or yellow to dark brown. The color should be stable for at least 1 min. This color change characterizes the Karl Fischer endpoint. The same endpoint (color) should be as close as possible for all the repetitions.
4. Weight 0.3 g of sodium tartrate dihydrate, previously dried at 150 °C for 2 h. A precise weight is very important at this step.
5. Add the Karl Fischer reagent into the burette and titrate the water in the sodium tartrate dihydrate sample. Use the volume (mL) of the KF reagent used during the titration to calculate the KF reagent water (moisture) equivalence (KF_{Req}) in mg H_2O/mL:

Karl Fischer Reagent Equivalence:

$$\frac{\left(36\frac{g}{mol} \times T \times 1000\right)}{\left(230.08\ g/mol \times K\right)}$$

T = weight of sodium tartrate dihydrate (g).

K = mL of Karl Fischer reagent used for titration of sodium tartrate dihydrate.

2.4.3 Sample Titration

The method might be adapted for powder or liquid samples.

- For powder samples, weigh approximately 0.3 g of sample using the analytical balance and record the exact sample weight (S).
- Transfer the sample to the reaction vessel through the sample port.
- The final color of the solution should switch to light yellow.
- For liquid samples, use a 1-mL syringe to draw approximately 0.1 mL of the sample. Weigh the syringe with the sample on an analytical balance and record the exact weight (S_1) to the nearest milligram.
- Inject 1–2 drops of sample into the reaction vessel and then weigh the syringe again (S_0), to the nearest milligram.
- Sample weight (S) is the difference of S_1 and S_0.

$$S = S_1 - S_0$$

1. The final color of the solution should switch to light yellow.
2. Add KF reagent to the burette and titrate the water in the sample. Annotate the volume to perform the calculations.
3. For new samples, use a new volume of methanol in a clean reaction vessel.

Calculate the moisture content of the sample as follows:

$$\%\text{water} = \frac{(\text{KFReq} \times Ks)}{S} \times 100$$

KFR_{eq} = water equivalence of KF reagent (mg H_2O/mL).

K_s = mL of KF reagent required for sample titration.

S = weight of sample (mg).

2.5 Distillation

This method (direct) is suitable for foods with a high volatile compound content. The technique involves co-distilling the moisture of a sample using a high-boiling solvent immiscible with water and measuring the volume of water collected. Determining the moisture content in foods can be done in two ways: direct distillation and reflux distillation.

In direct distillation, the food is heated in mineral oil or an immiscible solvent with a boiling point higher than water. Reflux distillation uses solvents less dense than water, such as toluene or xylene, or denser than water, such as tetrachlorethylene. The reflux distillation method uses toluene as a solvent, which is usually more used than direct distillation.

Distillation methods promote less thermal decomposition in specific foods, such as spices and cheese, than oven drying methods that use higher temperatures. Chemical reactions, such as oxidation, can be minimized using a lower boiling solvent. However, it must be considered that in this case, the distillation time will be increased.

Table 1 presents applications for food matrices.

2.5.1 Material

Glass distillation apparatus with ground-up joints (round bottom flask, short flask with gasket; condenser with drip tip with gasket and Bidwell-Sterling collector with gasket, with a capacity of 3 mL, graduated in 0.1 mL intervals); warming blanket; analytical balance, with a sensitivity of 0.1 mg.

2.5.2 Reagent

Xylene or toluene.

2.5.3 Procedure

1. Quickly grind the sample to avoid frictional heating.
2. Weight the sample so that 2 to 5 g of water is released. For example, 10 to 15 g of cheese or 40 g of spices. Transfer the sample quantitatively to the distillation bottle.
3. Add enough solvent to cover the sample, not less than 75 mL.

4. Mount the reflux device appropriately.
5. Adjust the heating to produce two drops per second until most of the water has been collected, then increase the reflux rate to four drops per second.
6. Continue the reflux until two consecutive readings are obtained, with an interval of 15 min, without changing. Rinse the condenser with approximately 5 mL of solvent and continue heating for another 3 to 5 min. The warm-up time is usually 60 to 90 min.
7. Turn off the heating and let the appliance cool down.
8. Record the volume of water collected.
9. Calculate the moisture content in the sample.

Calculations

$$\%\text{moisture} = \frac{\text{volume of water (mL)}}{\text{sample weight (g)}} \times 100$$

2.5.4 Attention

When handling toluene, wear gloves and safety glasses, and be in a suitable place with ventilation. It is a harmful and highly flammable liquid; it should not be inhaled. Your waste should be disposed of following the recommendations of good laboratory practices.

3 Indirect Methods for Moisture Evaluation in Food

3.1 Microwave Method

The microwave method is an indirect and rapid method for determining moisture in food samples. Despite the traditional oven being used to evaluate the moisture content, microwave drying needs less time to evaluate this parameter, approximately 3–8 min, compared to hours in the drying oven method, which is an advantage in industrial processes. A disadvantage of this method is that testing one sample at a time [1, 12] is only possible. Table 1 presents applications for food matrices.

Microwave energy is electromagnetic radiation (3 MHz–3000 GHz). Dipole rotation and ionic polarization are responsible for the heating in this technique. The heat is generated from inside the sample when it is exposed to microwave radiation, and molecules with dipolar electrical charges (water) rotate to align their dipoles with the changing electric field [13].

Moisture determination by microwave is based on water's ability to absorb microwave energy, which is significantly higher than that of dry substances. The moisture is directly proportional to the increase in microwave energy at 2450 GHz. Factors such as sample temperature, particle size, soluble salts, and sample polarization might affect the measurements [12, 13].

3.1.1 Materials

Microwave moisture analyzer, rectangular glass pads, balance, drying chamber.

3.1.2 Procedure

1. Insert the glass pads into the microwave moisture analyzer and tare.
2. Remove the glass pads from the microwave, add 4 g of sample to one glass pad, and cover it with the second one.
3. Insert the glass pads into the microwave and dry the sample for 3–5 min at a power of 80–100%, according to the product.
4. At the end of the process, the moisture will be displayed on the equipment's digital display.
5. Different types of pads, possibly made from fiberglass and quartz fiber, could be used in the analysis. The pads should not absorb the microwave energy; otherwise, the sample could burn. Loss of weight is used to calculate the moisture content.

Calculations:

$$\%\text{moisture} = [(\text{weight of wet sample} + \text{crucible}) - (\text{weight of dried sample} + \text{crucible})] / [(\text{weight of wet sample} + \text{crucible}) - (\text{weight of crucible})] \times 100$$

3.2 Dielectric Capacitance

The measurement of dielectric properties is a technique used to determine the moisture content in foods in different forms, such as solid, liquid, or semisolid samples. Several instruments are commercially available, and their calibrations are usually for specific food types. This means the food of interest needs its calibration curve of dielectric properties versus moisture content provided by an official method [1, 12, 14].

The principle of the method is based on the difference in the dielectric constant of water compared to other organic compounds. Water has a dielectric constant of 80.37, while most organic compounds are in the range 2–5 [11, 15]. The advantages of dielectric methods can be pointed out, such as their high sensitivity due to the large dielectric constant of water, convenience for industrial operations with continuous monitoring, and versatility for various products and food ingredients because their system is adaptable. On the other hand, its disadvantages include the influence of many variables, such as sample texture, electrolytes, temperature, and moisture distribution. Other disadvantages are the hard calibration on samples with pH values outside the 2.7–6.7 range and the difficult measurement of "bound" water at high frequencies [11]. All this makes its application in food chemistry slightly limited from an analytical point of view.

Some studies have investigated how the dielectric properties of some foods are affected by temperature, humidity, frequency, and salt content in food products, such as potato starch [16], pistachio kernels [17], macadamia nut kernels [18], edible oils [19], and apples [20]. In another work, dielectric spectroscopy measurements were applied to moisture prediction in onions [21]. Each study has its particularities and adaptations, and there is no universal method based on dielectric properties to determine the moisture content of foodstuffs and ingredients to date.

Implementing a dielectric method for moisture prediction in foods is dictated by measuring the change in capacitance or resistance to an electric current passed through a sample; therefore, it is measured as an index of capacitance or resistance [1, 15]. The central component of the dielectric instrument is a capacitance cell consisting of two metal plates spaced apart with equal but opposite charges. These charges reverse at fixed frequencies to produce an alternating current. When a sample is placed between the two plates, it will polarize and change the capacitance of the plates. This capacitance change, which is affected by moisture content, is then measured by the instrument. As the amount of capacitance or resistance correlates with the amount of moisture in food, the moisture content of food samples can be determined by performing a calibration against reference values on the same samples [11]. A general description of the procedure is given below. Specific details can be found in the literature provided.

3.2.1 Procedure

1. If applicable, the sample must be reasonably homogenized by simply mixing or other preparation.
2. The sample can be liquid (e.g., edible oils) or solid with smooth surfaces (e.g., compressed legume flour, macadamia nut kernels, onions, pistachio kernels, dehydrated apples, potato starch), but semisolid measurements can also be evaluated in certain circumstances.
3. The procedure can be carried out at room temperature (23 °C) [21] or using different temperatures when it is desired to evaluate temperature-dependent dielectric properties [16–20, 22].
4. Direct analysis is possible for grain or bulk samples, but often, a grind is appropriate. For flours, pressing is appropriate.
5. The reference moisture values must be obtained by an official method, such as gravimetric, mentioned previously in this chapter.
6. Dielectric properties are typically measured using an open-ended coaxial probe connected to a vector network analyzer.
7. Prior to the measurement, the system is calibrated by measurements on air, short (a metal block to short the coaxial cable),

and triple-distilled water at each temperature or at room temperature, following the idea in each study's conception.

8. The measured frequency varies from study to study and instrument to instrument but often ranges between MHz and GHz.
9. Electrodes are inserted into the sample to be measured (system 1), or the sample is placed between the two plates of the sample (system 2).
10. An electric potential is applied, instantly detecting moisture variations in the sample.
11. If penetration or direct coupling into the sample is not possible or desirable, the sensor with the plates (system 2) is placed close to the food to be measured (e.g., granules, powder, etc.). In system 2, a defined electrical field is applied, and after some time, the moisture around the sensor comes into equilibrium with its environment.
12. The measured data are converted and displayed by a computer or processed for further data analysis.
13. The reference moisture content is calibrated against reference values on the same samples. The literature provides more details about this [23].
14. The literature also properly describes other mathematical steps, functions, and interpretations of the generated data since they vary depending on the objective under investigation [21].
15. Briefly, the frequencies are useful in predicting moisture content because the dielectric constant increases nearly linearly with moisture content. It is not our purpose to discuss these details here; we will just mention some references for the interested reader.

3.2.2 Notes

- An increase in moisture content of 1% can increase the dielectric constant by approximately 0.8.
- For substances containing water, the dielectric constant increases almost linearly up to approximately 30–35% moisture.
- The moisture content in the samples must adequately cover a wide range of variations.
- The development of each study and its subsequent application is ideally valid only for those samples under investigation that originated it and within the estimated ranges.
- Consecutive model updates create a more representative model.

3.3 Conductivity Method

In conductivity methods, conductivity and resistance parameters are measured in an electrical circuit containing a food sample. There is a distinct relationship between the moisture content of materials and their electrical properties. Electrical resistance decreases, and conductivity increases as the moisture content of a food increases. However, measuring resistance appears to be most useful in food applications. The logarithm of resistance is plotted as a function of moisture content.

The logarithm of resistance is basically linear as a function of humidity but shows curvature as a function of moisture content. Various instruments exist for measuring either resistance or conductance of foods, and adequate calibration is necessary to achieve accuracy. Advantages include immediate analysis, no destructiveness, and precision. Disadvantages include those measures only requiring free water, difficulty maintaining equipment calibration, and conversion charts being needed to obtain total moisture values. Like the previously presented dielectric capacitance method, the accuracy and precision of the conductivity method are affected by temperature, electrolyte content, and contact between electrodes and samples. An official procedure based on AOAC to measure the conductivity of raisins and prunes is described below.

3.3.1 Procedure

1. Place the food between two electrodes of the instrument.
2. A current flow through the sample and is measured by the change in electrical resistance.
3. Conductance readings are converted into moisture contents using a table that corrects for sample temperature.
4. Conductivity methods measure only the "free" water in the sample; the amount of "bound" water needs to be added to approximate the total moisture content more closely.
5. Conversion charts may be required due to the variation of "bound" water among samples.

Today, developments in electronics have resulted in relatively low-cost, easy-to-use, portable moisture meters for assessing the moisture content and dryness of foods using electrical conductivity. Although not officially recognized by international agencies, these methods are relatively accurate and can be used for field and online measurements. An example commercially available is the "penlike" battery-powered moisture meter. It can be used to measure the approximate water content of dried fruit. This handheld device measures the moisture content using a DC resistance circuit, which measures the total current that moisture causes to flow between the needle electrodes. The automatic display directly shows the percent moisture content of the fruit [15].

3.4 Refractometry

Refractometry is based on the refraction that a beam of light suffers when it passes through from one medium to another with distinct densities. All chemical compounds have a specific index of refraction, but it can change according to the temperature, concentration, and wavelength of light [2].

The fact that the refractive index changes according to the concentration of the solution is the principle of the applicability of this technique to achieve the moisture content in food samples with high sugar content, such as condensed milk, syrups, or honey. It is possible to use a refractometer to measure the change of direction (refraction) of the light in those food samples. This equipment can be applied not only for moisture measurement but also for total soluble solids. Usually, the refractometers are commercially available in portable (handheld) and benchtop versions, and with them, tables are provided to adjust and/or correct values according to temperature [24].

If correctly performed and applied to adequate food matrices, this method is quick and accurate. Consequently, it is the official method to determine soluble solids in fruits and fruit products (methods 932.12, 976.20, 983.17) [6] and moisture in honey (IAL 2009).

3.4.1 Material

Digital refractometer or analogical refractometer, water bath, metal spatula, hydrophilic cotton, glass vial (10 mL), and thermometer from 0 °C to 130 °C.

3.4.2 Procedure

To crystallized samples:

1. Add approximately 10 mL of the sample into the glass vial and close it appropriately.
2. Set the water bath at a constant temperature, preferably at 50 °C ± 0.2 °C.
3. The glass vial should be placed into the water bath, which should cycle until the sample achieves the same temperature as the water and all the crystals have dissolved.
4. Using the thermometer, the temperature of the sample should be determined.
5. When the sample reaches room temperature, transfer three or four drops into the refractometer's prism and measure the value on the scale.
6. If the sample is read at a temperature above 20 °C, the value of the refraction index should be corrected according to the sample's temperature at the moment of the measurement, using the tables normally provided by the supplier of the equipment used.
7. Calculate the moisture percentage using the appropriate table correlating the refraction index and the moisture in the sample analyzed.

To liquid samples:

1. Add approximately 5 mL of the sample into the hydrophilic cotton.
2. Squeeze the cotton with the sample, let three or four drops into the prism of the refractometer, and measure the value on the scale.
3. Using the thermometer, the temperature of the sample should be determined.
4. If the sample is read at a temperature above 20 °C, the value of the refraction index should be corrected according to the sample's temperature at the moment of the measurement, using the tables normally provided by the supplier of the equipment used.
5. Calculate the moisture percentage using the appropriate table correlating the refraction index and the moisture in the sample analyzed.

3.5 Nuclear Methods

Nuclear magnetic resonance (NMR) spectroscopy is a quick, accurate, and nondestructive technique that applies to many foods; however, its application for moisture determination is challenging for nonexperts and food scientists. In NMR, the challenge is to differentiate the water signal from the other hydrogen-containing substances in the food, especially fat. Based on this issue, other NMR approaches have been developed for moisture determination, such as time-domain NMR (applied mainly to meat) and ^{1}H low-field NMR (applied to nonfood materials). In addition, both techniques are widely used to evaluate the behavior of water and its distribution in complex matrices. Here, some advantages versus disadvantages based on NMR studies can be mentioned. It can differentiate between free and bound water, and the particle size and packing of granular samples do not affect signal absorption. However, an NMR instrument is expensive, and the signals are hard to interpret and extract information from. In addition, separate calibration curves are required for different substances. It is also necessary to constantly and correctly sample weight, which is not applicable for foods with variable lipid contents [11].

3.5.1 Time-Domain NMR

T-D NMR has become a feasible option for monitoring several physical and chemical properties in food products, but the number of studies in moisture prediction is limited. The technique must be calibrated by a moisture reference method to make future predictions. In addition, multivariate analytical validation becomes essential to prove the performance and capability of the prediction model. Therefore, combining chemometric data handling in time domain NMR applications to predict quality parameters such as moisture is imperative.

T-D NMR studies are more dedicated to assessing the distribution and behavior of water in food than to predicting its total value. T-D NMR has been applied to meat to predict its moisture content by Pereira et al. [25, 26] Their group investigated the technique's capability for moisture prediction in beef samples using T-D NMR and chemometric multivariate calibration models and achieved feasible results. There was no significant difference between the reference values and those predicted by the models. This indicates that the moisture content could be predicted in the samples even without weighing and drying, as in the reference method or traditional NMR approaches. The procedure described below is based on methods from Pereira and their group [25, 26] for moisture prediction in meat samples.

Procedure

1. The procedure begins with a simple sample preparation, which, in the case of meat samples, involves cutting them into cylindrical slices with a 2 cm diameter.
2. Acquire T-D NMR signals in the spectrometer. The instrument settings can be verified in the papers of Pereira et al. (2013) and Pereira and Colnago (2012).
3. Obtain reference values for moisture content in the samples using an official method.
4. Built predictive models correlating T-D NMR data with multivariate calibration.
5. For building the predictive models using partial least squares regression, the T-D NMR signals are organized in a data matrix format $\mathbf{X}$ ($N \times J$), where N is the number of samples and J is the variable number. The T-D-NMR signals are independent variables ($\mathbf{X}$ matrix) and the moisture content values determined by the official method are the dependent variable ($\mathbf{Y}$ matrix). More details about the PLS tutorial and the step-by-step algorithm can be found in the literature [9, 27, 28].
6. The model is built and must be validated against a number of parameters of merit to establish its predictive capability. Some examples in the literature can be useful to guide the validation of PLS models in food applications, including moisture prediction [9, 25, 26, 29–31].
7. An advantage in the T-D NMR data treatments is that no pretreatments are needed on the spectra and the models are built from the original data.

3.5.2 Hydrogen Nuclear Magnetic Resonance

NMR is a versatile method. In addition to the technique's applicability using T-D NMR, the ^{1}H low-field NMR is also applied for moisture prediction but is more restricted for nonfood materials [32, 33]. For food and agro-food products, the applications of ^{1}H low-field NMR are more directed at studying the absorption, distribution, and mobility of water/moisture [34]. Despite its limitations in the food field, it is possible to extend its use to new applications, and this chapter encourages this. Moisture content is determined based on the effect of electromagnetic radiation on the angular momentum or spin of atomic nuclei (hydrogen proton being the most used) with odd numbers of protons and neutrons. As water molecules become charged by an external magnetic field, a magnetic dipole is created in the direction of the spin axis, producing a nuclear magnetic moment with a magnitude that is equal to the magnitude of the created dipole. This magnitude is excited by a low radio frequency alternating field, producing a detectable high-frequency resonance pulse, often between MHz and GHz. The moisture content of the material is directly proportional to the intensity of the detected signal [15, 34].

3.6 Cryoscopy Method

Cryoscopy is another indirect option for determining moisture, especially for measuring the water added to liquid milk. The method uses the colligative property known as freezing point depression. The freezing point of water decreases as the concentration of dissolved solutes increases. Thus, the freezing point is inversely correlated with the amount of water present for liquids containing a constant type of solute. The freezing point of "natural" milk is approximately 0.54 °C; a freezing point closer to 0 °C indicates water has been added to the milk (i.e., has become adulterated) [11]. The test is performed in a cryoscope device, in which the milk sample is frozen, and the freezing point is read on a very accurate thermometer. Usually, the procedure is described according to manufacturing, but a simple description using Brazilian legislation is provided below [35].

3.6.1 Procedure

1. It is necessary to use standard solutions with known freezing points to adjust the cryoscope and distilled water.
2. Prepare solutions of sodium chloride (NaCl) at 0.8646% (w/v) (freezing point about −0.512 °C) and 1.0155% (m/v) (freezing point about −0.600 °C).
3. Prepare three cuvettes of distilled water.
4. Start the equipment check with the first tube of distilled water. If the reading is adequate, i.e., 0.000 ± 2 mg, proceed to the next step. Otherwise, repeat the reading with the second cuvette, and so on.
5. Check the equipment with the first cuvette of the standard solution −0.621°H (or −0.600 °C). If the reading is adequate, go to the next step. Otherwise, repeat the reading with the second cuvette and so on.
6. Check the reading with the intermediate standard −0.530°H (or −0.512 °C). If the reading is adequate, perform the assay with the samples. Otherwise, repeat the verification and

adjustment procedures with distilled water and standard solution −0.621°H.

7. After calibrating the equipment, it is read in duplicate or triplicate 2.5 mL of fluid milk sample.
8. The result is obtained directly from the average of the readings.
9. The maximum freezing point accepted is around −0.530°H, equivalent to 0.512 °C for cow's milk.
10. After each reading, carefully clean the sensor and stirrer with water and dry gently with fine absorbent paper.

3.7 Infrared Spectroscopy

This method is based on the electromagnetic spectrum's infrared region (IR), which amplifies the molecular vibrational levels when absorbed by an organic molecule. Then, this process is registered as a spectrum [36]. Firstly, the spectra data obtained in the infrared region were studied through tables that presented stretches and deformations of molecular groups related to the chemical structure of the evaluated samples that helped to point to the structural information. More recently, the popularization of computers has allowed the development of multivariate methods to improve this process and interpretation [37].

The infrared region in the electromagnetic spectrum corresponds to radiation with wavenumbers ranging from 12,800 cm^{-1} to 10 cm^{-1}. This region is subdivided into near-infrared (NIR), comprising the portion between 14,000 cm^{-1} and 4000 cm^{-1}; mid-infrared (MIR), from 4000 cm^{-1} to 400 cm^{-1}; and far-infrared, from 400 cm^{-1} to 100 cm^{-1} [37].

From a chemical analytical point of view, only the NIR and MIR are used. The main difference between these regions is that the absorption in the NIR region corresponds to overtones and combinations due to the increase in the vibrational level of organic molecules, while in the MID region, the spectrum is registered due to the fundamental vibration of organic molecules [38].

In the food industry, infrared spectroscopy is commonly used to monitor the composition of food products during the process chain, from the reception of raw materials during the processing steps to final products. The many types of available equipment follow the versatility featured by infrared technology, with traditional benchtop devices or handheld devices until sensors allow the possibility of online monitoring [39, 40]. This technique has frequently been chosen by users who target new, green, and quick analytical methods capable of reporting reliable results that fit in the context of the fourth industrial revolution or industry 4.0 [39].

IR spectroscopy presents a strong signal of water bands in the NIR region and is well suited for moisture content determination [41]. It has been successfully applied in diverse cases, such as moisture in quinoa, fruit pulp, and in diverse plant matrices [42–44].

3.7.1 Sample Preparation

For solid samples:

1. If the samples present heterogeneous particle sizes, it is recommended that they be ground and sieved beforehand to avoid the light scattering phenomena caused by the particle size difference.

For liquid or pasty samples:

1. If the samples contain a high content of insoluble solids, a homogenization process is recommended to ensure representative sampling.
2. If the sample is an oil, the measurement must be conducted under temperature control once it is a critical parameter for this kind of sample.

For NIR measurements:

1. If the sample is liquid, weight a mass of the sample enough to cover the beam of infrared light of the equipment. This parameter could vary once, depending on the NIR equipment, there is a different size of cup to use during the measurements. Obtain the infrared spectra in absorbance or transflection mode.
2. If the sample is an oil, the measurement must be conducted under temperature control once it is a critical parameter for this kind of sample. Obtain the infrared spectra in absorbance or transflection mode.
3. If the sample is solid, weigh a mass of the sample enough to cover the beam of infrared light of the equipment. Obtain the infrared spectra in reflection mode.

For MIR measurements:

1. For liquid samples, use the attenuated total reflection accessory coupled to MIR to put a drop of each sample directly on the crystal. Obtain the spectra data in absorbance.
2. For solid samples, add enough sample to completely cover the crystal place, and compact the solid particles with the squeezer to avoid light passage through the sample. Obtain the spectra data in reflection mode.

For the determination and validation of the predictive method:

1. Obtain the reference values of moisture in each sample using an official method.
2. Build predictive multivariate methods correlating the spectra obtained to the official method results.
3. To determine the predictive method, the NIR/MIR signals must be organized in matrix format X ($N \times J$), where N is the samples and J is a variable number (Ferreira 2015).

4. To preprocess the NIR/MIR spectral data, there are many options depending on the demand of the profile of spectral data that should be applied. Some examples of existing preprocessing methods can be found in the literature [45, 46].
5. The model should be validated and evaluated according to the merit parameters to determine its predictive performance. Some examples of this process can be found in the literature [42].

References

1. Mauer LJ, Bradley RL Jr (2017) Moisture and total solids analysis. In: Nielsen SS (ed) Food analysis, pp 257–286
2. Nielsen SS (2017) Food analysis. Springer
3. Karmas E (1980) Techniques for measurement of moisture content of foods. Food Technol 34:52–59
4. Nollet LML, Toldrá F (2015) Handbook of food analysis. CRC Press
5. Pomeranz Y, Meloa CE (1994) Food analysis: theory and practice
6. AOAC (2016) Association of Official Analytical Chemists International. Appendix F: Guidelines for Standard Method Performance Requirements
7. Nielsen SS (2017) Moisture content determination. In: Nielsen SS (ed) Food analysis laboratory manual. Springer, pp 105–115. https://doi.org/10.1007/978-3-319-44127-6
8. Weinberg S (2011) Cost-contained regulatory compliance, cost-contained regulatory compliance https://doi.org/10.1002/9780470933510
9. Baqueta MR, Coqueiro A, Março PH, Valderrama P (2020) Quality control parameters in the roasted coffee industry: a proposal by using MicroNIR spectroscopy and multivariate calibration. Food Anal Methods 13:50–60
10. Smiderle OJ, Cezar NA, de Sousa RDCP (2003) Padronização do método infravermelho para determinação do teor de água em grãos de Zea mays. 31
11. Park YW, Bell LN (2004) Determination of moisture and ash contents of foods. In: Nollet LML (ed) Handbook of food analysis, pp 55–82. https://doi.org/10.5860/choice.194412
12. Park YW, Bell LN (2004) Determination of moisture and ash contents of foods. In: Handbook of food analysis second edition: physical characterization and nutrient analysis, vol 1
13. Bouraoui M, Richard P, Fichtali J (1993) A review of moisture content determination in foods using microwave oven drying. Food Res Int 26:49–57
14. Wernecke R, Wernecke J (2011) Industrial moisture and humidity measurement – moisture and humidity measurement methods in solid, liquid, and gaseous substances. In: Weinberg S (ed) Cost-contained regulatory compliance. Wiley-VCH. https://doi.org/10.1002/9780470933510
15. Zambrano MV, Dutta B, Mercer DG, MacLean HL, Touchie MF (2019) Assessment of moisture content measurement methods of dried food products in small-scale operations in developing countries: a review. Trends Food Sci Technol 88:484–496
16. Zhu Z, Guo W (2017) Frequency, moisture content, and temperature dependent dielectric properties of potato starch related to drying with radio-frequency/microwave energy. Sci Rep 7:1–11
17. Ling B, Wang S (2017) Dielectric properties of pistachio kernels as influenced by frequency, temperature, moisture and salt content. 2017 ASABE Annu Int Meet:1–15. https://doi.org/10.13031/aim.201700372
18. Wang Y, Zhang L, Gao M, Tang J, Wang S (2013) Temperature- and moisture-dependent dielectric properties of macadamia nut kernels. Food Bioprocess Technol 6:2165–2176
19. Lizhi H, Toyoda K, Ihara I (2008) Dielectric properties of edible oils and fatty acids as a function of frequency, temperature, moisture and composition. J Food Eng 88:151–158
20. Feng H, Tang J, Cavalieri RP (2002) Dielectric properties of dehydrated apples as affected by moisture and temperature. 45:129–135
21. McKeown MS, Trabelsi S, Tollner EW, Nelson SO (2012) Dielectric spectroscopy measurements for moisture prediction in Vidalia onions. J Food Eng 111:505–510

22. Guo W, Wang S, Tiwari G, Johnson JA, Tang J (2010) Temperature and moisture dependent dielectric properties of legume flour associated with dielectric heating. LWT Food Sci Technol 43:193–201
23. Wernecke R, Wernecke J (2011) Industrial moisture and humidity measurement – moisture and humidity measurement methods in solid, liquid, and gaseous substances. Cost-Contained Regulatory Compliance
24. da Silva Oliveira W, Neves DA, Ballus CA (2019) Mature chemical analysis methods for food chemical properties evaluation. Eval Technol Food Qual:63–90. https://doi.org/10.1016/B978-0-12-814217-2.00005-6
25. Pereira FMV et al (2013) Fast determination of beef quality parameters with time-domain nuclear magnetic resonance spectroscopy and chemometrics. Talanta 108:88–91
26. Pereira FMV, Colnago LA (2012) Determination of the moisture content in beef without weighing using benchtop time-domain nuclear magnetic resonance spectrometer and chemometrics. Food Anal Methods 5:1349–1353
27. Geladi P, Kowalski BR (1986) Partial least-squares regression: a tutorial. Anal Chim Acta 185:1–17
28. Ferreira MMC (2015) Quimiometria: conceitos, métodos e aplicações. Editora da Unicamp
29. Valderrama P, Braga JWB, Poppi RJ (2007) Validation of multivariate calibration models in the determination of sugar cane quality parameters by near infrared spectroscopy. J Braz Chem Soc 18:259–266
30. Baqueta MR, Coqueiro A, Valderrama P (2019) Brazilian coffee blends: a simple and fast method by near-infrared spectroscopy for the determination of the sensory attributes elicited in professional coffee cupping. J Food Sci 84:1247
31. Hashimoto JC et al (2018) Quality control of commercial cocoa beans (Theobroma cacao L.) by near-infrared spectroscopy. Food Anal Methods 11:1510–1517
32. Mao Y, Xia W, Xie G, Peng Y (2020) Rapid detection of the total moisture content of coal fine by low-field nuclear magnetic resonance. Meas J Int Meas Confed 155:107564
33. Xu K, Lu J, Gao Y, Wu Y, Li X (2017) Determination of moisture content and moisture content profiles in wood during drying by low-field nuclear magnetic resonance. Dry Technol 35:1909–1918
34. Patel KK, Khan MA, Kar A (2015) Recent developments in applications of MRI techniques for foods and agricultural produce – an overview. J Food Sci Technol 52:1–26
35. Brazil (2006) Ministério da Agricultura. Secretaria de Defesa Agropecuária. Métodos Analíticos Físico-Químicos para Controle de Leite e Produtos Lácteos. Instrução Normativa 68, 12/12/06. Ministério da Agricultura, Brasília
36. Silverstein RM, Webster FX, Kiemle DJ (2005) Spectrometric identification of organic compounds. J Mol Struct 30:424. https://doi.org/10.1016/0022-2860(76)87024-X
37. Skoog DA, Holler FJ, Crouch SR (2009) Princípios de Análise Instrumental
38. Bellon-Maurel V, McBratney A (2011) Near-infrared (NIR) and mid-infrared (MIR) spectroscopic techniques for assessing the amount of carbon stock in soils – critical review and research perspectives. Soil Biol Biochem 43: 1398–1410
39. Kakani V, Nguyen VH, Kumar BP, Kim H, Pasupuleti VR (2020) A critical review on computer vision and artificial intelligence in food industry. J Agric Food Res 2:100033
40. Pallone JAL, dos Santos Caramês ET, Alamar PD (2018) Green analytical chemistry applied in food analysis: alternative techniques. Curr Opin Food Sci 22:115–121
41. Achata EM, Esquerre C, Ojha KS, Tiwari BK, O'Donnell CP (2021) Development of NIR-HSI and chemometrics process analytical technology for drying of beef jerky. Innov Food Sci Emerg Technol 69:102611
42. Alamar PD, Caramês ETS, Poppi RJ, Pallone JAL (2016) Quality evaluation of frozen guava and yellow passion fruit pulps by NIR spectroscopy and chemometrics. Food Res Int 85:209–214
43. Ferreira DS, Poppi RJ, Lima Pallone JA (2015) Evaluation of dietary fiber of Brazilian soybean (Glycine max) using near-infrared spectroscopy and chemometrics. J Cereal Sci 64:43–47
44. Mayr S et al (2021) Challenging handheld NIR spectrometers with moisture analysis in plant matrices: performance of PLSR vs. GPR vs. ANN modelling. Spectrochim Acta Part A Mol Biomol Spectrosc 249: 119342
45. Rinnan Å (2014) Pre-processing in vibrational spectroscopy-when, why and how. Anal Methods 6:7124–7129
46. Rinnan Å, van den Berg F, Engelsen SB (2009) Review of the most common pre-processing techniques for near-infrared spectra. TrAC Trends Anal Chem 28:1201–1222

Chapter 2

Water Activity

Adriano Lucena de Araújo, Mayara Galvão Martins, and Rosinelson da Silva Pena

Abstract

Water activity (a_w) is a property of great relevance in the food area for both research and industry because it is directly related to food stability due to degrading processes of physical, chemical, biochemical, and microbiological nature. The most practical method for determining the a_w of food is the direct measurement in the water activity meter. However, many research laboratories and industries do not have such equipment due to financial limitations. In these cases, using a moisture sorption isotherm (MSI) of the product of interest is an alternative for indirectly determining a_w. In this sense, this chapter presents an alternative method: simple, low-cost, and fast for determining MSI.

Key words Alternative method, Indirect measurement, Accessible technique, Food stability

1 Introduction

Water plays a key role in the stability and preservation of foods. Several foods with the same water content differ significantly in perishability. The measure of water activity (a_w), instead of moisture content (MC), provides additional information and helps to control and establish specific food characteristics, such as nutritional, textural, microbiological, and sensorial [1].

From a historical perspective, a_w is almost a 70-year-old concept, which correlates more to the microbial, biochemical, chemical, and physical properties of food and other natural products than the MC itself. Studies from William James Scott demonstrated that microorganisms have a limiting a_w level for growth [2]. After Scott's initial statement, experimental and theoretical studies associating water with foods have been intensified in the endeavor to understand and interpret water behavior [3]. Likewise, research (especially from the food and microbiological sciences) acknowledges that a_w is a crucial parameter affecting food stability [4].

Lilian Mariutti and Frank Vriesekoop (eds.), *Food Chemistry: Basic Protocols*, Methods and Protocols in Food Science, https://doi.org/10.1007/978-1-0716-4678-6_2,

The term "water activity" in a food system reflects the thermodynamic capacity (energetic state) or the effective concentration of water in a food that can participate as a chemical agent in several chemical and biological processes. This thermodynamic property is related to fugacity, defined as the ratio of the partial vapor pressure of water in the system to the partial vapor pressure of pure water at the same temperature. At equilibrium, a_w is also equal to the relative humidity (RH) of the air surrounding the food [5].

The relationship between a_w and the MC, over a range of values and at a constant temperature, expresses the food's moisture sorption isotherm (MSI) [6]. The MSI gives excellent information about the water sorption mechanism and the interactions between food components and water, demonstrating how the food behaves under different RH to simulate the storage conditions for a given product.

Each food has its unique MSI, and its knowledge is essential to estimate the limiting MC a food would support at a particular a_w. Thus, MSI is a precious tool for scientists to predict the reactions that will increase stability and allow changing the a_w for ingredient selection. It can project moisture gain/loss in a package with a known water vapor permeability [7].

As stated earlier, it is fundamental to understand the sorption mechanism to establish critical a_w of biological materials and estimate potential changes during the handling, transporting, and storage stages. Therefore, several mathematical models fit the MSI data from different food matrices. Once the MSI is constructed, it is more viable (in a food industry or research laboratory) to perform MC routine essays and use the MSI data to predict the corresponding a_w of the material than performing a_w measures and predicting MC since the measure of a_w demands an water activity meter, equipment that is not accessible to many industries and research laboratories.

Likewise, the a_w concept should be considered when preserving and developing new food products. The knowledge of its interaction mechanisms with foods plays a critical role in achieving nutritional and economic goals. Furthermore, using this concept by the food industry can help manage ingredients and increase the quality of the final food products [8].

Several procedures are used to measure the a_w of a food matrix, and each one presents advantages and disadvantages. No methods are available to meter a_w directly; all procedures determine this by indirect measures. Methods vary in accuracy, reproducibility, measurement speed, calibration, stability, linearity, and convenience [9]. For a long time, methods employing simple laboratory equipment or apparatus were used to determine a_w. However, the water activity meters (devices that use different thermodynamic principles to determine a_w) have been improved or developed as technology advances. Although these devices may represent a high initial acquisition investment and need constant calibration, they are more versatile and provide results with greater precision and accuracy in shorter time intervals [9]. Thus, water activity meters are the most frequently used equipment to determine the food's a_w.

Nowadays, hygrometers are the most utilized water activity meters, with dew point sensors by chilled mirror, electrical resistance, or capacitance. In the chilled mirror dew point hygrometers, a sensor measures the dew formation temperature of the air humidity, an infrared thermometer measures the sample temperature, and a fan speeds up the balancing process and controls the conductance of the sensor's boundary layer [10]. From psychrometric charts, the dew point is correlated with vapor pressure and, consequently, with a_w [11]. These devices perform measures in the range of 0.03–1.00 a_w, being easy to handle and providing fast results with great accuracy and precision. Nevertheless, the reliability of the results can be reduced when the sample has volatiles, which can condense together with water on the surface of the chilled mirror [9].

Hygrometers with resistance or capacitance sensors are made of hygroscopic material containing a salt, such as lithium chloride, to which electrical circuits are associated and emit a signal, depending on the equilibrium RH, converted in a_w. The devices available commercially measure the a_w over its entire range [10, 11]. Besides, capacitance sensor hygrometers have a simple design and low installation cost. In contrast, this type of equipment requires a more cautious calibration and a longer equilibrium time [12].

Despite the versatility of hygrometers, in cases where the research laboratory or the industry does not have an water activity meter, the product's a_w can be estimated by the graphical interpolation method proposed by Landrock and Proctor [13] or determined from the product's MSI. In this case, the isotherm would have to be obtained by an outsourced laboratory at the temperature of interest. From the MSI, it is possible to determine the a_w of the product by knowing its MC, which is easily determined.

The method of graphic interpolation consists of exposing samples of the product with known masses (m_1) to atmospheres with different RH in an airtight environment. Saturated saline solutions [14] are used most to assure these RH atmospheres. The set is kept in an environment at a constant temperature for 2–3 h. Depending on the RH, the sample will exchange moisture with the environment, which will cause an increase or loss of mass (m_2). Thus, a graph shows the differences between m_2 and m_1, depending on the RH. The graph points are then connected by a line, which will cross the abscissa (horizontal axis) in a RH value [13]. This RH value corresponds to the condition in which the sample does not gain or lose mass once it is in equilibrium with the environment. The product's a_w is then obtained by dividing this RH value by 100 [5].

The MSI is classically obtained by the static gravimetric method and can be divided into adsorption and desorption [15]. This method keeps samples with known masses in airtight environments containing saturated salt solutions that generate atmospheres with different RH [14]. The set is kept in an environment with controlled and constant temperature until the sample reaches the equilibrium mass in a thermodynamic equilibrium condition; subsequently, the MC of the sample is determined. The MSI is the graphic representation of the product's MC (on a dry basis) against the different RH levels. Dry samples must be used to obtain adsorption isotherms, while wet samples are necessary for desorption isotherms [16].

Despite the practicality of the static gravimetric method, it has some disadvantages. This method is laborious and requires a long time for the sample to reach equilibrium at some RH levels, thus favoring the growth of fungi, which can promote significant changes in the product's mass and a_w. Furthermore, some salts that generate the RH atmospheres present high commercial value. Therefore, an alternative method for the determination of MSI has been proposed by Souza et al. [17] and will be described later in this chapter. The method is faster and simpler than the classical static gravimetric method. In summary, to obtain the adsorption data, a glass desiccator containing water at the bottom is used, and in consequence, the samples start gaining moisture. On the other hand, for desorption processes, the data are obtained in a desiccator containing silica gel at the bottom, and hence, the samples start losing moisture. The MSI is obtained through the same procedure used by the classical method.

2 Material

To perform the experiments, the following are necessary: a sample to be analyzed, preferably in powder or crushed form; a benchtop water activity meter; a BOD or bacteriological incubator; a drying oven; an analytical balance; a vacuum pump; a stopwatch; a large glass desiccator (300 mm diameter); a stainless-steel spatula; 20 capsules with specific lids from the water activity meter; a desiccant (e.g., silica gel); and distilled water.

3 Methods

3.1 Preparation of Capsules

Before initiating the tests, the capsules from the water activity meter and their lids must be sanitized with water and neutral detergent to remove dirt and eventual grease from surfaces when necessary. For safety reasons, dry the set in an oven at 50 °C since the capsules and lids are made of plastic, which does not withstand high temperatures. After the previous step, it is recommended to sanitize the capsules and lids with 70% alcohol, thus minimizing the action of microorganisms and contamination of the sample. After applying alcohol, the capsules and lids must be dried appropriately.

3.2 Obtaining the Moisture Adsorption Data

Sample Pretreatment

To obtain the moisture adsorption data (water gain), samples of the product (preferably in powder or crushed form) must be weighed in the capsules of the water activity meter using an analytical balance (0.0001 g precision). The sample mass should be approximately the same (±0.01 g) in all capsules; the amount of product must be sufficient to fill the bottom of the capsules and not exceed half of their height to avoid contamination of the water activity meter sensors during a_w readings. Using at least 20 capsules is recommended to guarantee ten pairs of capsules (ten duplicates).

In addition, the samples must be weighed in triplicate to determine the dry matter of the product, which is needed to calculate the MC on a dry basis (db) at the different levels of a_w. A conventional oven or a vacuum oven should be used to determine the dry matter, depending on the product's characteristics. The samples contained in the capsules must be subjected to complementary dehydration for at least 24 h in a desiccator containing silica gel or another efficient desiccant at the bottom, at the temperature at which the adsorption data will be obtained. Perform the procedure under negative pressure (vacuum) conditions to accelerate the water loss process and ensure that the samples reach a_w values of 0.1 or less. To provide an environment with controlled temperature (±1 °C), a BOD or a bacteriological incubator can be used, depending on the working temperature.

Data Acquisition

To acquire water adsorption data, the capsules containing the previously dehydrated samples must be placed in a large glass desiccator with a thin layer of distilled water at the bottom. The desiccator with water must be put earlier in an environment with controlled temperature (a BOD or bacteriological incubator), at the working temperature (±1 °C), for at least 30 min. Place the lids on the last two capsules (duplicate), and leave them outside the desiccator but inside the environment with a controlled temperature. These samples will correspond to the time zero of measurements. Fit the cover on the desiccator, but in this case, it is not recommended to use a vacuum to accelerate the mass transfer process since lowering the pressure inside the desiccator may promote the formation of water vapor and the subsequent condensation of water inside the desiccator cover, thus favoring the dripping of water on some samples, invalidating their use.

To monitor the evolution of water gain and the increase of a_w in the product, samples must be removed from the desiccator at increasing times (dynamic equilibrium). At the beginning of the process, the samples have low MC and, consequently, low a_w levels. In turn, the water in the desiccator generates levels of RH in the order of 100%, which promotes high rates of water gain by the product. Therefore, at the beginning of the process, samples must be removed from the desiccator in short intervals. Until the tenth minute, the samples should be removed every 2 min to ensure that data are obtained at the lowest a_w levels, in ranges of 0.05–0.1 a_w. After the tenth minute, samples can be taken every 5 min, and after 30 min, samples can be taken every 10 min. When the product reaches a_w level of 0.6, samples can be taken every 30 min. When a_w levels are in the order of 0.8, samples can be taken in 1-h intervals because, at these a_w levels, the product's water gain decreases considerably. Remove samples until the product reaches the maximum desired a_w level (commonly between 0.9 and 0.95 a_w).

The sample removal must be done in duplicate, preferably in pairs of capsules that are closely arranged over the porcelain plate of the desiccator, because, with the constant desiccator openings, it is not possible to ensure a good homogenization of the RH inside the desiccator. Immediately after removing the samples, close the capsules with the appropriate lids and let them rest in the environment at a controlled temperature for at least 5 min to ensure a thermodynamic equilibrium condition before performing the a_w measurement. After the resting time, weigh the capsule-sample set without the lid on the same analytical balance where the initial weighing of the samples was performed. Then, measure the a_w of the sample in an water activity meter. The capsule lid should only be removed at the time of weighing to avoid the exchange of sample moisture with the environment. After weighing and a_w determination, the pair of capsules with the sample must be returned to the desiccator to continue the moisture adsorption process, ensuring that the two capsules are closely arranged over the porcelain plate of the desiccator. Make sure that the same pairs of capsules are placed and removed together from the desiccator; for this, the capsules must be identified.

3.3 Obtaining the Moisture Desorption Data

Sample Pretreatment

The following procedure for obtaining the moisture desorption data (water loss) should be done after obtaining the adsorption data. The capsules containing the samples used in the adsorption process, in which masses and dry matter are already known, must be placed over the porcelain plate of the same desiccator with the water used during the adsorption process. Ensure that the same pairs of capsules used in adsorption continue to be used during desorption. The set must remain in the environment at a controlled working temperature for at least 12 h to guarantee that the samples reach a_w levels above 0.9. It is not recommended to use a vacuum in the desiccator containing water since the formation of water vapor and its condensation may occur inside the system on the desiccator cover. This may promote water dripping on some samples, which will suddenly increase a_w, generating discrepant a_w values among the duplicates.

Suppose the objective is only to obtain the desorption data without the previous execution of the adsorption process. In that case, the sample weighing procedure must be the same as that used to obtain the adsorption data, described in subsection 3.2. Additionally, samples must be weighed in triplicate to determine the dry matter in a conventional oven or vacuum oven, depending on the product's characteristics. Besides, the dry matter is needed to calculate the product's MC (db) at different a_w levels. As evidenced in the previous case, the capsules containing the samples must be placed over the porcelain plate of the desiccator with water, and the set must be kept at the working temperature in an environment with controlled temperature for 24 h or for a longer time, sufficient to ensure that the samples will reach a_w levels of 0.9 or more, from which it is desired to start the moisture desorption process.

Data Acquisition

To obtain the moisture desorption data, the capsules containing the wet samples must be placed on the porcelain plate of a large glass desiccator containing silica gel or other desiccant of good quality. Before positioning the capsules over the porcelain plate, the desiccator (containing the desiccant) must be put separately in an environment with a controlled temperature (±1 °C) (a BOD or bacteriological incubator) at the working temperature for at least 30 min. Cover the desiccator and keep the system in the same environment for a controlled temperature so that the loss of water and the reduction of the product's a_w can be monitored by taking samples from the desiccator at different times (dynamic equilibrium). At the beginning of the process, the high a_w of the product and the low levels of RH inside the desiccator, ensured by the silica gel, favor high water loss rates from the samples. However, at high levels of a_w, the variation in a_w is less sensitive to variations in the product's MC. Thus, at the beginning of the desorption process, samples must be taken every 30 min until the a_w reaches values around 0.7. Between 0.7 and 0.4 a_w, the time taken to remove samples from the desiccator should be reduced to 10 min because, at these a_w levels, small reductions in MC can promote significant variations in the product's a_w. Below 0.4 a_w, sample removal can be done every 30 min until reaching the desired final a_w.

Following the recommendation made for the adsorption process, in the monitoring of the desorption process, the removal of samples must occur in duplicate, in pairs of capsules that are closely arranged within the desiccator due to the deficiency of RH homogeneity in the system. Immediately after removal from the desiccator, the capsules must be sealed with the appropriate lids and left at rest in a temperature-controlled environment for 5 min to ensure thermodynamic equilibrium before a_w measurement. After the resting period of the sample, the capsule-sample set without the lid must be weighed on the same analytical balance used for the initial weighing of the samples, followed by the determination of a_w in an water activity meter. The capsule lid should only be removed at the time of weighing to avoid the exchange of sample moisture with the environment. After weighing and a_w measurement, the pair of capsules with the sample must be returned to the desiccator to continue the process of water loss (desorption), ensuring that the two capsules are closely arranged over the porcelain plate of the desiccator. Ensure that the same pairs of capsules are placed into and removed from the desiccator together.

3.4 Considerations for the Acquisition of MSI Data

For most products, achieving the highest levels of a_w for the adsorption process and the lowest levels of a_w for the desorption process on a working day is not possible. Thus, when there is a need to interrupt the adsorption process or the desorption process, all capsules containing samples must be covered with properly dried lids and fitted well to the body of the capsules to avoid the exchange of sample moisture with the environment. The capsules can be kept inside the desiccator with distilled water (adsorption) or silica gel (desorption) and in the same environment at working temperature. To restart both processes, simply remove the lids, assuring that the pairs of capsules are the same, and restart the sample removal procedure (in duplicate) until the desired a_w levels are reached. In the desorption process, performed in a desiccator with silica gel, vacuum can be applied to the desiccator to accelerate the water loss process at the lowest a_w levels. In this case, the times for sample removal, recommended in subsection 3.3, should be reduced by half.

3.5 Obtaining the MSI and a_w Determination

To obtain the product's moisture adsorption and desorption isotherms, the MC (db) must be plotted against the corresponding a_w of the product. The mean values of a_w, obtained in the water activity meter, must be arranged in the abscissa axis (*x*-axis), and the corresponding mean MC values, calculated by the following equation, must be arranged in the ordinate axis (*y*-axis):

$$m_t = \frac{M_t - \mathrm{DM}}{\mathrm{DM}} \times 100$$

where m_t = moisture content of the product at time t (g H_2O/100 g db); M_t = mass of the product at time t; and DM = dry matter.

A statistical program can be used to define a curve that fits well with the experimental data behavior of the moisture adsorption and desorption isotherms. Using one of these isotherms, it is possible to determine the a_w values of the product at the same temperature at which the adsorption or desorption data were obtained. To this end, it is only necessary to determine the product's MC, which can be measured using common equipment found in most research laboratories and industries, such as a conventional oven, vacuum oven, infrared moisture balance, and microwave oven. Determining the a_w of a product from the adsorption or desorption isotherm can be very useful when the water activity meter is unavailable, which is not uncommon in many laboratories and industries due to the commercial value of this type of equipment.

References

1. Duckworth RB (1974) Water relation of foods. Academic Press, London
2. Scott WJ (1957) Water relations of food spoilage microorganisms. Adv Food Nutr Res. https://doi.org/10.1016/S0065-2628(08)60247-5
3. Rizvi SSH, Benado AL (1984) Thermodynamic properties of dehydrated foods. Food Technol 38:83–92
4. Reid D (2007) Water activity: fundamentals and relationships. In: Barbosa-Cánovas GV, Fontana AJ Jr, Schmidit SJ, Labuza T (eds) Water activity in foods: fundamentals and applications, 1st edn. Blackwell, Iowa
5. Damodaran S (2017) Water and ice relations in foods. In: Damodaran S, Parkin KL (eds) Fennema's food chemistry, 5th edn. CRC Press, Boca Raton
6. Al-Muhtaseb AH, McMinn WAM, Magee TRA (2002) Moisture sorption isotherm characteristics of food products: a review. Trans IChemE 80(2):118–128
7. Bell LN, Labuza TP (2000) Moisture sorption: practical aspects of isotherms measurements and use, 2nd edn. AACC Int Pub, St. Paul
8. Welti-Chanes J, Pérez E, Guerrero-Beltrán JA et al (2007) Applications of water activity management in the food industry. In: Barbosa-Cánovas GV, Fontana AJ Jr, Schmidit SJ, Labuza TP (eds) Water activity in foods: fundamentals and applications, 1st edn. Blackwell, Iowa
9. Fontana AJ, Carter BP (2020) Measurement of water activity, moisture sorption isotherm, and moisture content of foods. In: Barbosa-Cánovas GV, Fontana AJ Jr, Schmidt SJ, Labuza TP (eds) Water activity in foods: fundamentals and applications, 1st edn. Blackwell, Iowa
10. Fraden J (2004) Handbook of modern sensors – physics, designs and applications, 3st edn. Springer, New York
11. Troller JA, Christian JHB (1978) Water activity and food. Academic Press Inc
12. Mathlouthi M (2001) Water content, water activity, water structure and the stability of foodstuffs. Food Control 12:409–417
13. Landrock AH, Proctor BE (1951) A new graphic interpolation method for obtaining equilibrium data with special reference to its role in food packaging studies. Food Technol 5:332–336
14. Rockland LB (1960) Saturated salt solutions for static control of relative humidity between 5°C and 40°C. Anal Chem 32(10):1375–1376
15. Wolf W, Spiesses WEL, Jung G (1985) Standardization of isotherm measurements (Cost-Project 90 and 90 BIS). In: Simatos D, Multon JL (eds) Properties of water in foods in relation to quality and stability. Springer, New York
16. Labuza TP (1983) Standard procedure for isotherm determination. Cereal Foods World 28(4):258
17. Souza TCL, Souza HAL, Pena RS (2013) A rapid method to obtaining moisture sorption isotherms of a starchy product. Starch/Stärk 65:433–436

Part II

Lipids

Chapter 3

Extraction and Total Lipid Quantification

Adriele Hacke, Marcella Camargo Marques, and Lilian Regina Barros Mariutti

Abstract

Lipid extraction is used daily in food analysis laboratories and is considered a routine analysis. Classical methods involve nonpolar solvents such as chloroform, which is used in the Folch and Bligh & Dyer methods, and ethyl ether, used in the Soxhlet method. With increasing concern for the environment, efforts have been made to develop sustainable methods that reduce or replace toxic solvents, such as the microwave-assisted extraction (MAE), in which small amounts of ethyl acetate are used to extract lipids. Regardless of the method used, extraction provides essential information about the composition of the food in terms of total lipid content, which is vital for the health of individuals and the food and biofuel industries.

Key words Lipid extraction methods, Total lipids, Microwave-assisted extraction, Solvent extraction, Cold extraction, Hot extraction

1 Introduction

Lipids are essential components of the human diet. They are an important energy source, form the lipid bilayer of cells, are precursors of hormones, and their metabolites are important mediators in several reactions. Furthermore, the accumulation of lipids in adipocytes ensures the body's energy reserves, in addition to functioning as a thermal protection barrier and against shocks to internal organs.

Due to the wide distribution of lipids in the human body, any change in their metabolism can lead to dysregulation of membranes with changes in permeability and lipid oxidation [1, 2]. These changes are responsible for several diseases, mainly those related to chronic inflammation. This is the case of inflammatory bowel diseases, in which lipids are associated with dysregulation of the intestinal barrier and changes in the immune cells in the intestine [3]. In cancer, lipid metabolism is responsible for promoting tumor growth and survival, and lipid peroxidation can lead to the

Lilian Mariutti and Frank Vriesekoop (eds.), *Food Chemistry: Basic Protocols*, Methods and Protocols in Food Science, https://doi.org/10.1007/978-1-0716-4678-6_3,

development of tumors, such as brain tumors [4, 5]. In addition, changes in lipid metabolism have also been linked to the severity of asthma [6], skin diseases [7], neurodegenerative diseases [2, 5, 8], aging [9], and Alzheimer's [10].

This relationship between lipids and health and disease leads the food industry to challenges in developing healthier foods that maintain lipid metabolism in full operation without losing the expected characteristics of the food, such as texture, flavor, and appearance. To do so, it is necessary to know the composition of the food, focusing, in this case, on lipids.

Several methods exist for quantifying lipids in food; for each method, the lipids must be extracted from the food matrix. The most used methods for extraction and quantification, the so-called classic methods, are the Folch method [11], the Bligh & Dyer method [12], and the Soxhlet method [13].

The Folch and the Bligh & Dyer methods are cold extraction methods, i.e., they do not use high temperature to accelerate extraction/solubilization. In both methods, the solvents used are chloroform and methanol, but in different proportions, 2:1 in the Folch method and 2:2 in the Bligh & Dyer method. In the latter, water can also be added if the sample has less than 80% of moisture content, resulting in a final chloroform/methanol/water ratio of 2: 2:1.8. The common point between these two methods is the use of chloroform as the lipid carrier phase [11, 12]. Cold extraction methods are considered standard methods for lipid extraction and are widely used. The Bligh & Dyer method uses a smaller volume of solvent but still uses chloroform, which is highly toxic to the environment [14].

The Soxhlet method, on the other hand, consists of hot extraction. It is one of the most traditional lipid extraction methods, mainly due to the autonomous extraction equipment, the Soxhlet extractor. The extractor consists of three distinct compartments: the distillation flask, the extraction chamber, which contains the Soxhlet cartridge, and the condenser (Fig. 1). Although it is a typical analysis, it has some disadvantages, such as excessive analysis time, use of toxic solvents such as ethyl ether, the need to have the equipment in the laboratory, and the possibility of degradation of fatty acids due to the high temperatures used in the analysis [14].

Due to environmental concerns, methods have been developed to reduce the harm caused by using organic solvents. One of these methods is microwave-assisted extraction (MAE), in which a small volume of ethyl acetate and methanol solution (2:1) is added to the sample, followed by microwave extraction for 16 min. The advantages of this analysis are the small volume of solvent used, the reduction of analysis time from hours to minutes, and the possibility of performing up to 12 extractions simultaneously, depending on the equipment [15].

Fig. 1 Soxhlet apparatus for lipid extraction

We know that several other methods of lipid extraction exist, but they require the use of specific equipment that is not always available in laboratories, such as supercritical extraction with CO_2, or they are specific methods for certain food groups, such as the Gerber acid extraction method. In this chapter, we will discuss the classic methods of total lipid extraction and an eco-friendly method, the microwave-assisted extraction.

2 Material

All methods use distilled water and analytical-grade reagents. Solutions can be prepared and stored at room temperature.

2.1 Cold Extraction

2.1.1 Folch Method

1. Samples may be dry (lyophilized or oven-dried) or fresh (*see* **Note 1**).
2. Chloroform and methanol: Prepare a 2:1 (v/v) chloroform/methanol solution.
3. NaCl: Prepare a 0.90% NaCl solution in distilled water.
4. Rotary shaker.
5. Rotary evaporator with vacuum pump at 30 °C.
6. 30 mL tube with cap (*see* **Note 2**).
7. Funnel.

8. 50 mL round-bottom flask.
9. Separatory funnel.
10. Whatman no. 1 filter paper.

2.1.2 Bligh & Dyer Method

1. Samples may be dry (lyophilized or oven-dried) or fresh. If their moisture content is higher than 80%, adding water is not necessary.
2. Chloroform and methanol.
3. 1.5% anhydrous sodium sulfate in distilled water (m/v).
4. Rotary shaker.
5. Oven.
6. Centrifuge (optional).
7. 70 mL tube with cap (*see* **Note 2**).
8. Funnel.
9. 50 mL beaker (*see* **Note 3**).
10. 10 and 20 mL volumetric pipettes.
11. 70 mL centrifuge tubes.
12. Whatman no. 1 filter paper.

2.2 Hot Extraction

2.2.1 Soxhlet Method

1. Dry the samples in an oven at 105 °C until constant weight. Then, grind them using a pestle and mortar until a powder is obtained.
2. Soxhlet cartridge.
3. Soxhlet extractor.
4. Heating plate.
5. Ethyl ether, methanol, and anhydrous sodium sulfate.
6. Oven.
7. Desiccator.
8. 250- or 300-mL flat-bottomed flask (*see* **Note 4**).

2.3 Microwave-Assisted Extraction (MAE)

1. Samples may be dry (lyophilized or oven-dried) or fresh (*see* **Note 1**).
2. Teflon tubes are specifically designed for use in microwave extraction systems.
3. Anhydrous sodium sulfate.
4. Ethyl acetate and methanol: Prepare a 2:1 (v/v) ethyl acetate/methanol solution.
5. Start-E microwave extractor system (*see* **Note 5**).
6. Whatman no. 1 filter paper.
7. 15 mL screw-cap tube (*see* **Note 6**).

8. Nitrogen gas.
9. Desiccator.

3 Methods

3.1 Cold Extraction

3.1.1 Folch Method

1. Weight 1.0 g of sample (regardless of whether dry or fresh) in the 30 mL tube. Record the weight with 3 decimals (0.001 g).
2. Add 20 mL of chloroform/methanol solution.
3. Homogenize the sample with the solvent using a rotary shaker for 15–20 min.
4. After shaking, filter the contents of the tube through qualitative filter paper.
5. Transfer the filtrate to a separatory funnel.
6. Add 4 mL of distilled water or 4 mL of 0.9% NaCl and allow to separate the phases.
7. Gently add 1.5 mL of methanol through the funnel wall using a pipette. This step only involves washing the tube's wall, taking care not to mix the methanol with the chloroform.
8. Remove the lower phase, containing the chloroform and lipids, to a previously weighed round-bottom flask and evaporate the contents in a rotary vacuum evaporator.
9. Weigh the flask with the dry lipids and record the result. This mass represents the total lipids in the sample, and the result can be expressed in mg/g of sample.

3.1.2 Bligh & Dyer

1. Weight 3–3.5 g of sample (regardless of whether dry or fresh) in the 70 mL tube. Record the weight with 3 decimals (0.001 g).
2. Using a volumetric pipette, add 10 mL of chloroform, 20 mL of methanol, and 8 mL of distilled water, the latter being added only if the sample has a moisture content lower than 80%.
3. Homogenize the sample with the solvents in a vortex for 20 s and shake constantly for 30 min on a rotary shaker.
4. Using a volumetric pipette, add 10 mL of chloroform and 10 mL of 1.5% sodium sulfate. Shake vigorously for 2 min.
5. Centrifuge the mixture at 1000 rpm for 2 min (opcional) to separate the phases or let it rest for 1 h.
6. After separating the phases, discard the upper layer containing methanol in a suitable container.
7. Filter the lower layer, containing the chloroform and lipids, through a filter paper containing 1 g of anhydrous sodium sulfate (*see* **Note 7**).

8. Pipette 5 mL of the filtrate into a previously weighed 50 mL beaker.
9. Dry the chloroform in an oven at 100 °C for approximately 1 h.
10. Weigh the beaker containing the lipids and record the result.
11. To calculate the percentage of lipids in the sample, use the formula below, where the weight of the lipids and the weight of the sample must be in grams (Equation 1).

Equation 1 Calculation of the percentage of total lipids by the Bligh & Dyer method (1959)

$$\%\text{Total Lipid} = \frac{\text{Weight of lipid} \times 4}{\text{Weight of sample}} \times 100$$

3.2 Hot Extraction

3.2.1 Soxhlet Method

1. Weigh 2–5 g of dry sample in a Soxhlet cartridge.
2. Place the cartridge in the Soxhlet extractor.
3. Attach the extractor to the flat-bottomed flask.
4. Fill the extractor with ethyl ether until siphoning occurs. Then, add ethyl ether until it reaches half of the cartridge.
5. Connect the system to a condenser, with the flat-bottomed flask resting on the heating plate, and turn on the heating plate between 50 and 70 °C.
6. Keep heating for 8 h (dripping 4–5 drops per s) or 16 h (dripping 2–3 drops per s).
7. After the extraction period is over, uncouple the Soxhlet extractor from the condenser and flat-bottomed flask, remove the Soxhlet cartridge, and discard the ether present in the extractor.
8. Reattach the extractor to the flat-bottomed flask, take it to the condenser and heated plate, and distill the ether content in the flat-bottomed flask (*see* **Note 8**).
9. Disconnect the extractor from the condenser and flat-bottomed flask and discard the ether in the extractor.
10. Place the flat-bottomed flask in an oven at 105 °C for approximately 1 h to evaporate any remaining ether that may have remained in the flat-bottomed flask along with the lipids.
11. Remove the flat-bottomed flask from the oven and cool it in a desiccator to room temperature (20–30 min).
12. Weigh the flat-bottomed flask containing the lipids and record the result.
13. To calculate the percentage of lipids in the sample, use the formula below, using values in grams (Equation 2).

Equation 2 Calculation of the percentage of total lipids using the Soxhlet method

$$\%\text{Total Lipid} = \frac{\text{Weight of lipid}}{\text{Weight of sample}} \times 100$$

3.3 Microwave-Assisted Extraction (MAE)

1. Weigh 0.65 g of sample in a Teflon tube and homogenize with 0.75 g of sodium sulfate.
2. Add 9 mL of ethyl acetate/methanol solution (2:1 v/v).
3. Place the tube in a microwave extractor system under magnetic stirring at 1200 rpm, with a maximum power of 400 W, and a 1-min heating ramp to reach the desired temperature and 15 min at 54 °C, totaling 16 min of irradiation. Then, cool the tubes naturally for 10 min in the equipment itself.
4. Filter the contents of the Teflon tubes into a previously weighed 15 mL bottle using qualitative filter paper containing 1 g of sodium sulfate.
5. Dry the filtrate under a nitrogen flow. Store the dried extract in a desiccator for 30 min.
6. Weight the tubes and record the result. The lipid mass represents the total lipid content of the sample, and the result can be expressed in mg/100 g of sample.

4 Conclusions

The choice of total lipid extraction method should consider the sample to be analyzed and equipment availability in the laboratory. Cold extractions do not require sophisticated equipment and glassware. They can be performed with simple glassware and standard equipment or even without equipment since the agitation in the Bligh & Dyer method can be done manually. Therefore, they are preferred extractions, even when using solvents that are toxic to the environment and require proper disposal. In addition, methods such as the Folch method are reliable and are widely used for analyzing lipids in food.

Soxhlet extraction requires specialized glassware for the analysis. Still, since it is a widely used method, most laboratories have a Soxhlet extractor, and the analysis can be performed without significant investments. However, this analysis is time-consuming, and since it is a hot extraction, it cannot be used in all samples, given the risk of oxidation of the lipids in the sample. Therefore, it is important to define the purpose of the analysis. For instance, this method is not recommended if the lipid extract will be used to determine fatty acids. On the other hand, MAE requires specific equipment that is not found in all laboratories. However, it is a quick analysis that uses a small sample and less solvent than the others, thus less harmful to the environment.

5 Notes

1. If the samples are dried before analysis, they must be thoroughly ground to form a powder. If they are moist, they must be thoroughly homogenized to ensure representativeness and correct extraction of all the contents in the sample.
2. Chloroform is a highly volatile reagent. If the vial is not well sealed, chloroform may leak, consequently causing loss of lipids. Therefore, the vials must be tested with a bit of chloroform before analysis to ensure no leakage during the procedure. Add a small amount of chloroform to the vial, cap, and shake the vial by inversion, and check if the outside is wet, indicating leakage.
3. Wash the beakers, dry them in an oven at 105 °C, and cool them in a desiccator. Weigh the beakers, note their weight, and store them in a desiccator until analysis. Always handle the beaker using tweezers to avoid contact with your hands.
4. Wash the flat-bottomed flasks, dry in an oven at 105 °C, and cool in a desiccator. Weigh the flasks and note the weight. Store in a desiccator until analysis. Always handle the flasks using tweezers to avoid contact with your hands.
5. The instrumentation is specifically developed for laboratory use, where the extraction occurs in a closed vessel and at a controlled temperature [16].
6. Wash the tubes, dry them in an oven at 105 °C, and cool them in a desiccator. Weigh the tubes and note the weight. Store them in a desiccator until analysis. Always handle the tubes using tweezers to avoid contact with your hands.
7. Perform this step quickly to avoid evaporation of the chloroform.
8. At this stage, after discarding the ethyl ether in the extractor, a small amount of ethyl ether remains in the flat-bottomed flask. This must be connected to the extractor again so that the ethyl ether is distilled into the extractor until there is no ether in the flat-bottomed flask or a small amount remains (approximately 20 mL) that can be dried in an oven.

Acknowledgments

This study was financed in part by the Coordenação de Aperfeiçoamento de Pessoal de Nível Superior—Brasil (CAPES)—Finance Code 001. LRBM thanks CNPq for the fellowship (305778/2022-6).

References

1. Orsavova J, Misurcova L, Vavra Ambrozova J et al (2015) Fatty acids composition of vegetable oils and its contribution to dietary energy intake and dependence of cardiovascular mortality on dietary intake of fatty acids. Int J Mol Sci 16:12871–12890
2. Gunay A, Shin HH, Gozutok O et al (2021) Importance of lipids for upper motor neuron health and disease. Semin Cell Dev Biol 112:92
3. Alhouayek M, Ameraoui H, Muccioli GG (2021) Bioactive lipids in inflammatory bowel diseases – from pathophysiological alterations to therapeutic opportunities. Biochim Biophys Acta Mol Cell Biol Lipids 1866:158854
4. Broadfield LA, Pane AA, Talebi A et al (2021) Lipid metabolism in cancer: new perspectives and emerging mechanisms. Dev Cell 56(10): 1363–1393
5. Jaganjac M, Cindrić M, Jakovčević A et al (2021) Lipid peroxidation in brain tumors. Neurochem Int 149:105118
6. Jiang T, Dai L, Li P et al (2021) Lipid metabolism and identification of biomarkers in asthma by lipidomic analysis. Biochim Biophys Acta Mol Cell Biol Lipids 1866:158853
7. Knox S, O'Boyle NM (2021) Skin lipids in health and disease: a review. Chem Phys Lipids 236:105055
8. Chausse B, Kakimoto PA, Kann O (2021) Microglia and lipids: how metabolism controls brain innate immunity. Semin Cell Dev Biol 112:137–144
9. Mutlu AS, Duffy J, Wang MC (2021) Lipid metabolism and lipid signals in aging and longevity. Dev Cell 56:1394–1407
10. Ooi KLM, Vacy K, Boon WC (2021) Fatty acids and beyond: age and Alzheimer's disease related changes in lipids reveal the neuro-nutraceutical potential of lipids in cognition. Neurochem Int 149:105143
11. Folch J, Lees M, Sloane GH (1957) A simple method for the isolation and purification of total lipides from animal tissues. J Biol Chem 226:497–509
12. Bligh EG, Dyer WJ (1959) A rapid method of total lipid extraction and purification. Can J Biochem Physiol 37:911–917
13. Instituto Adolfo Lutz (2008) Métodos Físico-Químicos para Análise de Alimentos
14. Hewavitharana GG, Perera DN, Navaratne SB et al (2020) Extraction methods of fat from food samples and preparation of fatty acid methyl esters for gas chromatography: a review. J Anim Sci 13:6865
15. dos Costa DSV, Bragagnolo N (2017) Development and validation of a novel microwave assisted extraction method for fish lipids. Eur J Lipid Sci Technol 119:1600108
16. Kaufmann B, Christen P (2002) Recent extraction techniques for natural products: microwave-assisted extraction and pressurised solvent extraction. Phytochem Anal 13:105–113

Chapter 4

Identification and Quantification of Fatty Acids in Meat and Fish

Celso Andrade Camargo, Lilian Regina Barros Mariutti, and Neura Bragagnolo

Abstract

The fatty acids in meat and fish, especially the n-3 and n-6 polyunsaturated fatty acids, are important from a nutritional point of view. In this sense, correctly identifying and quantifying fatty acids can be a tricky task. Nowadays, it is still quite common to find papers about food composition reporting fatty acid quantification, which does not reflect the actual food composition. This occurs when quantification is carried out using only the area percentage without considering important factors such as the lipid content of the food sample, detector response for the different fatty acids, and the use of an appropriate internal standard. Misidentification is also a common problem due to lack of standard availability or confirmation by mass spectrometry. This chapter will review the main methods for fatty acid analyses in meat and fish, emphasizing identification and calculation methods to generate reliable and accurate results.

Key words Fatty acids, Extraction, Methylation, Calculation

1 Introduction

Fatty acids are aliphatic monocarboxylic acids with different chemical structures, chain length, carbon number, position of double bonds, and additional functional groups along the carbon chain. Over 1,000 fatty acids are known, but only around 20 are found in significant amounts in oils and fats. In meat and fish, the main saturated fatty acids are palmitic, stearic, and myristic acids, while oleic and linoleic acids are the main unsaturated acids in meat, and linolenic, eicosapentaenoic (EPA), and docosahexaenoic (DHA) acids are predominant in fish (Fig. 1). The fatty acids found in animals are mainly esterified (~98%) with glycerol [1] because, in free form, they can disrupt the cell membrane organization and, therefore, are not common in living tissues.

Fatty acids in foods are analyzed mainly as methyl esters of fatty acids by gas chromatography coupled to flame ionization

Lilian Mariutti and Frank Vriesekoop (eds.), *Food Chemistry: Basic Protocols*, Methods and Protocols in Food Science, https://doi.org/10.1007/978-1-0716-4678-6_4,

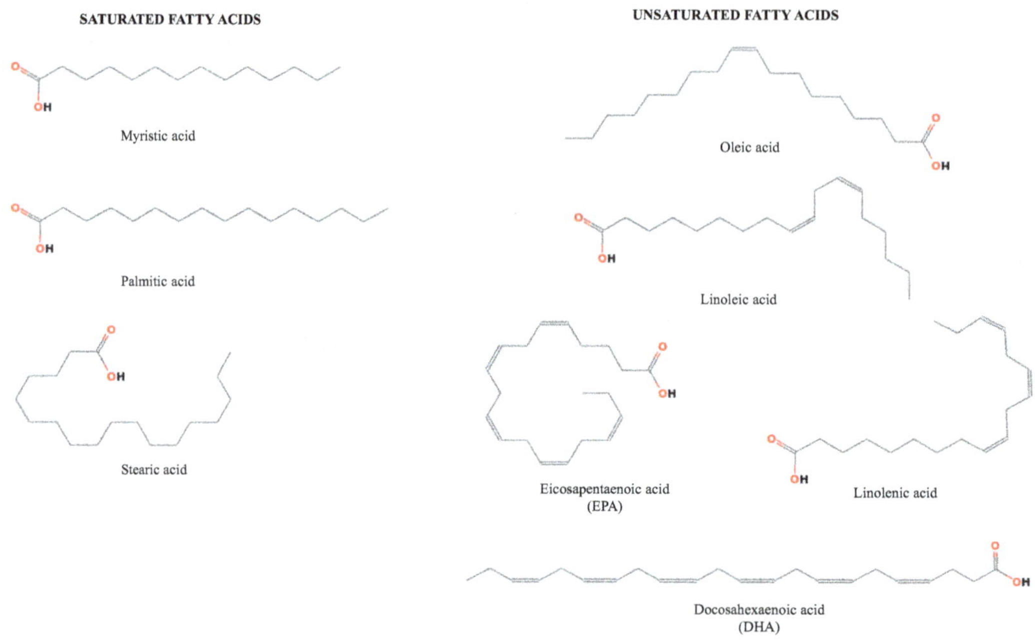

Fig. 1 Main fatty acids in meat and fish

(GC-FID) or mass spectrometry (GC-MS) detectors. The most common method of sample preparation used to determine fatty acids in meat and fish is the extraction of lipids, followed by converting fatty acids into methyl esters [2]. The lipid extraction can be done by conventional techniques, for example [3–5], or new technologies such as supercritical fluid extraction [6] pressurized liquid extraction [7], focused microwave-assisted Soxhlet extraction [4], microwave-assisted extraction [8–10], and dynamic ultrasound-assisted extraction [11], since they are more versatile, alternative, and sustainable techniques for the total extraction of lipids. The conversion of fatty acids into methyl esters involves two steps; first, the lipid is heated with NaOH in methanol, forming glycerol and free fatty acids (saponification), and secondly, the free fatty acids are then esterified with methanolic boron trifluoride (BF_3) (Fig. 2). Many procedures can be used to prepare methyl esters, and each method has advantages and limitations depending on the oil or fat to be analyzed. However, the methylation methods using methanolic BF_3 present the best results [12]. The quantification of fatty acid methyl esters (FAME) may be affected by incomplete conversion of lipids to FAME, changes in fatty acid composition during esterification, and the formation of compounds that may be mistakenly identified as fatty acids [13, 14].

To separate positional and configurational isomers of the methyl esters, fused silica capillary columns of 100 m coated with highly polar stationary phases, such as cyanoalkyl polysiloxane columns, are the most recommended [15].

Saponification

$$R-(CH_2)_n-COOH \ + \ NaOH \longrightarrow \underset{\text{Salt of free fatty acid}}{R-(CH_2)_n-COO^-Na^+} \ + \ H_2O$$

Methylation

$$R-(CH_2)_n-COOH \ + \ R'-OH \xrightarrow{BF_3} \underset{\text{FAME}}{R-(CH_2)_n-COOR'} \ + \ H_2O$$

Fig. 2 Conversion of fatty acids into methyl esters

Identification of the fatty acids in foods is carried out by comparing the retention time of the FAME in the sample with those of the standards, spiking, equivalent chain length (ECL) values [15–17], and mass spectrum. Structural information obtained by mass spectrometry, combined with parameters such as retention time, is one of the most accurate methods for identifying complex organic compounds [18–20].

The quantification of fatty acids in mg/100 g of sample is carried out using internal standards, and for accuracy, the application of correction factors for the FID response is necessary [21]. The main internal standards used are tricosanoic acid (23:0), nervonic acid (24:1n-9), nonadecanoic acid (19:0), and heneicosanoic acid (21:0). Tricosanoic acid is stable; however, it is less soluble than nervonic acid, which is an unstable compound and may suffer losses due to self-oxidation [21–23]. On the other hand, 19:0 and 21:0 improved the accuracy and precision of the analysis of PUFA compared to 23:0 [24]. In addition, other saturated fatty acids of odd carbon number in the chain, such as 11:0 and 13:0, can be used together with 19:0 or 23:0 if more than one internal standard is needed.

This chapter surveys current knowledge of fatty acids analysis in meat and fish, emphasizing identification and calculation methods to generate reliable and accurate results.

2 Material

2.1 Reagents

Analytical grade chloroform, methanol, 0.5 N sodium hydroxide in methanol, and saturated sodium chloride solution. 12% boron trifluoride (BF_3) in methanol (Sigma-Aldrich, St. Louis, MO, USA), HPLC grade isooctane, and hexane. A mixture of 37 fatty acid methyl ester standards from 4:0 to 24:0 (FAME MIX, ref. 47885-U, Supelco Co., Bellefonte, CA, USA), methyl esters of tridecanoic (13:0, 99% pure), tricosanoic (23:0, 99% pure) acids, or other appropriate FAME as internal standard.

2.2 Apparatus

1. Gas chromatograph equipped with an auto-injector system (2 μL) in split mode, operated at a split ratio of approximately 1:50, and flame ionization detector (FID), injection port temperature at 250 °C, detector temperature at 260 °C, oven temperature following temperature program: Start at 120 °C during 8 min, increasing to 160 °C at 20 °C/min, increasing from 160 to 195 °C at 3 °C/min, remaining at 195 °C for 10 min, increasing from 195 to 210 °C at 3 °C/min, and increasing from 210 to 220 °C at 35 °C/min, and after 3 min at 220 °C, it went up to 240 °C at 20 °C/min and remained for 5 min, totaling 46 min [9]. Figure 3 shows and example of a typical chromatogram obtained under the described conditions (Fig. 3).
2. Column: a fused capillary CP-SIL 88 column, 100 m × 0.25 mm × 0.20 μm (Chromopack, Middleburg, The Netherlands).
3. Carrier gas: H_2 with a minimum of 99.9% purity at 1 mL/min and linear velocity at 34 cm/s.
4. FID gas: N_2 with a minimum of 99.9% purity as the make-up gas at a flow rate of 30 mL/min, H_2 with a minimum of 99.9%

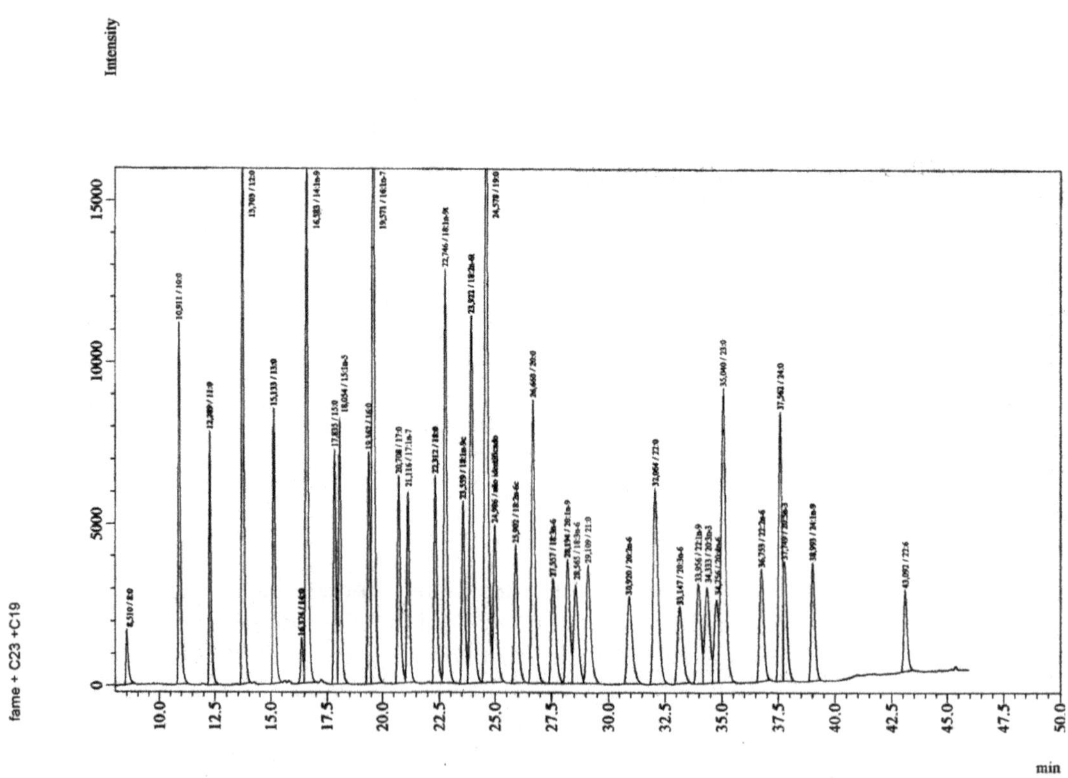

Fig. 3 Typical chromatogram of FAME obtained under the conditions described in Subheading 2.2, **item 1**

purity, and the linear velocity at 34 cm/s and synthetic air at 300 mL/min with a minimum of 99.9% purity.

5. Vials for autosampler: 2 mL, with screw cap or crimp cap.
6. Constant temperature water bath: maintained at 100 °C. Dry heater block may be used.
7. Analytical balance with an accuracy of ±0.0001 g.
8. Glass tubes: 16 × 125 mm, with leak-tight, Teflon-lined screw caps.
9. Dry nitrogen source.
10. Glassware: volumetric flasks of 25 and 100 mL; volumetric pipets of 1 and 2 mL; Pasteur pipets.

2.3 Preparation of Solutions

1. 0.5 N alcoholic sodium hydroxide: Weigh 2.0 g of NaOH in methanol and dilute to 100 mL of methanol.
2. Sodium chloride saturated solution. Weigh 36 g NaCl and dissolve in 100 mL H_2O.

2.4 Preparation of Standards

Accurately weigh 1 mg (±0.1 mg) of 13:0 and 23:0 methyl ester, internal standard (IS), and add 1 mL of isooctane. Pipette 50 μL portion into screw-cap glass tubes and evaporate the solvent under a gentle stream of nitrogen. Put the tubes in one plastic box and store them in the freezer if they are not to be used immediately.

3 Methods

3.1 Sampling and Sample Preparation

After removing some parts, such as the skin, cartilage, and connective tissue in fish or the superficial fat in meat, the samples are homogenized for 1 min in a multiprocessor until obtaining an homogenous paste.

3.2 Lipid Extraction

Lipids can be extracted by conventional methods, such as Folch, Lees, and Stanley (1957) or Bligh & Dyer (1959), which use a mixture of chloroform and methanol as the main extracting solvents. Lipid content can be gravimetrically determined. Lipids can also be extracted by microwave-assisted extraction [8, 9]. For more details, *see* Chapter 3.

3.3 Saponification, Methylation, and Analysis

Accurately weigh 25 mg of the lipids in a 25 mL glass tube, add 1.5 mL 0.5 N methanolic NaOH, flux with nitrogen for 1 min, and cap. Vortex for 1 min and heat for 5 min at 100 °C. Cool in an ice bath, add 2 mL BF_3 in methanol, flux with nitrogen gas for 1 min, cap tightly, vortex for 1 min, and heat for 30 min at 100 °C [21]. Cool to 30–40 °C, add 1 mL isooctane, flux with nitrogen gas for 1 min, cap, and shake vigorously for 30 s while still warm. Immediately add 5 mL saturated NaCl solution, flux with nitrogen gas for 1 min, cap, vortex for 2 min, and cool to room temperature. When the isooctane layer separates from the aqueous lower phase, transfer the isooctane to a glass tube containing the internal standard, flux with nitrogen gas for 1 min, and cap.

Extract the aqueous lower phase a second time with 1 mL isooctane. Combine isooctane extracts, concentrate in a stream of dry nitrogen, and add 1 mL of hexane.

Inject 2 μL into the GC apparatus.

3.4 Identification of Peaks

The identification of the fatty acids is made by the retention times of a known composition of fatty acid methyl ester standards peaks compared with the fatty acid methyl ester peaks of the samples under the same operating conditions.

3.5 Correction Factors

Each fatty acid methyl ester presents a differential response in the FID; therefore, it is necessary to use a correction factor for the methyl esters in relation to a specific internal standard that should be a methyl ester. Two factors are needed to correct the differential response of the detector: (1) the experimental or empirical correction factor, the value of which is determined experimentally, and (2) the theoretical correction factor, determined theoretically considering the number of active carbons (C*).

3.5.1 Experimental Correction Factor (ECF)

A solution must contain a known quantity of the methyl ester and the internal standard to determine the experimental or empirical correction factor. The solution is injected into the GC-FID, and considering the area and mass of the methyl esters, the experimental factor for the different esters is determined according to the equation:

$$\mathrm{ECF} = \frac{\mathrm{MX} \times \mathrm{AIS}}{\mathrm{MIS} \times \mathrm{AX}}$$

where:

ECF = experimental correction factor

MX = mass of methyl ester X

AX = area of methyl ester X

MIS = mass of internal standard

AIS = area of internal standard

For example, the methyl ester of palmitic acid (16:0) was used to verify the detector response for a saturated fatty acid in relation to the internal standard (23:0). The result obtained was 1.004 ± 0.03 [23, 24], similar to the theoretical correction factor (1.00, Table 1), demonstrating that the GC is perfectly working under the set conditions. Suppose the experimental correction factor differs from the theoretical correction factor. In that case, it indicates that the detector is not operating at optimum linearity or that the

Table 1
Correction factor and conversion factor for methyl ester to fatty acid

FAME	MW	Correction factor	Conversion factor
4:0	102.13	1.51	1.137
6:0	130.19	1.28	1.107
8:0	158.24	1.17	1.088
9:0	172.27	1.13	1.081
10:0	186.30	1.10	1.075
11:0	200.32	1.08	1.069
12:0	214.35	1.06	1.065
13:0	228.37	1.04	1.061
14:0	242.40	1.02	1.057
15:0	156.42	1.01	1.089
16:0[a]	270.46	1.00	1.051
17:0	284.49	0.99	1.049
18:0	298.52	0.97	1.046
19:0	312.52	0.97	1.044
20:0	326.57	0.97	1.042
21:0	340.57	0.96	1.041
22:0	354.62	0.95	1.039
23:0	368.62	0.95	1.037
24:0	382.68	0.94	1.036

[a]Reference: all other correction factors used in the calculation are relative to 16:0

methyl ester preparation or injection technique was faulty [25]. In this case, optimizing the equipment before using the theoretical correction factor will be necessary.

3.5.2 Theoretical Correction Factor (TCF)

The theoretical correction factor is calculated based on the fact that the FID responds to ions generated by the combustion of the C-H components of the molecule and does not respond to the C=O component.

Thus, a theoretical factor equal to unity is assigned for a given methyl ester, for example, for methyl palmitate, and the correction factors for other methyl esters can be calculated. Short-chain length fatty acids will require a relatively higher response factor than the unit, considering that they have fewer C-H numbers in the molecule [25]. For example, methyl palmitate (MW = 270.46, Table 1) was used as a reference to calculate the theoretical correction factor for methyl decanoate (MW = 186.30, Table 1). Therefore, it will be necessary to calculate the relative percentage of active carbon atoms (C*) for each methyl ester, remembering that the COO group (MW = 44.00) does not present a response in the detector, so only C* atoms will be considered. The C* atoms in the methyl palmitate are 192.16 (16 carbon atoms × MW of C = 12.01), and the C* atoms of methyl decanoate are 120.10 (10 carbon atoms × MW of C = 12.01). Calculating the percentage of C* of the methyl palmitate, we obtain the value of 71.04 and the methyl decanoate of 64.46. Dividing the percentage of C* of the methyl palmitate (71.04, reference, correction factor = 1.00) by the percentage of the methyl decanoate (64.46), we obtain the value of 1.10 (Table 1). This value will be used in the quantitative determination of decanoic acid. The correction factor for the other methyl esters will be calculated similarly. Although the values in Table 1 have been calculated for methyl esters, these values will be used to calculate the respective fatty acids since the methyl esters will be converted into fatty acids using the methyl ester to fatty acid conversion factor (Subheading 3.6). The theoretical correction factor can be obtained using other fatty acid methyl esters as a reference, for example, methyl stearate (18:0), with this methyl ester being assigned the unit value, and the calculations are carried out in the same way.

3.6 Conversion Factor from Methyl Ester to Fatty Acid

In the analysis of fatty acids by gas chromatography, the components injected into the GC are the methyl esters of fatty acids since they have a lower boiling point than the respective fatty acids. Thus, it is necessary to use a conversion factor to correct the differential response of the detector between a methyl ester and its corresponding fatty acid. The difference in the molecular structure of a methyl ester of a given fatty acid is that it has a methyl group (CH_3). Consequently, the methyl ester will present a greater response in the detector in relation to the corresponding fatty acid since the methyl group of the methyl ester will contribute to an increase in the number of active carbon (C*) in relation to the fatty acid. The conversion factor is calculated by dividing the methyl ester's molecular weight by the fatty acid's molecular weight.

3.7 Calculation

The quantification is made by internal standardization using the methyl esters of tricosanoic and tridecanoic acids as the internal standards, the theoretical correction factor, and the conversion factor from methyl ester to fatty acid. The results obtained are expressed in mg/100 g of the sample.

$$\text{FAME}\left(\frac{\text{mg}}{\text{g}}\right) = \frac{\text{mass of IS (mg)} \times \text{area of FAME}}{\text{area IS} \times \text{mass of sample (g)}}$$

Multiplying FAME by a correction factor (theoretical detector response factor for each fatty acid) and conversion factor (for methyl ester to fatty acid), according to Table 1, results in the concentration of fatty acids in mg/g of oil. The results can also be calculated for 100 g (just multiply by 100).

$$\mathrm{FA}\left(\frac{\mathrm{mg}}{\mathrm{g}}\ \text{of oil}\right) = \mathrm{FAME} \times \mathrm{TCF} \times \text{conversion factor}$$

If this result is multiplied by the lipid content, the FA can be expressed as mg/g of sample.

References

1. Scringeour C (2005) Chemistry of fatty acids. In: Bailey's industrial oil and fat products, 6th edn. Wiley, Hoboken
2. Metcalfe LD, Schmitz AA, Pelka JR (1966) Rapid preparation of fatty acid esters from lipids for gas chromatographic analysis. Anal Chem 38(3):514–515
3. Bligh EG, Dyer WJ (1959) A rapid method of total lipid extraction and purification. Can J Physiol Pharm 37:911–917
4. AOCS (1997) Official methods and recommended practices of the American Oil Chemists' Society, 5th edn (Firestone D ed). American Oil Chemists' Society Press, Champaign. Official Method Ce 1b-89
5. Priego-Capote F, Luque de Castro MD (2005) Focused microwave-assisted Soxhlet extraction: a convincing alternative for total fat isolation from bakery products. Talanta 65:98–103
6. Folch J, Lees M, Stanley S (1957) A simple method for the isolation and purification of total lipids from animal tissues. J Biophys Chem 226:497–509
7. Servaes K, Maesen M, Prandi B et al (2015) Polar lipid profile of nannochloropsis oculata determined using a variety of lipid extraction procedures. J Agric Food Chem 63:3931–3941
8. Ramalhosa MJ, Paıga P, Morais S et al (2012) Lipid content of frozen fish: comparison of different extraction methods and variability during freezing storage. Food Chem 131: 328–336
9. Medina AL et al (2015) Rapid microwave assisted extraction of meat lipids. Food Res Int 78:124–130
10. Costa DSV, Bragagnolo N (2017) Development and validation of a novel microwave assisted extraction method for fish lipids. Eur J Lipid Sci Technol 119:1–10. https://doi.org/10.1002/ejlt.201600108
11. Ruiz-Jimenez J, Luque de Castro MD (2004) Forward-and back dynamic ultrasound-assisted extraction of fat from bakery products. Anal Chim Acta 502:75–82
12. Craske JD (1993) Separation of instrumental and chemical errors in the analysis of oils by gas chromatography—a collaborative evaluation. J Am Oil Chem Soc 70:325–334
13. Shanta NC, Napolitano GE (1992) Gas chromatography of fatty acids. J Chromatogr B 624:37–51
14. Brondz I (2002) Development of fatty acid analysis by high-performance liquid chromatography, gas chromatography, and related techniques. Anal Chim Acta 465:1
15. Kramer JKG, Brian Blackadar C, Zhou J (2002) Evaluation of two GC columns (60-m SUPELCOWAX 10 and 100-m CP sil 88) for analysis of milkfat with emphasis on CLA, 18: 1, 18: 2 and 18: 3 isomers, and short-and long-chain FA. Lipids 37(8):823–835
16. Bragagnolo N, Rodriguez-Amaya DB (2001) Total lipid, cholesterol, and fatty acids of farmed freshwater prawn (Macrobrachium rosenbergii) and wild marine shrimp (Penaeus brasiliensis, Penaeus schmitti, Xiphopenaeus kroyeri). J Food Compos Anal 14:359–369
17. Bragagnolo N, Rodriguez-Amaya DB (2002) Simultaneous determination of total lipid, cholesterol and fatty acids in meat and back fat of suckling and adult pigs. Food Chem 79:255–260
18. Gutnikov G (1995) Fatty acid profiles of lipid samples. J Chromatogr B 671:71–89
19. Woollard PM, Mallet AI (1984) Lipoxygenase products—a novel gas—chromatography mass—spectrometric assay for monohydroxy fatty acids. J Chromatogr 306:1–21

20. Kuksis A, Myher JJ (1995) Application of tandem mass—spectrometry for the analysis of long-chain carboxylic acids. J Chromatogr B 671:35–70
21. Joseph JD, Ackman RG (1992) Capillary column gas chromatographic method for analysis of encapsulated fish oils and fish oil ethyl esters: collaborative study. J AOAC Int 75:488–506
22. Shantha NC, Ackman RG (1990) Nervonic acid versus tricosanoic acid as internal standards in quantitative gas chromatographic analyses of fish oil longer-chain n—3 polyunsaturated fatty acid methyl esters. J Chromatogr B Biomed Sci Appl 533:1–10
23. Schreiner M (2005) Quantification of long chain polyunsaturated fatty acids by gas chromatography—evaluation of factors affecting accuracy. J Chromatogr A 1095:126–130
24. Mazalli MR, Bragagnolo N (2007) Validation of two methods for fatty acids analysis in eggs. Lipids 42:483–490
25. Craske JD, Bannon CD (1987) Gas liquid chromatography analysis of the fatty acid composition of fats and oils; a total system for high accuracy. J Am Oil Chem Soc 64:1413–1417

Chapter 5

Analysis of Food Sterols by Gas Chromatography

Mara Mandrioli, Dario Mercatante, and Maria Teresa Rodriguez-Estrada

Abstract

Sterols are minor components of food lipids, often used as markers for confirming the authenticity of oils and food, so their correct characterization and quantification are very relevant for detecting adulterations. In addition, plant sterols can be used for the formulation of functional foods due to their cholesterol-lowering properties, so an accurate analytical determination of this compound class is also required for food labeling. Here, we describe the analytical determination of food sterols, including the various preparative steps (extraction, cold saponification, purification) and the gas chromatographic (GC) analysis with different detectors (flame ionization detector (FID) and mass spectrometer (MS)).

Key words Sterol, Lipid extraction, Unsaponifiable extraction, Thin-layer chromatography fractionation, Gas chromatography, Flame ionization detector, Mass spectrometer

1 Introduction

Sterols are isoprenoid-derived lipids that have an important role in both animal and plant regulation processes, particularly the control of fluidity, permeability, and functionality of cell membranes [1]. Sterols have a common structure based on the cyclopentanoperhydrophenanthrene ring system with a hydroxyl group at position C3 and a branched or unbranched side chain of variable length and unsaturation degree at C17. Cholesterol is the major animal sterol, whereas β-sitosterol, campesterol, and stigmasterol are the most abundant plant sterols. The main dietary sources of sterols are oils, animal and vegetable fats, spreads, nuts, and seeds. To determine sterols in food, it is first necessary to choose the most suitable method for extracting the lipid fraction according to the chemical characteristics of the food matrix. It is advisable to opt for a nondestructive method that is able to provide a representative amount of the food lipid fraction. Still, at the same time, it should ensure the lack of selectivity toward specific components and the production of unwanted artifacts or reactions.

Lilian Mariutti and Frank Vriesekoop (eds.), *Food Chemistry: Basic Protocols*, Methods and Protocols in Food Science,
https://doi.org/10.1007/978-1-0716-4678-6_5,

The method described below foresees the extraction of food lipids with a chloroform-methanol mixture, followed by cold saponification, extraction of the unsaponifiable matter, and purification of the sterol fraction. Finally, the quantitative determination of sterols is performed by gas chromatography (GC), coupled to a flame ionization detector (FID) or a mass spectrometry (MS) detector.

2 Materials

2.1 Laboratory Equipment

1. Laboratory balance: analytical balance with sensitivity ±0.1 mg and technical balance with sensitivity ±0.1 g.
2. Common laboratory glassware, bottles with sealing caps, cylinders, glass separating funnels, glass funnels, vacuum flasks, Buchner funnels, rotavapor flasks, Erlenmeyer flasks, volumetric flasks, spatulas, micro syringes, pipettes, test tubes, elution chamber for thin-layer chromatography (TLC).
3. Turbine homogenizer.
4. Oven with temperature control.
5. Refrigerator set at a temperature of 4 °C.
6. Vacuum pump.
7. Rotary evaporator.
8. Heating block.
9. Vortex stirrer.
10. Tilting agitator.
11. Centrifuge.
12. Ultraviolet (UV) lamp for TLC measurements at 254 nm.
13. Gas chromatograph with split-spitless injection system and various detectors (FID or MS).

2.2 Solvents, Standards, and Materials

Prepare all solutions using analytical grade solvents and reagents. Prepare and store all reagents at room temperature (unless otherwise stated). Rigorously stick to general safety instructions for using solvents and reagents (work under a fume hood and follow technical data sheets), using the required personal protection equipment (PPE). Carefully follow regulations for waste disposal.

1. Chloroform:methanol 1:1 (v/v) mixture. Prepare a mixture of chloroform (CAS No. 67-66-3) and methanol (CAS No. 67-56-1) in a 1:1 (v/v) ratio using graduated cylinders. Store the solution in a closed and correctly labeled bottle.
2. 1 M potassium chloride solution. Weigh 74.55 g of potassium chloride (CAS No. 7447-40-7), dissolve it in distilled water

(CAS No. 7732-18-5), and make up to volume in a 1000 mL volumetric flask.

3. Granular anhydrous sodium sulfate (CAS No 7757-82-6).
4. Qualitative filter paper for rapid filtration (Grade 1).
5. *n*-Hexane:isopropanol 4:1 (v/v) mixture. Prepare a mixture of *n*-hexane (CAS No. 110-54-3) and isopropanol (CAS No. 67-63-0) in a 4:1 (v/v) ratio using graduated cylinders. Store the solution in a closed and correctly labeled bottle.
6. *n*-Hexane:isopropanol 3:2 (v/v) mixture. Prepare a mixture of *n*-hexane (CAS No. 110-54-3) and isopropanol (CAS No. 67-63-0) in a 3:2 (v/v) ratio using graduated cylinders. Store the solution in a closed and correctly labeled bottle.
7. *n*-Hexane:diethyl ether 65:35 (v/v) mixture. Prepare a mixture of *n*-hexane (CAS No. 110-54-3) and diethyl ether (CAS No. 60-29-7) in a 65:35 (v/v) ratio using graduated cylinders. Store the solution in a closed and correctly labeled bottle.
8. 2 N potassium hydroxide solution in methanol. Weigh 112.2 g of potassium hydroxide (ACS reagent, ≥85%, pellets, CAS No. 1310-58-3), dissolve it in methanol (CAS No. 67-56-1), and make up to volume in a 1000 mL volumetric flask (*see* **Note 1**).
9. 0.5% potassium hydroxide solution. Weigh 5 g of potassium hydroxide (ACS reagent, ≥85%, pellets, CAS No. 1310-58-3), dissolve it in distilled water, and make up to volume in a 1000 mL volumetric flask.
10. 0.2 M potassium hydroxide solution in methanol. Weigh 11.22 g of potassium hydroxide (ACS reagent, ≥85%, pellets, CAS No. 1310-58-3), dissolve it in methanol, and make up to volume in a 1000 mL volumetric flask (*see* **Note 1**).
11. 20% (w/v) sodium chloride solution. Weigh 200 g of sodium chloride (CAS No. 7647-14-5), dissolve it in distilled water, and make up to volume in a 1000 mL volumetric flask.
12. Silica gel 60 TLC plates (20 cm × 20 cm × 0.25 mm film thickness).
13. 0.2% 2′,7′-dichlorofluorescein (w/v) solution in absolute ethanol. Weigh 2 g of 2′,7′-dichlorofluorescein (CAS No. 76-54), dissolve it in absolute ethanol (CAS Number 64-17-5), and make up to volume in a 1000 mL flask (*see* **Note 2**).
14. Silylation mixture. Prepare the derivatization mixture with pyridine (CAS No. 110-86-1): hexamethyldisilazane (CAS No. 999-97-3): trimethylchlorosilane (CAS No. 75-77-4) in a 5:2:1 (v/v/v) ratio, adding the reagents in this decreasing concentration order. Store the solution in a fridge at 4 °C in a closed, correctly labeled bottle (*see* **Note 3**). Before use, bring

the solution to room temperature and aliquot it into the individual samples, about 500 μL, using a glass Pasteur pipette.

15. Internal standard (IS) solution. Prepare a 2 mg/mL solution of 5α-cholestan-3β-ol (CAS No. 80-97-7) (v/v). Weigh 100 mg of 5α-cholestan-3β-ol in a 50 mL volumetric flask and make up to volume with chloroform (CAS No. 67-66-3). Store the solution in a closed and correctly labeled bottle in the freezer at −18 °C until use. Before use, thaw the solution and bring it to room temperature, aliquot it into the individual samples using a glass microsyringe.
16. Sterol standard solutions for TLC. Prepare a 5 mg/mL β-sitosterol (CAS No. 83-46-5) (v/v) solution. Weigh 25 mg of β-sitosterol in a 5 mL volumetric flask and make up the volume with chloroform (CAS No. 67-66-3). Store the solution in a closed and correctly labeled bottle in the freezer at − 18 °C until use. Before use, thaw the solution and bring it to room temperature (*see* **Note 4**).
17. Sterol standard solutions for GC analysis. The following sterol standards (all commercially available) can be used for their quantification depending on the sterol composition of the food matrix: β-sitosterol (CAS no. 83-46-5), β-sitostanol (CAS no. 83-45-4), Δ^5-avenasterol (CAS no. 18472-36-1), brassicasterol (CAS no. 474-67-9), campesterol (CAS no. 474-62-4), campestanol (CAS no. 474-60-2), stigmasterol (CAS no. 83-48-7), cholesterol (CAS no. 57-88-5), and Δ^7-avenasterol (CAS no. 23290-26-8). For each sterol, prepare a 1 mg/mL (v/v) solution in a volumetric flask and make up the volume with chloroform (CAS No. 67-66-3). Store the solution in a closed and correctly labeled bottle in the freezer at − 18 °C until use. Before use, thaw the solution and bring it to room temperature (*see* **Note 5**).
18. Nitrogen (CAS No. 07727-37-9) (*see* **Note 6**).

3 Methods

3.1 Extraction of the Lipid Fraction

The choice of the lipid extraction method depends on the type of matrix and must be able to provide a representative lipid fraction. In the case of milk, where the fat is organized as globules protected by a lipoprotein membrane, it is difficult for solvents to permeate therein, so the Rose-Gottlieb et al. method [2] is usually recommended, as it uses NH_4 and ethanol, which help denaturing the lipoprotein membrane and releasing fat from the globules. To extract the lipid fraction from vegetable samples (such as seeds), the Hara and Radin method [3] is recommended since isopropanol minimizes the enzymatic degradation of lipids in plant tissues.

Furthermore, the solvents used in the Hara and Radin [3] method are less toxic than those used with the Folch et al. method [4].

On the other hand, the Folch et al. method [4] is suggested for the extraction of food minor lipid components by exploiting the combined action of solvents with diverse polarity, homogenization, and heating; it is able to extract the intracellular fat as well, thus providing a more representative sample of the lipid fraction. The method described as follows is a modified version [5] of the Folch et al. method [4]:

1. Weigh a suitable amount of sample (10–25 g for food in general) according to the food matrix and the estimated fat to be present therein. The sample must be as homogeneous as possible (previously ground if necessary) and representative of the food to be analyzed.
2. Transfer the sample to a bottle with a tight-fitting cap and add 200 mL of a chloroform:methanol solution (1:1, v/v).
3. Mix it with a dispersion homogenizer for 3 min at a speed that does not cause excessive overheating.
4. Place the bottle in a thermostated lab oven at 60 °C for 20 min. Once the first 10 min have passed, vent the bottle under the suction hood and reposition it in the oven.
5. Remove the bottle from the oven, allow it to cool down, and add 100 mL of chloroform before homogenizing it again for 2 min with the turbine homogenizer.
6. Filter the bottle content with a Buchner funnel under vacuum using a fast filtration paper filter.
7. Transfer the filtrate to a bottle with a tight-fitting cap and add 100 mL of a 1 M aqueous solution of KCl.
8. Shake the bottle very vigorously for 1 min and keep it overnight in a refrigerator at 4 °C.
9. Transfer the contents of the bottle to a separating funnel, allow the two phases to separate very well, and transfer the organic phase (lower one) to a clean bottle.
10. Add 30 g of anhydrous sodium sulfate and place the bottle in the refrigerator at 4 °C for 2 h to allow removal of residual water from the organic phase.
11. Filter through a folded paper filter containing 10–20 g of anhydrous sodium sulfate and collect the filtrate in a rotavapor flask.
12. Evaporate the excess solvent with a rotary evaporator, keeping the temperature at 40 °C in the water bath.
13. Eliminate traces of residual solvent inside the fat extract by placing the balloon under a gentle nitrogen stream.

14. Transfer the fat obtained to a test tube using small aliquots of a mixture of *n*-hexane:isopropanol (4:1, v/v) and store it in the freezer at −20 °C until analysis.

3.2 Cold Saponification

To extract sterols, the lipid extract is first subjected to cold saponification to breakdown the ester bond of esterified sterols and thus release free sterols without artifact formation [6]:

1. Weigh 250 mg of perfectly homogeneous anhydrous fat (*see* **Note 7**) into a 20 mL glass tube.
2. Add 100 μL of internal std. solution (5α-cholestan-3β-ol, corresponding to 0.2 mg) (*see* **Note 8**) and evaporate the solvent with a slight nitrogen flow.
3. Add 10 mL of 2 N methanolic solution of KOH (*see* **Note 9**), and stir it very vigorously in the vortex until a light emulsion appears (*see* **Note 10**).
4. Place the test tube on a horizontal shaker for 18–24 h to allow the saponification reaction to occur.

3.3 Extraction of the Unsaponifiable Fraction

Once the lipid saponification is completed, it is necessary to extract the unsaponifiable matter, which includes the sterols [6]:

1. Place the content of the saponified solution in a 250 mL separating funnel, add 10 mL of distilled water and 20 mL of diethyl ether, stir it with alternated venting of the funnel, and then allow the two phases to separate.
2. Collect the aqueous phase (lower one) in a second separating funnel, perform an extraction with 10 mL of diethyl ether, remove the ether phase, and transfer it to a clean separated funnel together with the first ether extract. This procedure should be done twice to extract all the unsaponifiable matter that may have remained in the aqueous phase.
3. Pull the three ether fractions together in a separating funnel and carry out two washes with 10 mL of 0.5 N KOH aqueous solution, followed by two more washes with 10 mL of distilled water, until a pH close to neutrality (7–8) is reached.
4. The ethereal extract thus obtained is anhydrified by adding 5 g of anhydrous sodium sulfate and leaving it standing for 2 h in a refrigerator at 4 °C to allow the removal of residual water from the organic phase.
5. Filter through a folded paper filter containing a small amount of anhydrous sodium sulfate and collect the filtrate in a rotavapor flask.
6. Remove the diethyl ether using a rotary evaporator with a water bath set at 40 °C.

7. Transfer the unsaponifiable fraction into a test tube using 1 mL of *n*-hexane:isopropanol (4:1, v/v), and store it in the freezer at −20 °C until analysis.

3.4 Purification by Preparative Silica Thin-Layer Chromatography (TLC)

Since the cold saponification treatment followed by liquid/liquid extraction is able to pull out all the unsaponifiable matter, it is often necessary to run a TLC step to purify the sterol fraction prior to the chromatographic analysis, especially if the unsaponifiable fraction is complex and abundant and/or the sterol analysis is performed by GC-FID. If the analytical determination of sterols is performed by GC-MS, the TLC purification step can be avoided.

1. Immerse the silica TLC silica gel 60 in a 0.2 M KOH methanolic solution for 20 s, leave it drying under a suction hood for 2 h, and then dry it in an oven at 100 °C for 1 h. Remove the TLC from the oven, cover it with aluminum foil, and place it in a lab dryer containing granular silica gel with cobalt-free indicator (CAS No. 7631-86-9), protecting the TLC from light and humidity (*see* **Note 11**).
2. Prepare the TLC development chamber with 100 mL of an *n*-hexane:diethyl ether solution (65:35, v/v), and keep it for 1 h in the refrigerator at 4 °C to allow the liquid/vapor equilibrium to be reached within the chamber.
3. Dry the unsaponifiable matter, dissolve it in 100 μL of chloroform, and load it as a band along the starting line of the TLC plate (*see* **Note 12**). A spot of sterol standard solution (2–5 μL of a 5 mg/mL solution) is usually loaded in the same TLC plate as a reference point for the qualitative identification of the sterol band.
4. Place the TLC plate inside the development chamber, rapidly close it, and put it back in the fridge at 4 °C to allow vertical development of the TLC plate for 10 cm from the fixed starting baseline.
5. Once the solvent mixture has reached the predetermined TLC frontline (10 cm of full development), remove the TLC plate from the chamber, let it dry under a fume hood, and spray it with the 2′,7′-dichlorofluorescein ethanolic solution (0.2%, p/v).
6. Place the sprayed TLC plate under a UV light lamp at a wavelength of 254 nm to visualize and identify the sterol band.
7. Collect the sterol band by carefully scraping it off with a spatula and put it into a vial.
8. Add 1 mL of chloroform to extract sterols from the silica, stir it in the vortex for 30 s, and pick the extract up with a Pasteur pipette (*see* **Note 13**). Carry out the whole extraction procedure three times.

9. Evaporate the chloroform under nitrogen flow at room temperature, dissolve the extracted sterols in a solution of *n*-hexane:isopropanol (3:2, v/v), and store it in the freezer at −20 °C until the next analysis.

3.4.1 Various TLC Applications

To carry out a qualitative and quantitative assessment of the sterol composition of the matrix of interest, it could be useful to obtain a preliminary picture of the total lipid profile by using TLC.

Figure 1 shows a TLC of three samples of tomato seed oil, where band 4 corresponds to free sterols. In this case, the TLC plate was sprayed with a saturated solution of $K_2Cr_2O_7$ in H_2SO_4 (80%) and then charred at 130 °C/25 min in an oven. This is a destructive visualization method, which could be very helpful for obtaining an overall picture of the sample as it allows to evidence all organic compounds present therein.

Figure 2 shows a TLC of industrial sterol-based ingredients used to formulate functional foods; these ingredients were provided by different industrial suppliers, which used diverse sterol sources, concentrations and formulations (water or lipid soluble).

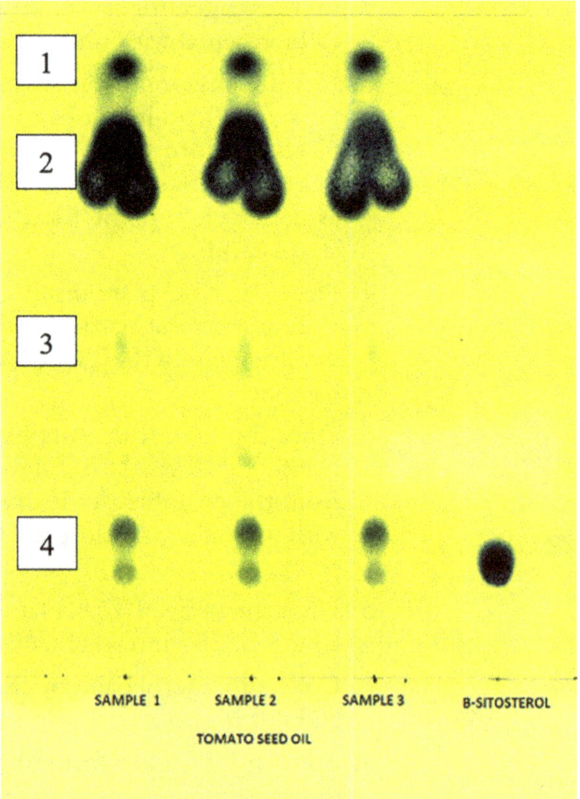

Fig. 1 Image of a TLC plate of three different samples of tomato seed oils (indicated as samples 1, 2 and 3 at the bottom of the TLC plate) and a spot of a standard solution of β-sitosterol. Four main bands were detected in the samples, which corresponded to: **1**- carotenes, **2**- triglycerides, **3**-tocopherols, **4**- free sterols

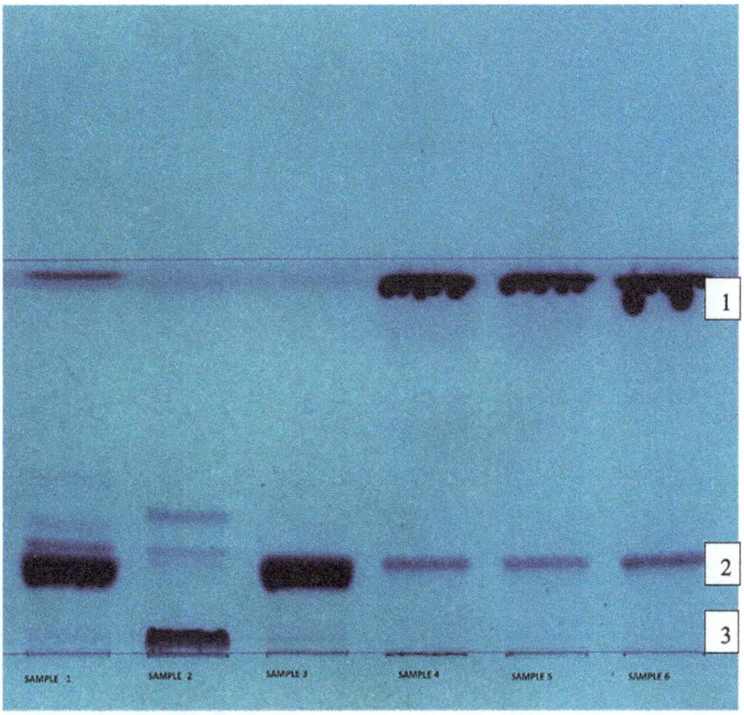

Fig. 2 Image of a TLC plate of sterol-based ingredients (labeled as samples 1-6 at the bottom of the TLC plate), used as pre-screening method of their lipid composition before further analyses. Three main bands were detected in the samples, which corresponded to: 1- esterified sterols, 2- free sterols, 3- monoglycerides

In this case, the TLC helps depicting the lipid composition of such ingredients before proceeding with further analyses (*see* **Note 10**). An aliquot of the ingredient is dissolved in chloroform, loaded onto the silica TLC plate, and eluted with a mixture of *n*-hexane:diethyl ether (50:50, v/v). As in Figure 1, the TLC plate was sprayed with a saturated sulfochromic solution and then charred. The TLC plate evidences that ingredient samples 1 and 3 contain only free sterols (band 2), that ingredient sample 2 does not have sterols, and that the other ingredients (samples 4–6) are mostly formulated with esterified sterols (band 1). Therefore, it can be deduced that to determine the total sterol content of the ingredients, it is not necessary to perform a cold saponification for samples 1 and 3. In contrast, it is indispensable for samples 4–6 to carry out the saponification treatment to free all the sterols from the esterified molecules.

TLC elution of the unsaponifiable matter could help verifing the yield of the saponification reaction (*see* **Note 10**). An aliquot of the unsaponifiable fraction is dissolved in chloroform, loaded onto the silica TLC plate, eluted with a mixture of *n*-hexane:diethyl ether

Fig. 3 Image of a TLC plate of the unsaponifiable fractions from two sterol-based ingredients (indicated as samples 1 and 2 at the bottom of the TLC plate). Two main bands were detected in the samples, which corresponded to: 1-esterified sterols, 2-free sterols

(50:50, v/v), sprayed with a saturated sulfochromic solution and then charred. Figure 3 shows that, under the same treatment conditions, sample 1 is completely saponified as only free sterols (band 2) are detected, while sample 2 is partially saponified since esterified sterols (band 1) are still present.

3.5 Derivatization

To make sterols more volatile and improve their peak resolution and detectability, they are converted into trimethylsilyl derivatives as follows:

1. Dry the purified sterol band with a weak nitrogen flow, add 1 mL of the derivatization solution (pyridine:hexamethyldisilazane:trimethylchlorosilane, 5:2:1, v/v/v) [7], and close the vial.
2. Place the vial in a thermoblock at 40 °C for 20 min, and stir it with a vortex after 10 min.

3. Once the reaction is completed, dry the derivatized sample with a weak nitrogen flow, add 500 μL of *n*-hexane (*see* **Note 14**), centrifuge, and inject 1 μL of the supernatant into the GC-FID or GC-MS.

3.6 Gas Chromatographic Analysis

Sterol determination can be performed either by GC-FID (Figs. 4 and 5) or GC-MS.

3.6.1 GC-FID Analysis

For the separation and detection of sterols, inject 1 μL of the sample solution into a GC-FID under the following conditions:

- Capillary GC column: apolar (100% dimethylpolysiloxane or 5% phenyl-95% dimethylpolysiloxane), 30 m length × 0.32 mm internal diameter (I.D.) × 0.25 μm film thickness.
- Oven program: from 265 to 280 °C at 0.5 °C/min and then from 280 to 325 °C at 4 °C/min, with final isotherm at 325 °C for 30 min.
- Carrier gas: Helium (He).
- Pressure: 75 kPa.
- Flow: 2.88 mL/min.
- Split: 43.7 mL/min.
- Split ratio: 1:15.
- Injector temperature: 325 °C.
- Detector temperature: 325 °C.

GC-FID Identification and Quantification

The single sterol peaks are identified by comparing their retention times with those of the corresponding commercial sterol standards. In case of partial overlapping with other matrix components, it is also possible to spike the sample with solutions of commercial sterol standards.

The content of each single sterol (in mg/100 mg of fat) is calculated by applying the internal standard method (*see* **Note 8**) according to the following formula:

$$C_x = \frac{A_x}{A_{\text{is}}} \cdot \frac{C_{\text{is}}}{P} \cdot 100$$

where:

A_x = the peak area of sterol x

A_{is} = the area of the internal standard (5α-cholestan-3β-ol or betulinol)

C_{is} = the amount of the internal standard added (5α-cholestan-3β-ol or betulinol) in mg

P = sample weight in mg

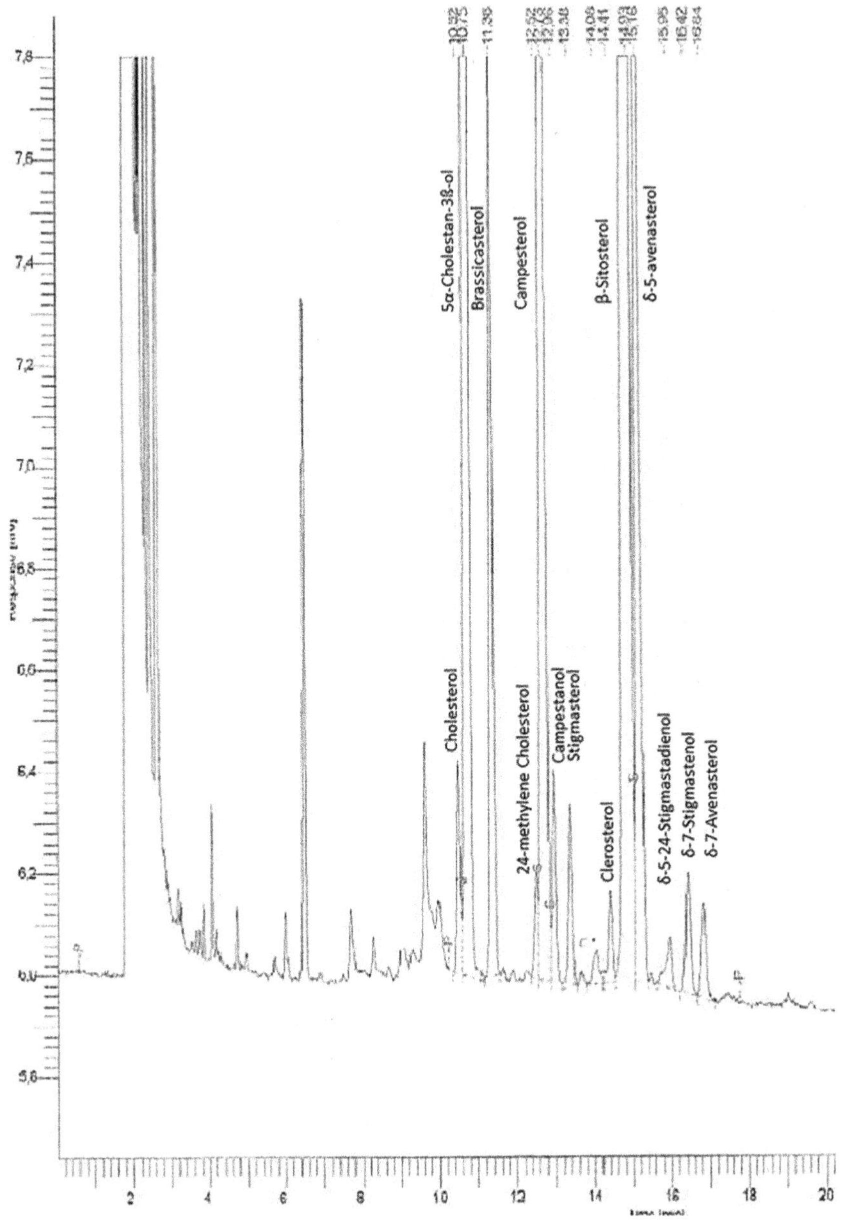

Fig. 4 GC-FID trace of TMS derivatives of sterols from an industrial sterol-based ingredient obtained with a 30-m CP-Sil 5 CB column (100% dimethylpolysiloxane) under the analytical conditions indicated in Subheading 3.6.1. Peaks are annotated with sterol names, indicating their presence and relative abundance in the sample

Generally, the FID provides a fairly linear response for sterols. However, to perform a quantification that also considers the differences in chromatographic response for the single sterol molecules with respect to the internal standard, it is possible to introduce a

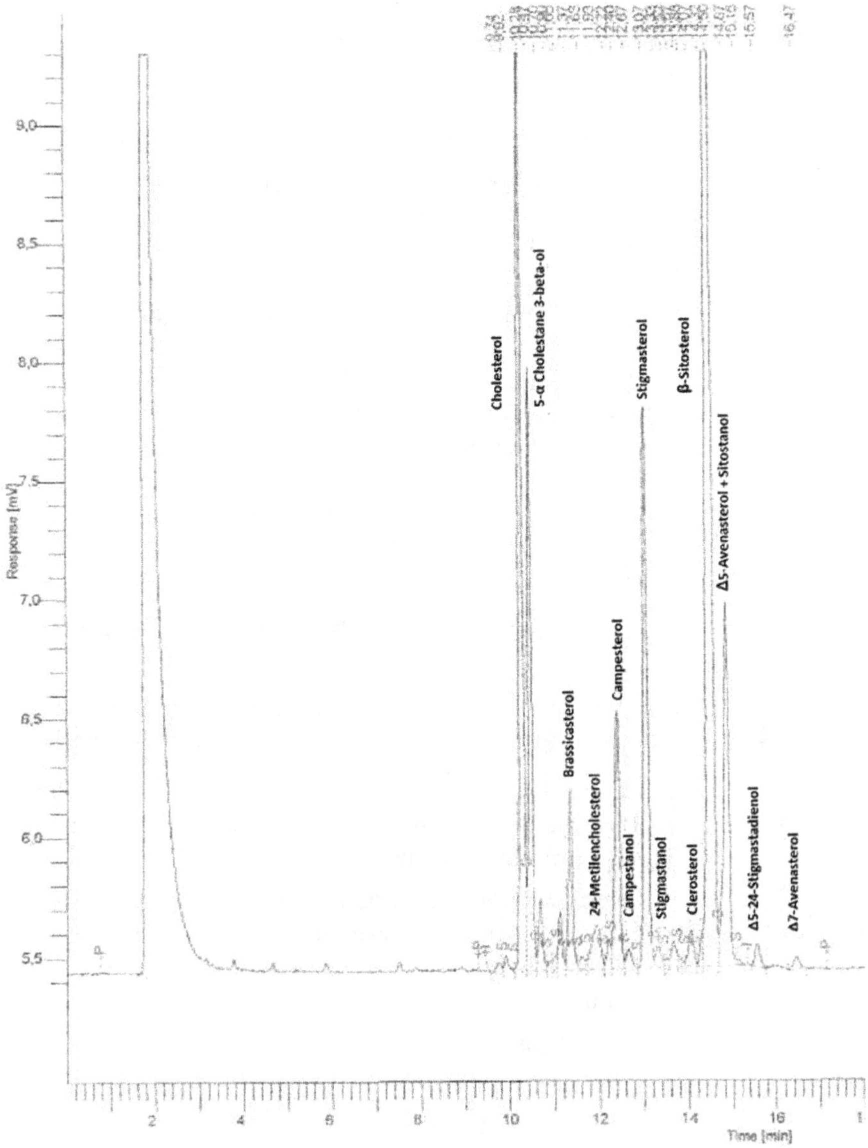

Fig. 5 GC-FID trace of TMS derivatives of sterols from tomato seed oil, obtained with a 30-m CP-Sil 5 CB column (100% dimethylpolysiloxane) under the analytical conditions indicated in Subheading 3.6.1. Peaks are annotated with sterol names, indicating their presence and relative abundance in the sample

response constant calculated for each individual sterol according to the following formula:

$$K = \frac{A_x}{A_{is}} \cdot \frac{C_{is}}{C_x}$$

where:

A_x = the peak area of sterol x

A_{is} = the area of the internal standard

C_x = the concentration of sterol x in mg

C_{is} = the concentration of the internal standard in mg

Therefore, the corrected calculation of the content of each single sterol (in mg/100 mg of fat) using the response constant is the following:

$$C_x = \frac{A_x}{A_{is}} \cdot \frac{C_{is}}{P} \cdot \frac{1}{K} \cdot 100$$

To calculate the sterol content expressed in a food basis (mg sterol/100 g of food), it is necessary to consider the lipid content of the food matrix according to the following formula:

$$x = \frac{C_x \cdot \mathrm{LC}}{0.1}$$

where:

LC = the lipid content of the sample (g in 100 g food)

3.6.2 GC-MS Analysis

For the separation and detection of sterols, inject 1 μL of the sample solution into a GC-MS under the following conditions:

- Capillary GC column: apolar (5% phenyl-95% dimethylpolysiloxane), 30 m length × 0.25 mm internal diameter (I.D.) × 0.25 μm film thickness.
- Oven program: isotherm at 250 °C for 3 min, from 250 to 280 °C at 2 °C/min, isotherm at 280 °C for 12 min, and then from 280 to 320 °C at 1.7 °C/min, with final isotherm at 320 °C for 2 min.
- Carrier gas: Helium (He).
- Pressure:75 kPa.
- Flow rate: 0.41 mL/min.
- Injector temperature: 325 °C.
- Ion source temperature: 230 °C.
- Transfer line temperature: 210 °C.
- Electron energy: 70 eV.
- Mass range *m/z*: 50–550.
- Split ratio 1:15.

GC-MS Identification and Quantification

GC-MS allows to acquire spectra in the TIC (Total Ion Current) and SIM (single ion monitoring) modes, which makes it possible to carry out a qualitative and quantitative determination of sterols, eliminating problems related to coelution with matrix interferences.

Table 1
Main fragmentation ions (*m/z*) of each sterol. The selected highlighted ions are used for sterol quantification

Sterols	Main fragmentation ions, *m/z*
5α-cholestan-3β-ol **(IS)** (*see* **Note 15**)	**460**, 445, 403, 370, 355, 306, 215, 75
β-sitosterol	**486**, 471, 396, 381, 357, 255, 129
β-sitostanol	**488**, 473, 431, 398, 383, 305, 230, 215, 187, 161
Δ^5-avenasterol	**386**, 371, 281, 257, 255, 253, 213, 211, 129, 73, 55
Brassicasterol	**470**, 380, 255, 129, 69
Campesterol	**472**, 457, 382, 367, 343, 255, 129, 73
Campestanol	**472**, 457, 382, 343, 367, 255, 129, 73
Stigmasterol	**484**, 394, 379, 355, 255, 213, 185, 159, 129, 83
Cholesterol	**458**, 443, 368, 353, 329, 255, 247, 129
Δ^7-avenasterol	**386**, 343, 281, 253, 213

Sterols are identified by comparing their retention times and mass spectra with those of the corresponding commercial standards. It is also advisable to compare the sterol mass spectra with those reported in mass spectral databases (i.e., NIST®). Table 1 shows the fragmentation ions (*m/z*) used for sterol identification.

For the GC-MS quantitative determination of sterols, it is necessary to build a calibration curve for each sterol, including the internal standard for every curve point (*see* **Note 16**). To calculate the amount of sterol (in mg), it is necessary to extrapolate from the calibration curve the sterol quantity (in the *x*-axis) corresponding to the detected sterol area during the sample analysis.

4 Notes

1. The reaction is strongly endothermic and produces toxic vapors, so working under a fume hood is important. It is reccomended to wait until the solution reaches room temperature to correctly bring it up to volume. This solution is highly aggressive for glass, so it is suggested to store it in PET containers. If the solution becomes yellow, it must be re-prepared.
2. It is suggested to add 1–2 pellets of KOH (which approximately corresponds to 0.2–0.4 g), to make the solution slightly basic. Store the solution in a dark bottle.
3. The analyst must verify that pyridine is anhydrous since water interferes with the derivatization reaction, hindering sterol

silylation. If pyridine appears cloudy, it is necessary to anhydrify it by adding 3 g of anhydrous sodium sulfate, leaving it at 4 °C in the refrigerator for 2 h, and filtering it through a folded paper filter. The derivatization mixture should be prepared 2–3 days before it is used so that it becomes clear.

4. β-sitosterol is often used as the most representative phytosterol in vegetable matrices. For qualitative purposes, it is possible to use a commercial standard with purity ≥70% for the solution preparation.
5. Select the appropriate standards based on the type of matrix you will analyze. GC-grade commercial standards with high purity levels should be used. It is important to check and determine their effective purity before using them by analyzing the standard solutions with the chosen chromatographic method and eventually applying a correction coefficient in the quantification formula.
6. Nitrogen is used in some steps to remove small quantities of solvent, carefully avoiding sample oxidation or generating unwanted artifacts.
7. Margarine, butter, and their mixtures contain water, so it is necessary to anhydrify them before subjecting them to cold saponification. The fat should be first melted in the oven (at a temperature that is sufficient to reach liquefaction), and when the fat is liquefied and transparent, it should be added with anhydrous sodium sulfate (3 g of anhydrous sodium sulfate/ 10 g of fat), filtered, and collected in a suitable glass container. The sample's anhydrification treatment should be carried out in the lab oven.
8. Adding an internal standard helps overcoming and correcting eventual sterol losses during preparation, purification, and inlet sample steps. In case of samples particularly rich in sterols (such as sterol-enriched functional foods) or co-elution of the internal standard with other matrix compounds, it is possible to use betulinol (CAS no. 473-98-3) as the internal standard since it elutes after all compounds of interest. However, sometimes betulinol is difficult to derivatize in sterol-rich samples; in this case, it is recommended to add first 100 μL of pyridine to the sample before adding the silylation mixture, thus improving its solubilization and the derivatization efficacy.
9. In the case of sterol-enriched products or sterol-based ingredients that have high quantities of esterified sterols, it is necessary to calculate and add a higher stoichiometric quantity of 2 N KOH methanolic solution as a function of the percentage of esterified sterols estimated to be present, to guarantee the complete saponification yield of all esterified sterols.

10. It is recommended to check the efficacy of the saponification reaction (especially in ingredients or matrices rich in sterol esters) by carrying out a qualitative TLC of the unsaponifiable matter using *n*-hexane:diethyl ether solution (80:20, v/v) as mobile phase or a semi-quantitative GC-FID determination of the main lipid classes according to the method reported by Gallina Toschi et al. [8].
11. The pre-impregnated TLC must be used within 15 days of the treatment. If the plate is not well preserved, it turns yellow or loses silica, which will no longer be usable.
12. The TLC development starting line is usually fixed at 2.5 cm from the TLC plate bottom edge to avoid sample dissolution in the mobile phase. The TLC frontline is set at 10 cm from the starting line, that is, 7.5 cm from the TLC plate top edge. These lines are usually drawn with pencil, so the analyst should avoid damaging the layer surface. The border effect should be prevented by loading the samples about 0.5–1 cm from the TLC external borders. To obtain a narrow band, the sample should be carefully loaded with a TLC syringe in the starting line, letting it dry little by little (or with a gentle nitrogen flow) as the sample is loaded. Regarding the reference standard spot for the correct identification of the band, about 2–5 μL of the standard solution should be loaded in the TLC starting line in alignment with the loaded sample band. The reference standard spot should be loaded by keeping the needle perpendicular to the TLC surface to obtain a circular spot. If a jagged edge forms, the point is saturated, which could lead to an anomalous elution and separation.
13. To withdraw the chloroformic extract, it is advisable to cover the Pasteur pipette tip with a bit of pure cellulose paper to avoid silica suction.
14. During sterol derivatization with the silylating mixture, ammonium salts are formed; this resulting white precipitate can hinder the correct sample dilution and subsequent withdrawal steps, so it is advisable to add *n*-hexane, centrifuge it, and transfer it to a clean vial. Regarding the amount of solvent to be added to solubilize the sample, it is recommended that run some preliminary tests to ensure that the quantity of solvent used for sample solubilization allows the correct identification and quantification of the analytes.
15. If you decide to use betulinol as an internal standard for sterols quantification, it is necessary to consider its fragmentation ions (*m/z*): **203**, 279, 393, 483, 496.
16. To prepare the calibration curves, make a stock solution of a known concentration for each sterol standard and then make at least five dilutions starting from the stock solution. These

solutions will be used for building the calibration curve for each sterol. Each solution corresponding to a concentration point of the curve should be injected at least 3 times, and the average data for the 3 injections should be employed to create the curves.

Acknowledgments

This work was supported by the *Alma Mater Studiorum*-Università di Bologna (Basic Research Funding RFO).

References

1. Demel RA, De Kruyff B (1976) The function of sterols in membranes. Biochim Biophys Acta-Biomembr 457(2):109–132
2. A.O.A.C. (1961) Official method 905.02 Fat in milk. Methods of test for dairy industry. 1479 (Part II) –17th edn
3. Hara A, Radin NS (1978) Lipid extraction of tissues with a low-toxicity solvent. Anal Biochem 90(1):420–426
4. Folch J, Lees M, Sloane-Stanley GH (1957) A simple method for the isolation and purification of total lipids from animal tissues. J Biol Chem 226:497–509
5. Boselli E, Velazco V, Caboni MF, Lercker G (2001) Pressurized liquid extraction of lipids for the determination of oxysterols in egg-containing food. J Chromatogr A 917: 239–244
6. Sander BD, Addis PB, Park SW, Smith DE (1989) Quantification of cholesterol oxidation products in a variety of foods. J Food Prot 52: 109–114
7. Sweeley CC, Bentley R, Makita M, Wells WW (1963) Gas–liquid chromatography of trimethylsilyl derivatives of sugars and related substances. J Am Oil Chem Soc 85:2497–2507
8. Gallina Toschi T, Cardenia V, Bonaga G, Mandrioli M, Rodriguez-Estrada MT (2014) Coffee silverskin: characterization, possible uses, and safety aspects. J Agric Food Chem 62:10836–10844

Chapter 6

Analysis of Sterols Oxidation Products (SOPs) in Food, by Gas Chromatography

Mara Mandrioli, Dario Mercatante, and Maria Teresa Rodriguez-Estrada

Abstract

Sterols have a chemical structure that contains at least one double bond and one tertiary carbon, which make them particularly prone to oxidation during food processing and storage. Sterol oxidation products (SOPs) are structurally very similar and are often present at trace levels, so high-resolution and sensitive analytical methods are required to accurately determine them. A correct identification and quantification of these compounds is necessary, as several adverse biological effects have been attributed to them, especially to cholesterol oxidation products (COPs). The accurate analytical control of sterol-enriched food is also of outmost importance, as plant sterols added therein can easily oxidize, which may contrast their functional cholesterol-lowering properties. This chapter faces the analytical determination of SOPs in food, including the various preparative steps (extraction, cold saponification, purification by solid-phase extraction (SPE)), and the gas chromatographic (GC) analysis with different detectors (flame ionization detector (FID) and mass spectrometer (MS)).

Key words Sterols oxidation products, Cholesterol oxidation products, Phytosterols oxidation products, Lipid extraction, Unsaponifiable extraction, SPE purification, Gas chromatography, Flame ionization detector, Mass spectrometer

1 Introduction

Lipid oxidation is mainly initiated in the highly unsaturated phospholipid fraction of cell membranes, which also contain other unsaturated liposoluble molecules, like sterols, that are also prone to oxidation. Cholesterol is an unsaturated lipophilic biomolecule that exposes the A ring to the outer side of the double phospholipid layer in the biological membrane. The side chain is located among the alkyl chains of phospholipids. However, cholesterol is prone to oxidation (chemical, photosensitized, and enzymatic), which generates cholesterol oxidation products (COPs), compounds with different characteristic functional groups (7-hydroxy, 5,6-epoxy, triol, and 7-keto derivatives). COPs are usually present in low amounts in raw animal-origin food (meat, milk, eggs, sea

Lilian Mariutti and Frank Vriesekoop (eds.), *Food Chemistry: Basic Protocols*, Methods and Protocols in Food Science,
https://doi.org/10.1007/978-1-0716-4678-6_6,

products), but their concentration significantly increases after exposure to light, heat, metals, and oxygen; in fact, processing conditions may promote a relevant COP formation in highly processed food products [1]. On the other hand, phytosterols are synthesized by plants and introduced into the human body through the diet. Plant sterols can also undergo oxidation in a similar way as cholesterol, giving rise to phytosterol oxidation products (POPs) that have analogous structures and functional groups to those of COPs; in particular, sterol oxidation could be relevant in functional sterol-enriched food and the sterol-based ingredients used for their formulation [2, 3]. While toxicological aspects of COPs have been well documented for their wide range of adverse biological effects (i.e., inflammation processes and cytotoxicity) [2] and their role in the induction of atherosclerosis, cancer, autoimmune diseases, and neurodegenerative diseases (i.e., Parkinson and Alzheimer diseases) [2], there is increasing evidence on both positive and harmful biological effects of POPs. Some investigations have highlighted the potential atherogenicity, cytotoxicity, and pro-inflammatory effects of POPs [2, 3]. The determination of sterol oxidation products (SOPs = COPs + POPs) often is quite arduous since, as mentioned before, SOPs have similar structures and are quite unstable trace components, which can easily form or decompose during sample preparation and analysis; in addition, the lipid fractions of some foods and biological samples are very complex, increasing the possibility of matrix interference and consequent peak overlapping. It is, therefore, necessary to carry out a purification of the unsaponifiable fraction by using solid phase extraction (SPE) cartridges with the aminopropyl stationary phase (SPE-NH_2) for COPs [4] and silica stationary phase (SPE-Si) for POPs [5], before the gas chromatographic analysis.

2 Material

2.1 Laboratory Equipment

1. Laboratory balance: analytical balance with sensitivity ±0.1 mg and technical balance with sensitivity ±0.1 g.
2. Common laboratory glassware, bottles with sealing caps, cylinders, glass separating funnels, glass funnels, vacuum flasks, Buchner funnels, rotavapor flasks, Erlenmeyer flask, volumetric flasks, spatulas, microsyringes, pipettes, test tubes, elution chamber for thin-layer chromatography (TLC).
3. Turbine homogenizer.
4. Oven with temperature control.
5. Refrigerator set at a temperature of 4 °C.
6. Vacuum pump.
7. Rotary evaporator.

8. Heating block.
9. Vortex stirrer.
10. Tilting agitator.
11. Centrifuge.
12. Tube vacuum manifold for SPE cartridges.
13. SPE- NH_2 cartridges 500 mg/3 mL.
14. SPE-Si cartridges 500 mg/3 mL.
15. Gas chromatograph with split/spitless injection system and various detectors (flame ionization detector (FID) or mass spectrometer (MS)).

2.2 Solvents, Standards, and Materials

Prepare all solutions using analytical-grade solvents and reagents. Prepare and store all reagents at room temperature (unless otherwise stated). Rigorously stick to general safety instructions for the utilization of solvents and reagents (work under a fume hood and follow technical data sheets), using the required personal protection equipment (PPE). Carefully follow regulations for waste disposal.

1. Chloroform:methanol 1:1 (v/v) mixture. Prepare a mixture of chloroform (CAS No. 67-66-3):methanol (CAS No. 67-56-1) in a 1:1 (v/v) ratio using graduated cylinders. Store the solution in a closed and correctly labeled bottle.
2. 1 M potassium chloride solution. Weigh 74.55 g of potassium chloride (CAS No. 7447-40-7), dissolve it in distilled water (CAS No. 7732-18-5), and make up to volume in a 1000 mL volumetric flask.
3. Granular anhydrous sodium sulfate (CAS No 7757-82-6).
4. Qualitative filter paper for rapid filtration (Grade 1).
5. *n*-Hexane:isopropanol 4:1 (v/v) mixture. Prepare a mixture of *n*-hexane (CAS No. 110-54-3):isopropanol (CAS No. 67-63-0) in a 4:1 (v/v) ratio using graduated cylinders. Store the solution in a closed and correctly labeled bottle.
6. 2 N potassium hydroxide solution in methanol. Weigh 112.2 g of potassium hydroxide (ACS reagent, ≥85%, pellets, CAS No. 1310-58-3), dissolve it in methanol (CAS No. 67-56-1), and make up to volume in a 1000 mL volumetric flask (*see* **Note 1**).
7. 0.5% potassium hydroxide solution. Weigh 5 g of potassium hydroxide (ACS reagent, ≥85%, pellets, CAS No. 1310-58-3), dissolve it in distilled water, and make up to volume in a 1000 mL volumetric flask.

8. 20% (w/v) sodium chloride solution. Weigh 200 g of sodium chloride (CAS No. 7647-14-5), dissolve it in distilled water, and make up to volume in a 1000 mL volumetric flask.
9. *n*-Hexane:ethyl acetate 95:5 (v/v) mixture. Prepare a mixture of *n*-hexane (CAS No. 110-54-3):ethyl acetate (CAS No. 141-78-6) in a 95:5 (v/v) ratio using graduated cylinders. Store the solution in a closed and correctly labeled bottle.
10. *n*-Hexane:ethyl acetate 90:10 (v/v) mixture. Prepare a mixture of *n*-hexane (CAS No. 110-54-3):ethyl acetate (CAS No. 141-78-6) in a 90:10 (v/v) ratio using graduated cylinders. Store the solution in a closed and correctly labeled bottle.
11. *n*-Hexane:diethyl ether 95:5 (v/v) mixture. Prepare a mixture of *n*-hexane (CAS No. 110-54-3):diethyl ether (CAS No. 60-29-7) in a 95:5 (v/v) ratio using graduated cylinders. Store the solution in a closed and correctly labeled bottle.
12. *n*-Hexane:diethyl ether 90:10 (v/v) mixture. Prepare a mixture of *n*-hexane (CAS No. 110-54-3):diethyl ether (CAS No. 60-29-7) in a 90:10 (v/v) ratio using graduated cylinders. Store the solution in a closed and correctly labeled bottle.
13. *n*-Hexane:diethyl ether 80:20 (v/v) mixture. Prepare a mixture of *n*-hexane (CAS No. 110-54-3):diethyl ether (CAS No. 60-29-7) in an 80:20 (v/v) ratio using graduated cylinders. Store the solution in a closed and correctly labeled bottle.
14. Silylation mixture. Prepare the derivatization mixture with pyridine (CAS No. 110-86-1): hexamethyldisilazane (CAS No. 999-97-3): trimethylchlorosilane (CAS No. 75-77-4) in a 5:2:1 (v/v/v) ratio, adding the reagents in this decreasing concentration order. Store the solution in a fridge at 4 °C in a closed, correctly labeled bottle (*see* **Note 2**). Before use, bring the solution to room temperature and aliquot it (about 500 μL) into the individual dried samples using a glass Pasteur pipette.
15. Sterols internal standard (IS) solution. Prepare a 2 mg/mL solution of 5α-cholestan-3β-ol (CAS No. 80-97-7) (v/v). Weigh 100 mg of 5α-cholestan-3β-ol in a 50 mL volumetric flask and make up to volume with chloroform (CAS No. 67-66-3). Store the solution in a closed and correctly labeled bottle in the freezer at −18 °C until use. Before use, thaw the solution and bring it to room temperature.
16. SOPs Internal standard (IS) solution. Prepare a 1 mg/mL solution of cholest-5-en-3β,19-diol (19-hydroxycholesterol; CAS No. 561-63-7) (v/v). Weigh 5 mg of 19-hydroxycholesterol in a 5 mL volumetric flask and make up to volume with *n*-hexane (CAS No. 110-54-3). Store the solution in a closed and correctly labeled bottle in the freezer at

−18 °C until use. Before use, thaw the solution and bring it to room temperature.

17. COPs standard solutions for GC analysis. The following COPs standards (all commercially available) can be used for their quantification: cholest-5-en-3β,7α-diol (7-α-hydroxycholesterol) (CAS No. 566-26-7), cholest-5-en-3β,7β-diol (7β-hydroxycholesterol) (CAS No. 566-27-8), 5α,6α-epoxy-cholestan-3β-ol (α-epoxycholesterol) (CAS No. 1250-95-9), 5β,6β-epoxy-cholestan-3β-ol (-β-epoxycholesterol) (CAS No. 4025-59-6), cholestan-3β,5α,6β-triol (cholestanetriol) (CAS No. 1253-84-5), cholest-5-en-3β-ol-7-one (7-ketocholesterol) (CAS No. 566-28-9), 5-cholestene-3β,25-diol (25-hydroxycholesterol) (CAS No. 2140-46-7), 5-cholestene-3β,20α-diol (20-α-hydroxycholesterol) (CAS No. 516-72-3). For each COP, prepare a 1 mg/mL (v/v) solution in a volumetric flask and make up to volume with *n*-hexane (CAS No. 110-54-3). Store the solution in a closed and correctly labeled bottle in the freezer at −18 °C until use. Before use, thaw the solution and bring it to room temperature (*see* **Note 3**).
18. POPs standard solutions for GC analysis. Only one POP standard is commercially available, the 7-keto-β-sitosterol (CAS n. 2034-74-4). For the POPs, prepare a 1 mg/mL (v/v) solution in a volumetric flask and make up to volume with *n*-hexane (CAS No. 110-54-3). Store the solution in a closed and correctly labeled bottle in the freezer at −18 °C until use. Before use, thaw the solution and bring it to room temperature (*see* **Note 3**).
19. Nitrogen (CAS No. 07727-37-9) (*see* **Note 4**).

3 Methods

Figure 1 shows the flow diagram of the methodology used for the sample preparation for the analysis of COPs and POPs in food, by GC-FID and GC-MS.

3.1 Extraction of the Lipid Fraction

The choice of the lipid extraction method depends on the matrix type and should provide a representative lipid fraction. In the case of milk and dairy products, the Rose-Gottlieb et al. method [6] is usually recommended, while the Hara and Radin method [7] is suggested for vegetable samples (such as seeds). The Folch et al. method [8] is particularly recommended for the extraction of food minor lipid components as it can extract the intracellular fat as well, thus providing a more representative sample of the lipid fraction. The method described as follows is a modified version [9] of the Folch et al. method [8]:

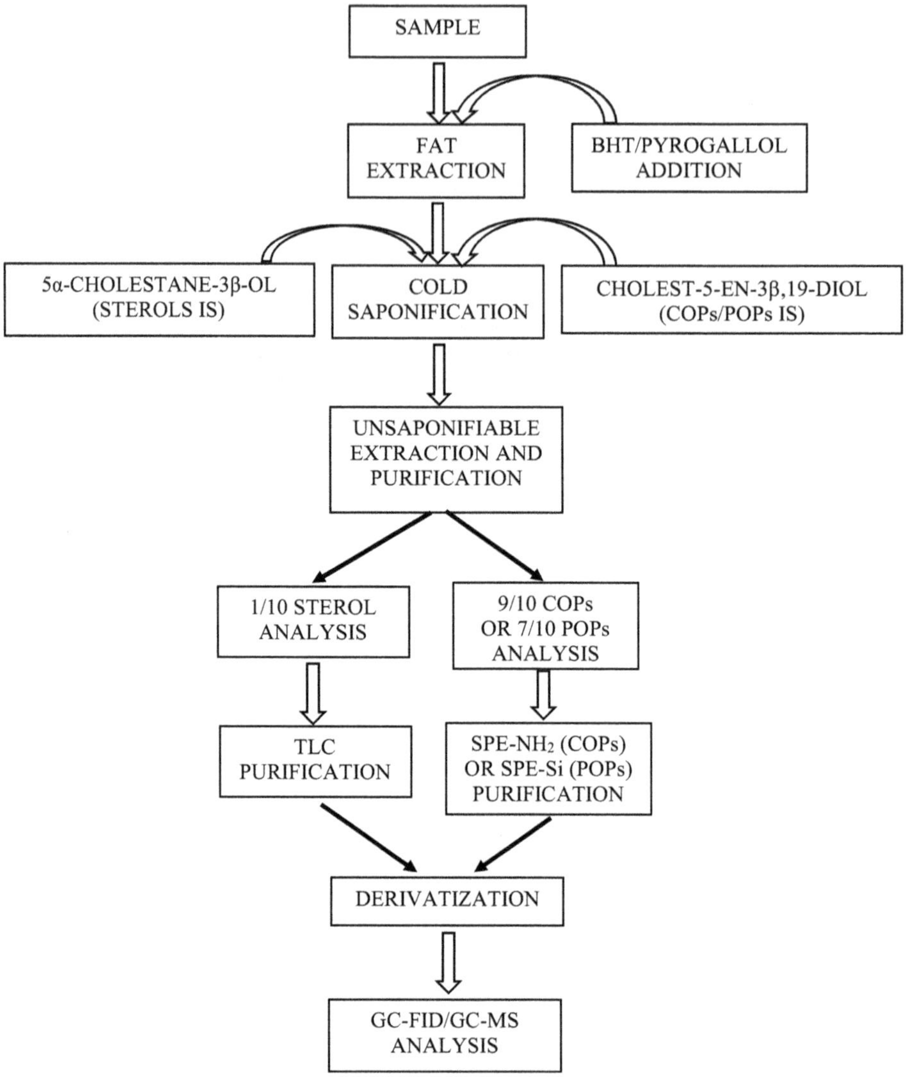

Fig. 1 Flow diagram of the sample preparation for the analysis of COPs and POPs in food, by GC-FID and GC-MS

1. Weigh a suitable amount of sample (10–25 g) according to the food matrix and the estimated fat present therein. The sample must be as homogeneous as possible (previously ground if necessary) and representative of the food to be analyzed (*see* **Note 5**).
2. Transfer the sample to a bottle with tight-fitting cap and add 200 mL of a chloroform:methanol solution (1:1, v/v).
3. Mix it with a dispersion homogenizer for 3 min, at a speed that does not cause excessive overheating.
4. Place the bottle in a thermostated lab oven at 60 °C for 20 min. Once the first 10 min have passed, vent the bottle under the suction hood and reposition it in the oven.

5. Remove the bottle from the oven, allow it to cool down, and add 100 mL of chloroform before homogenizing it again for 2 min with the turbine homogenizer.
6. Filter the bottle content with a Buchner funnel under vacuum, using a fast filtration paper filter.
7. Transfer the filtrate to a bottle with a tight-fitting cap and add 100 mL of a 1 M aqueous solution of KCl.
8. Shake the bottle very vigorously for 1 min and keep it overnight in a refrigerator at 4 °C.
9. Transfer the contents of the bottle to a separating funnel, allow the two phases to separate very well, and transfer the organic phase (lower one) to a clean bottle.
10. Add 30 g of anhydrous sodium sulfate and place the bottle in the refrigerator at 4 °C for 2 h to allow the removal of residual water from the organic phase.
11. Filter through a folded paper filter containing 10–20 g of anhydrous sodium sulfate and collect the filtrate in a rotavapor flask.
12. Evaporate the excess solvent with a rotary evaporator, keeping the temperature at 40 °C in the water bath.
13. Eliminate traces of residual solvent inside the fat extract by placing the balloon under a gentle nitrogen stream.
14. Transfer the fat obtained to a test tube by using small aliquots of a mixture of *n*-hexane:isopropanol (4:1, v/v), and store it in the freezer at −20 °C until analysis.

3.2 Cold Saponification

To extract SOPs, the lipid extract is first subjected to cold saponification to breakdown the ester bond of esterified sterols and thus release free sterols without artifact formation [10]:

1. Weigh 250 mg of perfectly homogeneous anhydrous fat (*see* **Note 6**) into a 20 mL glass tube.
2. Add 100 μL of the 5α-cholestan-3β-ol solution (corresponding to 0.2 mg; IS for sterols) and 25 μL of 19-hydroxycholesterol (corresponding to 0.025 mg; IS for SOPs) using glass microsyringes (*see* **Note 7**), and evaporate the solvent with a slight nitrogen flow. In any case, the quantities of standards must be modulated according to the amounts of analytes present in the sample.
3. Add 10 mL of 2 N methanolic solution of KOH (*see* **Note 8**), and stir it very vigorously in the vortex until a light emulsion appears (*see* **Note 9**).
4. Place the test tube on a horizontal shaker for 18–24 h to allow the saponification reaction to occur.

3.3 Extraction of the Unsaponifiable Fraction

Once the lipid saponification is completed, it is necessary to extract the unsaponifiable matter, which contains both sterols and SOPs [10]:

1. Place the content of the saponified solution in a 250 mL separating funnel, add 10 mL of distilled water and 20 mL of diethyl ether, stir it with alternated venting of the funnel, and then allow the two phases to separate.
2. Collect the aqueous phase (lower one) in a second separating funnel, perform an extraction with 10 mL of diethyl ether, remove the ether phase, and transfer it to a clean separated funnel together with the first ether extract. This procedure should be done twice to extract all the unsaponifiable matter that may have remained in the aqueous phase.
3. Pull the three ether fractions together in a separating funnel and carry out two washes with 10 mL of 0.5 N KOH aqueous solution, followed by two more washes with 10 mL of distilled water, until a pH close to neutrality (7–8) is reached.
4. The ethereal extract thus obtained is anhydrified by adding 5 g of anhydrous sodium sulfate, and leave standing for 2 h in a refrigerator at 4 °C to allow the removal of residual water from the organic phase.
5. Filter through a folded paper filter containing a small amount of anhydrous sodium sulfate, and collect the filtrate in a rotavapor flask.
6. Remove the diethyl ether using a rotary evaporator with a water bath set at 40 °C.
7. Transfer the unsaponifiable fraction into a test tube using 1 mL of *n*-hexane:isopropanol (4:1, v/v), and store it in the freezer at −20 °C until analysis (*see* **Note 10**).

3.4 SPE Purification

Since cold saponification followed by liquid/liquid separation can extract all the unsaponifiable matter, an SPE step to purify the SOPs fraction prior to the chromatographic analysis is often necessary (*see* **Note 11**).

3.4.1 COPs Purification

1. Place 2–3 mm of anhydrous sodium sulfate in the SPE-NH_2 cartridge and insert it in the tube vacuum manifold.
2. Elute 3 mL of *n*-hexane to activate the stationary phase of the SPE-NH_2 (*see* **Note 12**).
3. Once the SPE is activated, the sample can be loaded into the cartridge. Take a suitable aliquot (about 9/10, but it depends on the matrix nature) of unsaponifiable matter, place it in a vial, dry it under a nitrogen stream, dissolve it in 500 μL of a mixture of *n*-hexane:ethyl acetate in a 95:5 (v/v), and load it into the SPE cartridge.

4. Elute first 6 mL of *n*-hexane:ethyl acetate (95:5, v/v) and then 10 mL of *n*-hexane:ethyl acetate (9:1, v/v), discarding these eluate fractions.
5. Elute 10 mL of acetone (CAS n. 67-64-1) and collect the eluate in a clean test tube; this phase contains the compounds of interest (*see* **Note 13**).
6. Remove the acetone using a rotary evaporator with a water bath set at 40 °C.
7. Transfer the COPs fraction into a test tube by using 1 mL of *n*-hexane:isopropanol (4:1, v/v), and store it in the freezer at −20 °C until analysis.

3.4.2 POPs Purification

1. Place 2–3 mm of anhydrous sodium sulfate in the SPE-Si cartridge and insert it in the tube vacuum manifold.
2. Elute 5 mL of *n*-hexane to activate the stationary phase of the SPE-Si (*see* **Note 12**).
3. Once the SPE is activated, the sample can be loaded into the cartridge. Take a suitable aliquot (about 7/10, but it depends on the matrix nature) of unsaponifiable matter, place it in a vial, dry it under a nitrogen stream, dissolve it in 500 μL of a mixture of *n*-hexane:diethyl ether in a 95:5 (v/v), and load it into the SPE cartridge.
4. Elute first 10 mL of *n*-hexane:diethyl ether (95:5, v/v), then 30 mL of *n*-hexane:diethyl ether (90:10, v/v), and finally 10 mL of *n*-hexane:diethyl ether (80:20, v/v), discarding all these eluate fractions.
5. Elute 10 mL of acetone (CAS n. 67-64-1) and collect the eluate in a clean test tube; this phase contains the compounds of interest (*see* **Note 13**).
6. Remove the acetone using a rotary evaporator with a water bath set at 40 °C.
7. Transfer the POPs fraction into a test tube using 1 mL of *n*-hexane:isopropanol (4:1, v/v), and store it in the freezer at −20 °C until analysis.

3.5 Derivatization

To render SOPs more volatile and improve their peak resolution and detectability, they are converted into trimethylsilyl derivatives as follows:

1. Dry the purified SOPs fraction with a weak nitrogen flow, add 1 mL of the derivatization solution (pyridine:hexamethyldisilazane:trimethylchlorosilane, 5:2:1, v/v/v) [11], and close the vial.
2. Place the vial in a termoblock at 40 °C for 20 min and stir it with vortex after 10 min.

3. Once the reaction is completed, dry the derivatized sample with a weak nitrogen flow, add 100 μL of *n*-hexane (*see* **Note 14**), centrifuge, and inject 1 μL of the supernatant into the GC-FID or GC-MS.

3.6 Gas Chromatographic Analysis

The determination of COPs, POPs, or SOPs can be performed either by GC-FID (Figs. 2 and 3) or GC-MS.

3.6.1 GC-FID Analysis

For the separation and detection of COPs, POPs, or SOPs, inject 1 μL of the sample solution into a GC-FID under the following conditions:

- Capillary GC column: apolar (100% dimethylpolysiloxane or 5% phenyl-95% dimethylpolysiloxane), 30 m length × 0.32 mm internal diameter (I.D.) × 0.25 μm film thickness.
- Oven program: from 265 to 280 °C at 0.5 °C/min and then from 280 to 325 °C at 4 °C/min, with final isotherm at 325 °C for 30 min.
- Carrier gas: helium (He).

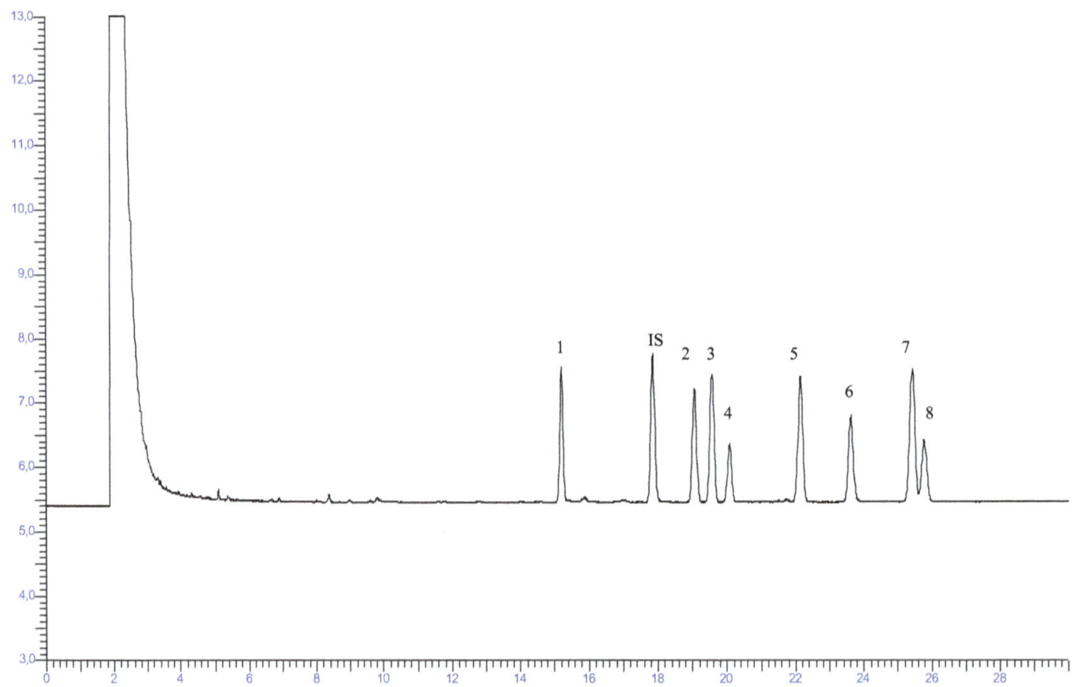

Fig. 2 GC-FID trace of TMS derivatives of COPs standards, obtained with a 30-m CP-Sil 5 CB column (100% dimethylpolysiloxane) under the analytical conditions indicated in Subheading 3.6.1. Peaks identification: **1**, 7α-hydroxycholesterol; **IS**, cholest-5-en-3β,19-diol (19-hydroxycholesterol); **2**, 7β-hydroxycholesterol; **3**, β-epoxycholesterol; **4**, α-epoxycholesterol; **5**, 20-hydroxycholesterol; **6**, cholestanetriol; **7**, 7-ketocholesterol; **8**, 25-hydroxycholesterol

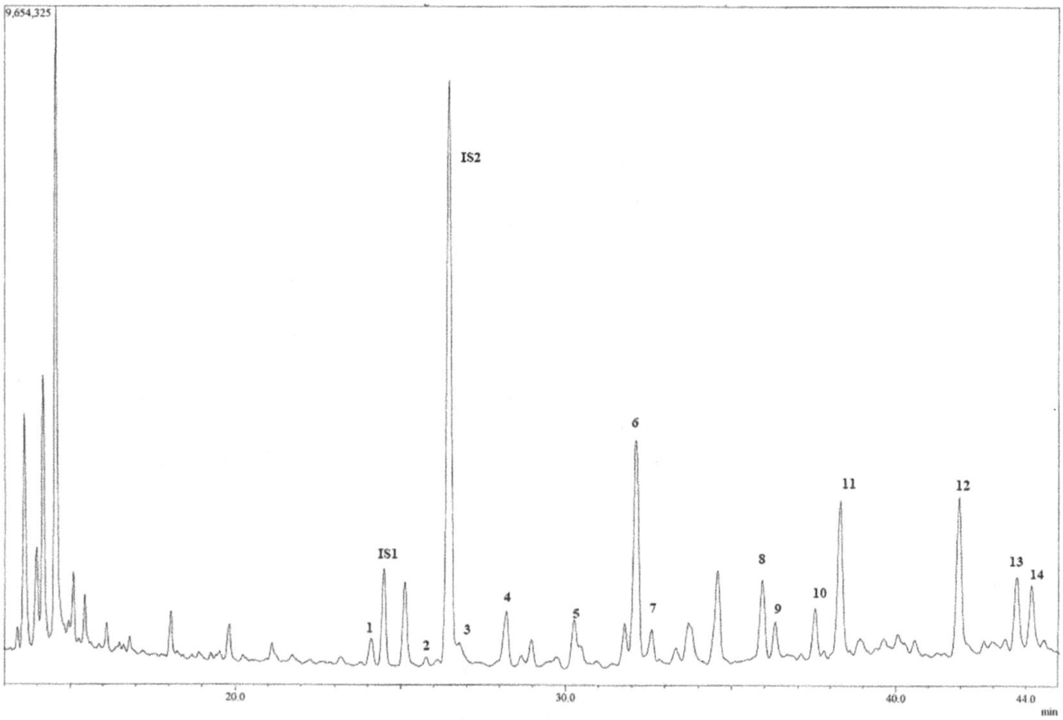

Fig. 3 GC-FID trace of TMS derivatives of POPs profile of a fermented milk enriched in phytosterols obtained with a 30-m CP-Sil 5 CB column (100% dimethylpolysiloxane) under the analytical conditions indicated in Subheading 3.6.1. Peaks identification: **1**, cholesterol; **IS1**, 5α-cholestane3-β-ol; **2**, brassicasterol; **IS2**, cholest-5-en-3β,19-diol (19-hydroxycholesterol); **3**, 7α-hydroxycampesterol; **4**, campesterol; **5**, 7α--hydroxy-β-sitosterol; **6**, β-sitosterol; **7**, sitostanol; **8**, 7β-hydroxy-β-sitosterol; **9**, 25-hydroxysitosterol; **10**, β-epoxysitosterol; **11**, 7-ketocampesterol; **12**, triol derivate of β-sitosterol; **13**, 6-ketositosterol; **14**, 7-ketositosterol

- Pressure: 75 kPa.
- Flow: 2.88 mL/min.
- Split: 43.7 mL/min.
- Split ratio: 1:15.
- Injector temperature: 325 °C.
- Detector temperature: 325 °C.

GC-FID Identification and Quantification

The single SOP peaks are identified by comparing their retention times with those of the corresponding commercial COPs and POPs standards. In case of partial overlapping with other matrix components, it is also possible to spike the sample with solutions of commercial standards.

The content of each single SOP (in mg/100 mg of fat) is calculated by applying the internal standard method (*see* **Note 9**) according to the following formula:

$$C_x = \frac{A_x}{A_{is}} \cdot \frac{C_{is}}{P} \cdot 100$$

where:

A_x = the peak area of SOP x

A_{is} = the area of the internal standard (cholest-5-en-3β,19-diol)

C_{is} = the amount of the internal standard added (cholest-5-en-3β,19-diol) in mg

P = sample weight in mg

Generally, the FID provides a fairly linear response for SOPs. However, to perform a quantification that also considers the differences in chromatographic response of the single SOP molecules with respect to the internal standard, it is possible to introduce a response constant calculated for each individual SOP according to the following formula:

$$K = \frac{A_x}{A_{is}} \cdot \frac{C_{is}}{C_x}$$

where:

A_x = the peak area of SOP x

A_{is} = the area of the internal standard

C_x = the concentration of SOP x in mg

C_{is} = the concentration of the internal standard in mg

Therefore, the corrected calculation of the content of each single SOP (in mg/100 mg of fat) using the response constant is the following:

$$C_x = \frac{A_x}{A_{is}} \cdot \frac{C_{is}}{P} \cdot \frac{1}{K} \cdot 100$$

To calculate the SOP content expressed in food basis (mg SOP/100 g of food), it is necessary to consider the lipid content of the food matrix according to the following formula:

$$x = \frac{C_x \cdot \mathrm{LC}}{0.1}$$

where:

LC = the lipid content of the sample (g in 100 g food)

3.6.2 GC-MS Analysis

For the separation and detection of SOPs, inject 1 μL of the sample solution into a GC-MS under the following conditions:

- Capillary GC column: apolar (5% phenyl-95% dimethylpolysiloxane), 30 m length × 0.25 mm internal diameter (I.D.) × 0.25 μm film thickness.

- Oven program: isotherm at 250 °C for 3 min, from 250 to 280 °C at 2 °C/min, isotherm at 280 °C for 12 min, and then from 280 to 320 °C at 1.7 °C/min, with final isotherm at 320 °C for 2 min.
- Carrier gas: helium (He).
- Pressure: 75 kPa.
- Flow rate: 0.41 mL/min.
- Split ratio 1:15.
- Injector temperature: 325 °C.
- Ion source temperature: 230 °C.
- Transfer line temperature: 210 °C.
- Electron energy: 70 eV.
- Mass range *m/z*: 50–550.

GC-MS Identification and Quantification

GC-MS allows to acquire spectra in the TIC (total ion current) and SIM (single ion monitoring) modes, which enables to carry out a qualitative and quantitative determination of SOPs, eliminating problems related to coelution with matrix interferences.

COPs Identification and Quantification

COPs are identified by comparing their retention times and mass spectra with those of the corresponding commercial standards. It is also advisable to compare the COPs' mass spectra with those reported in mass spectral databases (e.g., NIST®). Table 1 shows the fragmentation ions (*m/z*) used for COP identification.

For the GC-MS quantitative determination of COPs, it is necessary to build a calibration curve for each COP using the internal standard for every curve point (*see* **Note 15**). To calculate the amount of COP (in mg), it is necessary to extrapolate from the calibration curve the COP quantity (in the x-axis) that corresponds to the detected COP area during the sample analysis.

POPs Identification and Quantification

POPs are identified by comparing their retention times and mass spectra with those of the corresponding commercial standards. However, since there are very few commercial POPs standards (*see* **Note 16**), it is highly advisable to chemically synthesize POPs [12] for an accurate determination of the retention times and mass spectra, as well as to compare the POPs mass spectra with those reported in mass spectral databases (i.e., NIST®) and literature [13]. Table 2 shows the fragmentation ions (*m/z*) used for POPs identification.

For the quantitative determination of POPs by GC-MS, a calibration curve for each POP, including the internal standard for every curve point (*see* **Notes 15** and **16**), must be built. To calculate the amount of POP (in mg), the POP quantity (on the x-axis) that corresponds to the detected POP area during the sample analysis must be extrapolated from the calibration curve.

Table 1
Main fragmentation ions (*m/z*) of each COP

Cholesterol oxidation product (COP)	Main fragmentation ions, *m/z*
Cholest-5-en-3β,19-diol (19-hydroxy-cholesterol) (**IS**)	**353**, 456, 366, 253, 197, 145
7α-hydroxycholesterol	**456**, 233, 546, 351
7β-hydroxycholesterol	**456**, 233, 546, 367
α-epoxycholesterol	**384**,474, 135, 95
β-epoxycholesterol	**384**, 474, 135, 253
Cholestanetriol	**403**, 456, 546, 321
24-hydroxycholesterol	**145**, 413, 503, 323
25-hydroxycholesterol	**131**, 445, 546, 271
7-ketocholesterol	**472**, 367, 382, 296
27-hydroxycholesterol	**456**, 546, 417, 255

The selected highlighted ions are used for COP quantification

Table 2
Main fragmentation ions (*m/z*) of each POP

Phytosterols oxidation product (POP)	Main fragmentation ions, *m/z*
Cholest-5-en-3β,19-diol 1(9-hydroxy-cholesterol) (**IS**)	**353**, 456, 366, 253, 197, 145
7α-hydroxycampesterol	**470**, 129, 233, 455, 470, 471, 472, 560
7α-hydroxystigmasterol	**482**, 208, 233, 343, 483, 484, 467, 572
6β-hydroxycampesterol	**417**, 455, 470, 545
7α-hydroxysitosterol	**484**, 233, 253, 485, 486, 574
6β-hydroxystigmasterol	**429**, 467, 482, 543, 557, 572
7β-hydroxycampesterol	**470**, 129, 209, 233, 381, 455, 471, 472, 486, 560
7β-hydroxystigmasterol	**482**, 209, 233, 483, 484, 572
α-Epoxystigmasterol	**253**, 83, 349, 410, 500
β-Epoxystigmasterol	**253**, 83, 129, 157, 211, 349, 410, 442, 471, 500
7β-hydroxysitosterol	**484**, 233, 485, 486, 574
α-epoxysitosterol	**412**, 253, 384, 394, 431, 502, 503, 575
β-epoxysitosterol	**412**, 253, 384, 394, 474, 502, 503
7-ketocampesterol	**381**, 129, 396, 469, 486
7-ketostigmasterol	**386**, 129, 269, 357, 359, 455, 498

(continued)

Table 2
(continued)

Phytosterols oxidation product (POP)	Main fragmentation ions, *m/z*
6-ketositosterol	**473**, 474, 487
7-ketositosterol	**500**, 129, 161, 395, 410, 483, 501, 502
25-hydroxysitosterol	**131**, 129
Triol derivate of campesterol	**470**, 560, 500, 496, 485, 431, 405, 397, 369, 341, 309, 283, 253, 208, 149
Triol derivate of β-sitosterol	**431**, 574, 559, 546, 512, 484, 470, 395, 349, 253, 177, 157, 145
Triol derivate of stigmasterol	**590**, 253, 482, 349, 430,159

The selected highlighted ions are used for POP quantification

4 Sterols Oxidation Ratio

After having quantified the total SOPs (COPs + POPs) present in the sample under analysis, it is possible to calculate the oxidation ratio according to the following formula:

$$\mathrm{OR\%} = \frac{\mathrm{SOPs\ Tot} \cdot 100}{\mathrm{Sterols}}$$

A similar calculation approach can be also applied for the determination of the OR% of COPs and POPs, using the total amount of cholesterol or phytosterols in the formula denominator, respectively.

5 Some Considerations About GC Analysis: The Choice of GC/MS

As reported by Gachumi et al. [14], in the case of hybrid food matrices or sterol-enriched products, the chromatogram will display the simultaneous elution of COPs and POPs so, in these cases, the GC-MS analysis assures the correct identification of these trace compounds. In the example reported in Fig. 3, focused on the determination of phytosterols and POPs on a phytosterol-enriched fermented milk, only a small residual amount of cholesterol was found; considering this fact and that it was a fresh sample, no COPs formation was detected in this fresh dairy product. The determination carried out with GC-MS allows to acquire spectra with TIC (Total Ion Current) and SIM (Single Ion Monitoring) modes and to perform a qualitative and quantitative evaluation of SOPs, eliminating any interference due to other compounds that could coelute with the analytes of interest. The use of pure reference standards allows to quantify SOPs with calibration curves constructed for each individual analyte in the concentration range of interest. In the case of POPs, as stated before, there are few pure standards on the market, so the calibration curves used for COPs are often used for the quantification of these compounds. However, it is highly advisable to chemically synthesize POPs [12] and to use them for their GC/MS quantification.

5.1 Fast GC-MS

In recent years, GC/MS has become the preferred analytical technique for the determination of COPs, due to its ability to simultaneously provide sensitive identification and quantification while mitigating matrix interference issues. The benefits of GC/MS can be further enhanced by employing a fast GC system, which significantly reduces analysis time compared to conventional GC instruments. The fast GC system operates with specific components, including accelerated temperature programming rates (up to 120 °C/min), optimized sample introduction systems for narrower band loading onto the column, high-pressure pneumatic control of carrier gas flow, rapid oven cooldown times (from 350 to 50 °C in 1 min), and an injector capable of functioning under high pressures (<1000 kPa) with high split ratios (e.g., 1:1200). In addition, this system can be coupled with various detectors (e.g., FID, MS, etc.), provided that the detection is sufficiently fast and sensitive, with appropriate signal sampling rates (in millisec) and frequencies (in Hz).

Cardenia et al. [15] developed a rapid method for the determination of COPs using fast GC/MS, which demonstrated high sensitivity, speed, and repeatability in both food and biological samples (Fig. 4). The method's analytical performance, coupled with a substantial reduction in analysis time and resource consumption, suggests that fast GC/MS is a promising alternative to conventional GC/MS. However, it is important to note that fast GC/MS may not be suitable for sterol-enriched food or complex matrices, particularly those in which co-elution of COPs and POPs is significant. In such cases, it is recommended to utilize conventional GC/MS, with longer capillary columns (e.g., 30-m or 50-m length), to improve compound separation and minimize issues related to compound identification and quantification.

For the separation and detection of COPs, inject 1 μL of the sample solution into a Fast GC-MS under the following conditions:

- Capillary GC column: apolar (5% phenyl-95% dimethylpolysiloxane), 10 m length × 0.1 mm internal diameter (I.D.) × 0.1 μm film thickness.
- Oven program: isotherm at 230 °C, from 230 to 350 °C at 20 °C/min, isotherm at 350 °C for 5 min.

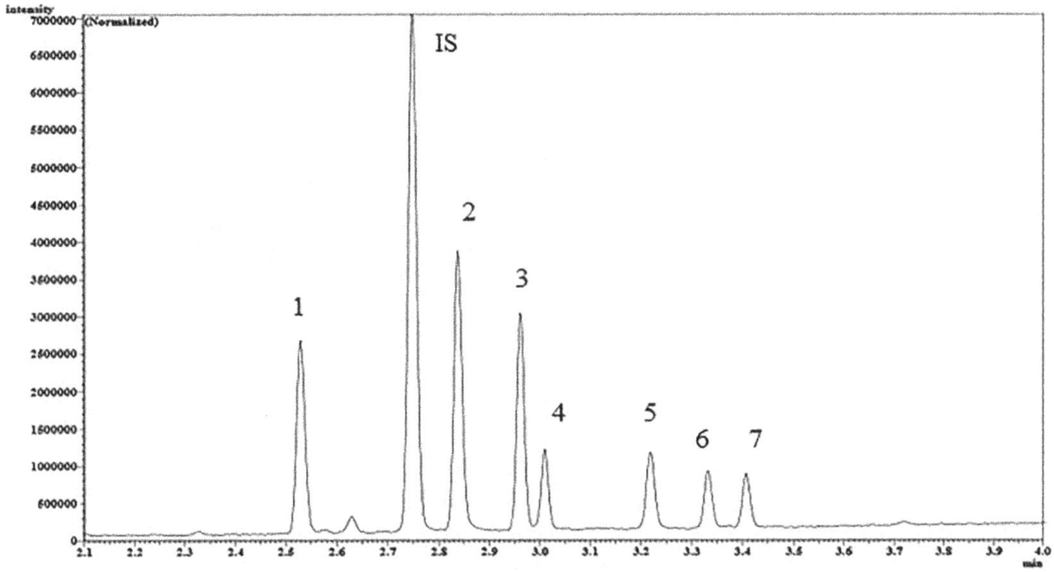

Fig. 4 Cholesterol oxidation products (COPs) profile of a standard mixture obtained with a 10 m Rtx-5 column (95% dimethylpolysiloxane/5% phenylpolysiloxane) under the analytical conditions indicated in Subheading 4.1 [15]. **1**, 7α-hydroxycholesterol; **IS**, cholest-5-en-3β,19-diol (19-hydroxycholesterol); **2**, 7β--hydroxycholesterol; **3**, 5,6β-epoxycholesterol; **4**, 5,6α-epoxycholesterol; **5**, cholestanetriol; **6**, 25-hydroxycholesterol; **7**, 7-ketocholesterol. (From Cardenia et al. [15], with permission.)

- Carrier gas: helium (He).
- Pressure: 412.6 kPa.
- Constant flow: 15.5 mL/min.
- Split ratio 1:50.
- Injector temperature: 325 °C.
- Ion source temperature: 230 °C.
- Transfer line temperature: 330 °C.
- Electron energy: 70 eV.
- Scan rate: 3333 amu/s.
- Mass range *m/z*: 50–900.

6 Notes

1. The reaction is strongly endothermic and produces toxic vapors, so it is important to work under a fume hood. It is recommended to wait until the solution reaches room temperature to correctly bring it up to volume. This solution is highly

aggressive for glass, so it is suggested to store it in PET containers. If the solution becomes yellow, it must be re-prepared.

2. The analyst must verify that pyridine is anhydrous, since water interferes with the derivatization reaction, hindering sterol silylation. In case pyridine appears cloudy, it is necessary to anhydrify it by adding 3 g of anhydrous sodium sulphate, leaving it standing at 4 °C in the refrigerator for 2 h, and filtering it through a folded paper filter. The derivatization mixture should be prepared 2–3 days before it is used, so that it becomes clear.
3. GC grade commercial standards with a high purity level should be used. It is important to check and determine their effective purity before using them by analyzing the standard solutions with the chosen chromatographic method and eventually applying a correction coefficient in the quantification formula.
4. Nitrogen is used in some steps to remove small quantities of solvent, being careful to avoid sample oxidation or the generation of unwanted artifacts.
5. It is advisable at this stage to insert an antioxidant BHT or pyrogallol (600 μg BHT or pyrogallol/mg of matrix), in order to avoid the formation of oxidation products (artifacts) during sample manipulation, extraction, and/or storage.
6. Margarine, butter, and their mixtures contain water, so it is necessary to anhydrify them before subjecting them to cold saponification. The fat should be first melted in an oven (at a temperature that is sufficient to reach liquefaction), and, when the fat is liquefied and transparent, it should be added with anhydrous sodium sulphate (3 g of anhydrous sodium sulphate/10 g of fat), filtered, and collected in a suitable glass container. All the anhydrification treatment of the sample should be carried out in the lab oven.
7. The addition of an internal standard helps overcoming and correcting eventual sterol losses during preparation, purification, and inlet sample steps. In case of samples particularly rich in sterols (such as sterol-enriched functional foods) or co-elution of the internal standard with other matrix compounds, it is possible to use betulinol (CAS no. 473-98-3) as internal standard, since it elutes after all compounds of interest. However, sometimes betulinol is difficult to derivatize in sterol-rich samples, so it is thus recommended to add first 100 μL of pyridine to the sample before adding the silylation mixture, which helps improving its solubilization and the derivatization efficacy.
8. In the case of sterol-enriched products or sterol-based ingredients that have high quantities of esterified sterols, it is necessary to calculate and add a higher stoichiometric quantity of

2 N KOH methanolic solution, considering the percentage of esterified sterols estimated to be present, to guarantee the complete saponification yield of all esterified sterols.

9. It is recommended to check the efficacy of the saponification reaction (especially in ingredients or matrices rich in sterol esters), by carrying out a qualitative TLC of the unsaponifiable matter using *n*-hexane:diethyl ether solution (80:20, v/v) as mobile phase or a semiquantitative GC-FID determination of the main lipid classes according to the method reported by Gallina Toschi et al. [16].
10. It is important to underline that, with a single saponification, it is possible to obtain the unsaponifiable fraction to be used for the analysis of sterols and their oxidation products (SOPs). To warrant a correct SOPs purification, it is necessary to dose the right amount of unsaponifiable fraction to be loaded into the SPE cartridge, which usually ranges between 7/10 and 9/10 of the total unsaponifiable matter, depending on the oxidation products to be analyzed (SOPs and COPs, respectively) and the type of matrix.
11. When the unsaponifiable matter of complex food matrices (containing the oxidation products of both cholesterol and phytosterols) is to be SPE purified, it is advisable to choose the SPE fractionation method according to the relative abundance of sterols (SPE-NH_2 for cholesterol-rich food, and SPE-Si for phytosterol-rich matrices).
12. During all the conditioning and elution phases, the SPE phase should never become dry; the meniscus of the solvent should always be kept at the sodium sulphate level.
13. The solvent elution during SPE purification must be carried out by gravity, with a regular rate of about 1 drop/s; if necessary, apply a weak vacuum to the tube vacuum manifold.
14. During SOPs derivatization with the silylating mixture, ammonium salts are formed; this results in a white precipitate that can hinder the correct sample dilution and subsequent withdrawal steps, so it is advisable to add *n*-hexane, centrifuge it, and transfer it to a clean vial. Regarding the amount of solvent to be added to solubilize the sample, it is recommended to carry out preliminary tests to ensure that the chosen quantity of solvent allows the correct identification and quantification of the analytes.
15. For the preparation of the calibration curves, make a stock solution of a known concentration for each COP standard and then make at least 5 dilutions starting from the stock solution. These solutions will be used to build the calibration curve of each COP. Each solution, corresponding to a concentration point of the curve, should be injected at least 3 times;

the average data of the three injections should be employed to create the curve. It is advisable to prepare separate solutions for each COP, but if you want to prepare a mix of standards, prepare the following three COPs mixtures: (a) 7-β-hydroxycholesterol, cholestanetriol and 7-ketocholesterol; (b) 7α-hydroxycholesterol and 5,6α-epoxycholesterol; (c) 25-hydroxycholesterol and 5,6β-epoxycholesterol. In this way, it is avoided an inaccurate quantification of COPs due to the presence of some cholesterol oxides generated by the oxidative degradation (conversion of epoxides to cholestanetriol) and/or isomerization of the commercial COP standards (i.e. isomerization between α and β isomers of 7-hydroxycholesterol, as well as between α and β isomers of epoxycholesterol).

16. For POPs, only one commercial standard is available on the market (7-keto-β-sitosterol, CAS n. 2034-74-4) so, to construct the calibration curves, it would therefore be advisable to chemically synthesize POPs [12]; upon customer's request, some chemical suppliers can synthesize POPs (deuterated and non-deuterated). Otherwise, the calibration curves prepared with the commercial standards of the corresponding COPs could be used.

Acknowledgments

This work was supported by the *Alma Mater Studiorum*-Università di Bologna (Basic Research Funding RFO).

References

1. Rodriguez-Estrada MT, Garcia-Llatas G, Lagarda MJ (2014) 7-Ketocholesterol as marker of cholesterol oxidation in model and food systems: when and how. Biochem Biophys Res Commun 446(3):792–797
2. Garcia-Llatas G, Mercatante D, López-García G, Rodriguez-Estrada MT (2021) Oxysterols—how much do we know about food occurrence, dietary intake and absorption? Curr Opin Food Sci 41:231–239
3. Wang M, Lu B (2018) How do oxyphytosterols affect human health? Trends Food Sci Technol 79:148–159
4. Rose-Sallin C, Huggett AC, Bosset JO, Tabacchi R, Fay LB (1995) Quantification of cholesterol oxidation products in milk powders using [$^2H_7$] cholesterol to monitor cholesterol autoxidation artifacts. J Agric Food Chem 43(4):935–941
5. Guardiola F, Codony R, Rafecas M, Boatella J (1995) Comparison of three methods for the determination of oxysterols in spray-dried egg. J Chromatogr A 705:289–304
6. A.O.A.C. (1961) Official method 905.02 Fat in milk. Methods of test for dairy industry. 1479 (Part II) – 17th edn
7. Hara A, Radin NS (1978) Lipid extraction of tissues with a low-toxicity solvent. Anal Biochem 90(1):420–426
8. Folch J, Lees M, Sloane-Stanley GH (1957) A simple method for the isolation and purification of total lipids from animal tissues. J Biol Chem 226:497–509
9. Boselli E, Velazco V, Caboni MF, Lercker G (2001) Pressurized liquid extraction of lipids for the determination of oxysterols in egg-containing food. J Chromatogr A 917: 239–244

10. Sander BD, Addis PB, Park SW, Smith DE (1989) Quantification of cholesterol oxidation products in a variety of foods. J Food Prot 52: 109–114
11. Sweeley CC, Bentley R, Makita M, Wells WW (1963) Gas–liquid chromatography of trimethylsilyl derivatives of sugars and related substances. J Am Oil Chem Soc 85:2497–2507
12. Cercaci L, Conchillo A, Rodriguez-Estrada MT, Ansorena D, Astiasarán I, Lercker G (2006) Preliminary study on health-related lipid components of bakery products. J Food Prot 69(6):1393–1401
13. Dutta PC (2002) Determination of phytosterol oxidation products in foods and biological samples. In: Guardiola F, Dutta PC, Codony R, Savage GP (eds) Cholesterol and phytosterol oxidation products: analysis, occurrence, and biological effects. AOCS Press, Champaign, pp 335–374
14. Gachumi G, Poudel A, Wasan KM, El-Aneed A (2021) Analytical strategies to analyze the oxidation products of phytosterols, and formulation-based approaches to reduce their generation. Pharmaceutics 13(2):268
15. Cardenia V, Rodriguez-Estrada MT, Baldacci E, Savioli S, Lercker G (2012) Analysis of cholesterol oxidation products by fast gas chromatography/mass spectrometry. J Sep Sci 35(3):424–430
16. Gallina Toschi T, Cardenia V, Bonaga G, Mandrioli M, Rodriguez-Estrada MT (2014) Coffee silverskin: characterization, possible uses, and safety aspects. J Agric Food Chem 62:10836–10844

Part III

Protein

Chapter 7

Total Protein

Maria Elisa Caetano-Silva and Flavia Maria Netto

Abstract

This chapter describes the most commonly used methods for total protein determination in different laboratories' routines, including nitrogen content determination and spectrophotometric methods. The methods are subdivided into protein-dye interaction, copper-protein interactions, and wavelength absorption (280 nm) by aromatic residues. The methods described here include the apparatus, material, and reagents related, protocol, notes, and troubleshooting.

The protein composition and the sample matrix might influence protein determination, and when selecting the method that best suits your needs, consider the method's different principles. Our goal is to provide the reader with relevant notes beside the main procedure, aiding in the better performance of each method.

Key words Nitrogen content, Protein-dye interaction, Protein-copper interaction, UV absorption

1 Introduction

Total protein analysis is crucial from a nutritional, legal, health, and economic perspective. It is also largely applied for quality control purposes in the food industry. In food chemistry, apart from food products, quantifying proteins is also important in other matrices derived from food research, such as supernatants after in vitro digestion, cell culture medium, and plasma from animal models or clinical trials [6].

A great variety of methods for protein quantification, with different mechanisms to assess the amount of protein material in a sample, are available. The assessment can be based on measuring the total nitrogen, a colored product formed by complexation with protein structure, a compound whose oxidation state changes upon reaction with proteins, or the aromatic groups of specific amino acids in protein structures [1]. It is of utmost importance to consider the mechanism involved in total protein determination to select the best method for each sample, the objective of the analysis, and the laboratory infrastructure. Therefore, to choose the method

Lilian Mariutti and Frank Vriesekoop (eds.), *Food Chemistry: Basic Protocols*, Methods and Protocols in Food Science, https://doi.org/10.1007/978-1-0716-4678-6_7,

and get reliable results, factors such as the presence of interferents, the available amount of sample, and the type of matrix should be considered.

It is worth highlighting that each method has a distinct principle, and they do not measure protein directly. Each principle relies on the formation of a measurable compound (e.g., a colored product), or it measures a specific component of the protein structure (e.g., nitrogen or aromatic residues). Thus, there is no an all-purpose method. The researchers must select the method that best suits their needs, considering speed, type of sample, protein composition, presence of interferents, and sensitivity required.

The most applied method for total protein quantification in food products is based on the quantification of nitrogen, and protein content is calculated by multiplying N content by a conversion factor. This method has been used for the calibration of other methods. The most used method for nitrogen determination is the Kjeldahl method, followed by the Dumas method, which requires specific equipment/apparatus that is less available and much more expensive than the one required by the Kjeldahl method. Dumas' method consists of the combustion of the sample material to release nitrogen gas, quantified by a thermal conductivity detector. As in the Kjeldahl method, total protein quantification is obtained by using a nitrogen-protein conversion factor [5, 6].

Methods based on nitrogen content do not depend on specific reactions with functional groups in protein structure; thus, the protein composition is not an issue. However, all the nitrogen is converted to protein, which means nitrogen from compounds other than protein is also measured. Therefore, getting information about the sample is advisable to avoid overestimating the protein content. This chapter presents only the Kjeldahl method as the representative method based on determining nitrogen content.

Although the Kjeldahl method is lengthy and labor-intensive, even accounting for the associated hazards, it is still considered an accurate and reliable method, widely used in laboratories worldwide.

The Kjeldahl method is appropriate for measuring complex matrices, with compounds bound to protein and entrapped in its structure, such as grains, cheese, meat, processed foods, and crude extracts from different food sources. However, the proper homogeneity of samples is still required for accurate results, and sample size and preparation issues are of critical importance [5, 8].

The Kjeldahl method is accomplished in three main steps:

1. Digestion: The sample is boiled in concentrated sulfuric acid (H_2SO_4) in the presence of a catalyst under heating, and the acid digested contains ammonium sulfate $(NH_4)_2SO_4$ and other by-products.

2. Neutralization and distillation: an excess base (NaOH) is added to the acid digestion mixture to convert NH_4^+ to NH_3, followed by boiling and condensation of the NH_3 gas in a receiving solution.
3. Titration: The receiving solution (H_2BO_3) is titrated using an acid solution to quantify ammonia.

Depending on the sample characteristics and available apparatus, the Kjeldahl method can be carried out in macro, micro, or semimicro variants. In this chapter, we address the semimicro Kjeldahl assay. The semimicro variant is a modification of the macro-Kjeldahl method designed to quantify nitrogenous compounds in relatively small quantities, in which the total nitrogen content is in the range of 1 to a few milligrams.

The ability of molecules to absorb light at specific wavelengths is the principle of the spectrophotometric methods used to quantify specific compounds. Protein can be quantified by spectrophotometric methods based on the absorption of (1) colored products formed by reactions between protein molecules and other compounds or (2) specific functional groups, such as aromatic rings or specific amino acids [1, 6]. All the methods described in this section are applied only to protein in solution.

Some methods might be affected by amino acid composition, which should be considered when selecting the best assay to analyze a specific sample. Also, some substances other than the analyte of interest can generate color, being known as interferents in such assays. One should consider all the possible interferents that might cause changes in the protein concentration measurement [6].

In the Bradford assay, acid and basic functional groups of amino acid residues can interact with oppositely charged organic dyes. This method is based on the absorbance shift of the Coomassie Brilliant Blue dye upon binding with protein structures. When the dye binds to the protein, the reaction stabilizes the protonated form of the dye that absorbs at 595 nm. The reaction occurs in acidic conditions, in which the dye assumes the blue color [3].

Colorimetric assays for proteins can be based on the binding of Cu^{2+} to proteins, which yields a colored product, such as the Biuret and Lowry methods. The color intensity and the protein concentration are linearly proportional in a specific range. In the Biuret method, the peptide bonds of protein structure react with Cu^{2+} under alkaline conditions, yielding a violet $Cu2^+$-protein complex that absorbs at 540 nm. The higher the number of peptide bonds (i.e., the protein concentration), the higher the absorption at this specific wavelength. The reaction does not depend on amino acid composition since Cu^{2+} ions react directly with the peptide backbone [2, 6, 10].

The Lowry method [4, 7] is derived from the Biuret method. The Lowry method is based on 2 reactions: (1) Amino acidic residues react with copper in alkali; (2) subsequently, the copper-treated protein reduces the Folin-Ciocalteu reagent (phosphomolybdic-phosphotungstic reagent), yielding a blue chromophore which absorbs at 750 nm. Using the Folin-Ciocalteu reagent increases color development and allows the detection of lower protein concentrations. This method is around 100 times more sensitive than the Biuret assay. This method is very useful for protein determination in serum, antigen-antibody precipitates, cell lysates, and samples with low protein content or available in small quantities.

The bicinchoninic acid (BCA) assay or Smith [9] assay relies on the formation of a Cu^{2+}-protein complex in a basic environment, followed by the reduction of the Cu^{2+} to Cu^{+}. The amount of Cu^{2+} that is reduced is proportional to the amount of protein in the solution. Every two molecules of BCA chelate to each Cu^{+} ion, causing a change of color from green to purple with strong absorbance at 562 nm.

Quantifying a protein in a solution is possible using only a spectrophotometer. Proteins in solution absorb ultraviolet light in the vicinity of 280 nm. The absorption at 280 nm (A_{280}) by proteins depends on the Tyr and Trp content and, to a small extent, on the amount of Phe and disulfide bonds [1]. The A_{280} varies significantly between different proteins and is mainly in the range of 0.5–1.5.

The advantages of this method are that it is simple and the sample is recoverable. The method has some disadvantages, including interference from other chromophores and the specific absorption value for a given protein must be determined. The extinction of nucleic acid in the 280-nm region may be as much as tenfold that of protein at the same wavelength, and hence, a few percent of nucleic acid can significantly influence absorption [1].

2 Material

2.1 Nitrogen Content: Kjeldahl Method

Apparatus

- Digestion block.
- Kjeldahl distiller (Fig. 1).
- Analytical balance.

Material and Reagents Prepare all solutions using deionized or distilled water and analytical grade reagents.

- Kjeldahl tubes, Erlenmeyer flasks (250 mL), pipettes, volumetric flasks, burette.
- 40% NaOH solution (m/m): Dissolve 400 g NaOH in 600 mL deionized water in an ice bath. Mix until completely dissolved.

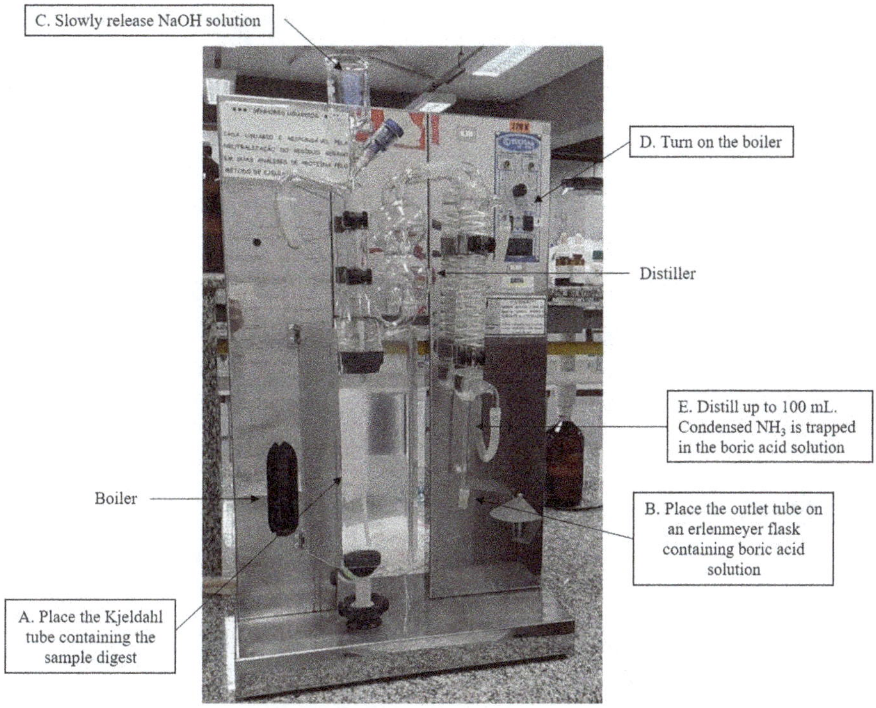

Fig. 1 Kjeldahl distiller and main steps in Kjeldahl assay

Store the solution in a plastic container. *Remind*: Work on a hood and use proper personal protective equipment (PPE).

- conc H_2SO_4: Always manipulate concentrated acid solution in the hood, using proper PPE.
- Catalyst: Mixture 99.5 g K_2SO_4 and 0.5 g $CuSO_4$. This mixture is stable and can be stored in a plastic container in a dry place indefinitely.
- Indicator solution: mixture 40 mL 0.1% methyl red (m/v) in ethanol and 10 mL 0.1% methylene blue (m/v) in 50% ethanol.
- Boric acid solution: Dissolve 40 g boric acid (H_3BO_3) in 800 mL deionized water. After completely dissolved, add 15 mL indicator solution and adjust the volume to 1 L with deionized water. Store under refrigeration.
- Standardized 0.02 eq/L HCl solution: Slowly add 3.3 mL of conc HCl (37%; $d = 1.19$ g/mL) to 500 mL deionized water. Adjust the final volume of the solution to 2000 mL with deionized water. To standardize this solution, titrate sodium carbonate (Na_2CO_3) using this HCl solution in a burette, following the steps:

  (a) Dry Na_2CO_3 for 2 h in a muffle furnace at 250 °C.

  (b) In triplicate, weigh out 21.2 mg of dried Na_2CO_3 in an Erlenmeyer flask, and add volumetrically 20 mL of decarbonated water. To decarbonate, the water should be previously boiled to remove the carbonic gas.

(c) Fill a burette with the HCl solution and proceed with the titration.

(d) Calculate the HCl concentration using Eq. 1. The standard deviation among the triplicate must be <1%.

$$NHCl = (mNa_2CO_3/eq\ Na_2CO_3)/V\ HCl \quad (1)$$

Where:
mNa_2CO_3 = sodium carbonate mass (mg).
Eq Na_2CO_3 = sodium carbonate equivalent (53).
V HCl = HCl volume (mL).

2.2 Bradford Assay

Apparatus

- Spectrophotometer.
- Analytical balance.

Material and Reagents

- Bradford reagent: Dissolve 10 mg of Coomassie Brilliant Blue G-250 in 10 mL of 95% ethanol. Add 10 mL 85% phosphoric acid and mix well. Adjust the volume to 100 mL with distilled water. Filter the solution using Whatman No. 1 filter paper. This solution is stable for 2 weeks stored in an amber flask at room temperature (20–25 °C).
- Standard solution: BSA (1.0 mg/ml): Weigh exactly 10 mg BSA and dissolve in roughly 8 mL of deionized water. Do not vortex to avoid foaming. When completely dissolved, transfer the solution to a 10 mL volumetric flask and complete the volume with deionized water. This solution can be aliquoted in microtubes, quickly frozen (preferably in dry ice or liquid nitrogen), and stored at—20 °C for 1 year. This solution is expected to have an absorbance value around 0.66 at 280 nm [3].

2.3 Biuret Method

Apparatus

- Spectrophotometer (VIS).
- Analytical balance.

Material and Reagents.

Prepare all solutions using deionized or distilled water and analytical grade reagents. Prepare and store all reagents at room temperature (25 ± 2 °C).

- Erlenmeyer flasks, pipettes, volumetric flasks, and glass or plastic (disposable) cuvettes.
- $CuSO_4.5H_2O$.

- Sodium and potassium tartrate.
- 10% NaOH solution (w/v): Dissolve 100 g NaOH in 800 mL deionized water in an ice bath. Mix it until completely dissolved. Transfer to a volumetric flask and complete the volume to 1000 mL. Store the solution in a plastic container. Always work on the hood and use proper PPE.
- Biuret reagent*:

  (a) Weigh 1.50 g $CuSO_4.5H_2O$ and 6.00 g sodium potassium tartrate.

  (b) Add 500 mL decarbonated water and stir. To decarbonate, the water should be previously boiled to remove the carbonic gas.

  (c) While stirring, slowly add 300 mL 10% NaOH; transfer the solution to a 1 L volumetric flask; complete the volume with decarbonated water.

  (d) Store the reagent in an amber plastic recipient at room temperature. This solution is stable and can be stored at room temperature. If the reagent shows any sign of precipitation, discard using a proper way.

  *The Biuret reagent is commercially available.
- Standard solution: BSA (10 mg/ml)—Weigh precisely 100 mg BSA and dissolve in roughly 8 mL deionized water. Do not vortex to prevent foaming. When completely dissolved, transfer the solution to a 10 mL volumetric flask and complete the volume with deionized water. This solution can be aliquoted in microtubes, quickly frozen (preferably in dry ice or liquid nitrogen), and stored at—20 °C for up to 1 year.

2.4 Lowry Method

Apparatus

- Spectrophotometer.
- Analytical balance.

Material and Reagents.

Prepare all solutions using deionized or distilled water and analytical grade reagents. Prepare and store all reagents at room temperature unless indicated otherwise.

- Reagent A: 2% Na_2CO_3 in 100 mmol/L NaOH.
- Reagent B: 0.5% $CuSO_4.5H_2O$ in 1% sodium and potassium tartrate. If using hydrate tartrate, make sure the concentration is corrected by the anhydrous salt.
- Folin-Ciocalteu reagent (phosphomolybdic-phosphotungstic reagent) 1 eq/L. Consider the original concentration of the commercial reagent. Most of the time, it is 2 eq/L. In this case, dilute the reagent 1:1 with distilled water right before use.

- Standard solution: BSA (300 μg/mL)—Weigh precisely 30 mg BSA and dissolve in roughly 80 mL of deionized water. Do not vortex to avoid foaming. When completely dissolved, transfer the solution to a 100 mL volumetric flask and complete the volume with deionized water. This solution can be aliquoted in microtubes, quickly frozen (preferably in dry ice or liquid nitrogen), and stored at—20 °C for 1 year.

2.5 Bicinchoninic Acid Assay

Apparatus

- Spectrophotometer.
- Analytical balance.

Material and Reagents.

Prepare all solutions using deionized or distilled water and analytical grade reagents. Prepare and store all reagents at room temperature (25 ± 2 °C) unless indicated otherwise.

- Reagent A: Weigh the following reagents: 10 g BCA, 20 g Na_2CO_3, 1.6 g Na_2 tartrate, 4 g NaOH, and 9.5 g $NaHCO_3$. Add distilled water and mix well. Complete the volume for 1 L. If needed, add NaOH or solid $NaHCO_3$ to adjust the pH to 11.25.
- Reagent B: 4% $CuSO_4.5H_2O$ in deionized water (2 g $CuSO_4.5H_2O$ in 50 mL distilled water). Reagents A and B are stable for at least 12 months at room temperature. Both reagents A and B are commercially available.
- BCA working solution: Mix reagents A and B in the proportion of 50:1 (the reagent is apple green in color). This solution can be stored stably for up to a week.
- Standard solution: BSA (1.0 mg/ml)—Weigh precisely 10 mg BSA and dissolve in roughly 8 mL of deionized water. Do not vortex to avoid foaming. When completely dissolved, transfer the solution to a 10 mL volumetric flask and complete the volume with deionized water. This solution can be aliquoted in microtubes, quickly frozen (preferably in dry ice or liquid nitrogen), and stored at—20 °C for 1 year.

2.6 Based on Absorption at 280 nm

Apparatus

- Spectrophotometer.
- Analytical balance.

3 Methods

3.1 Nitrogen Content: Kjeldahl Method

Remember, always wear appropriate PPE, gloves, glasses, a lab coat, and closed shoes!

Digestion

1. Weigh 20–300 mg of sample in Kjeldahl tubes. The amount depends on the protein concentration. The sample should have from 20 to 40 mg of protein to get good results. For example, a protein isolate usually contains 0.9 g of protein/g of sample; thus, around 20 mg of the sample would be sufficient for the analysis using the current protocol. Carry out a blank assay with no sample (*see* **Notes 1** and **2**).
2. Add 0.5–1.0 g catalyst to each tube.
3. Carefully add 5 mL conc H_2SO_4. Always work in a hood and wear adequate PPE to handle concentrated acid (*see* **Notes 2** and **3**). The acid volume might vary depending on the sample. The total volume should not exceed 1/4 of the tube.
4. Place the tubes in the heater block, and the digestion should be performed inside a hood appropriate/resistant to acid fumes. Start the digestion at 100 °C and gradually increase the temperature (50 °C every 30 min) up to 400 °C. Monitoring all the samples frequently is important to avoid overflow or splattering. Keep the samples at 400 °C for at least 2 h. If samples still present a brownish color or particles, turn the heater on again and digest it further at 400 °C for a couple of hours more. Standardize the digestion time for each type of sample.
5. To avoid salt caking and bumping during neutralization, add very carefully around 10 mL of ammonium-free water while the digestion mixture is still warm.
6. Let the tubes cool down to room temperature before proceeding.

Neutralization and Distillation

This step is performed in a Kjeldahl distiller (Fig. 1).

1. Place the tube containing the sample digest in the proper support (Fig. 1a).
2. Insert the outlet tube (Fig. 1b) into a 250 mL Erlenmeyer containing 50 mL of acid boric solution with indicator solution. The outlet tube must be completely immersed in the boric acid solution.
3. Add 25 mL of 40% NaOH to the proper compartment of the Kjeldahl distiller (Fig. 1c). Slowly release the NaOH solution to

neutralize the acid digest. Stop releasing NaOH when the digested sample reaches the turning point (brown color). Generally, for each 5 mL of H_2SO_4 used in the digestion, 25 mL of 40% NaOH is required to increase the pH to values >11.

4. Turn on the boiler (Fig. 1d).
5. Distill until a volume of 100 mL is collected in the Erlenmeyer (Fig. 1e). The boric acid captures the ammonia gas, forming an ammonium-borate complex. The boric acid solution changes from purple to light green. The receiving solution should remain below 45 ° C during distillation to prevent the loss of ammonia.
6. The residue remaining in the digestion tube must be stored in a proper plastic container and professionally disposed of. Make sure to follow the guidelines from your Laboratory for appropriate waste disposal.

Titration

1. Titrate the boric acid solution with a standardized HCl solution (0.02 eq/L) in a 5–50 mL burette. The adequate burette volume depends on the HCl volume needed per sample. The changing color from light green to light purple characterizes the titration endpoint. The original boric acid solution was diluted; thus, the purple color might be less intense than the original solution.
2. Calculate the nitrogen and protein concentrations according to the item "Conversion Factor and Calculation."

Conversion Factor and Calculation

The calculations for % nitrogen or % protein explained here are adequate for acid boric as the receiving solution. The nitrogen content obtained by the Kjeldahl method can be calculated according to Eq. 2.

$$\%N = \frac{(V\ HCl\ sample - V\ HCl\ blank)\ x\ N\ HCl\ x\ 14.007}{M\ sample} \times 100 \tag{2}$$

Where:

V HCl = volume (mL) of the standardized HCl solution used in the titration.

N HCl = concentration of the standardized HCL solution (eq/L).

M sample = mass (mg) of the sample used in the analyses.

14.007 is the atomic mass of the nitrogen

Table 1
Average conversion factors for some food proteins

Product	Conversion factor
Milk	5.85
Casein	6.15
Cheese	5.85
Fish	5.58
Whole egg	5.68
Egg white	5.74
Beef	5.48
Chicken	5.53
Wheat flour and products	5.52
Rice	5.34
Corn	5.62
Soybean/soy flour	5.50
Pea	5.36
Dry bean	5.28
Vegetables and leaf protein	5.6

The protein concentration is determined according to Eq. 3:

$$\%protein = \%N \times conversion\ factor \tag{3}$$

An adequate conversion factor is necessary to convert the nitrogen concentration into protein concentration (g protein/ 100 g sample). The conversion factor represents the concentration of N in the protein. Most animal proteins contain 16 g of nitrogen /100 g protein; thus, the general conversion factor is 6.25 (i.e., 100/16). The average nitrogen content for plant proteins is commonly higher, roughly 17.5 g/100 g. Thus, the general conversion factor is 5.7 (i.e., 100/17.5). Such values can be used if there is no specific conversion factor for the protein in the study (e.g., unknown protein) or a mixture of different proteins in the analyzed sample. The conversion factors of some proteins based on their average amount of nitrogen are shown in Table 1.

Adapted from [5]. The conversion factor was calculated from amino acid analysis.

If the amino acid composition is available, a more accurate conversion factor can be calculated [6].

Troubleshooting: Kjeldahl

Problem	Possible causes	Action
Protein concentration higher than expected	If a sample has nitrogen from sources other than protein, the protein content will be overestimated (*see* **Notes 1–5**)	Protein can be precipitated by TCA addition in a protein dispersion (final concentration TCA 8–10%). Determine nitrogen concentration before and after protein precipitation. The difference between results is equivalent to nonprotein nitrogen
Protein content lower than expected	1. Inadequate or excessive time for digestion. 2. Sample loss during digestion due to overflowing. 3. Leak in Kjeldahl distiller connections. 4. Inadequate amount of NaOH solution (pH must reach values >11). 5. Bumping or foaming of the digested solution during NaOH addition or distillation. 6. Inadequate or excessive distillation.	1. Verify if conc. H_2SO_4 amount and/or the time is adequate and sufficient to digest the sample. 2a. The total volume in the digestion tube should not exceed ¼ of the tube volume. 2b. Use glass beads to prevent foaming and overflow 3. Test all the connections before using the distiller 4. For more than 5 mL of conc H_2SO_4 adjust the amount of NaOH solution 5a. Dilute the digested solution with distilled water (cautiously) before starting the distillation step 5b. The NaOH should be added slowly to digestion 6. An adequate volume to collect is 100 mL. Remember that the outlet tube must be completely immersed in the acid solution to prevent gas escape

(continued)

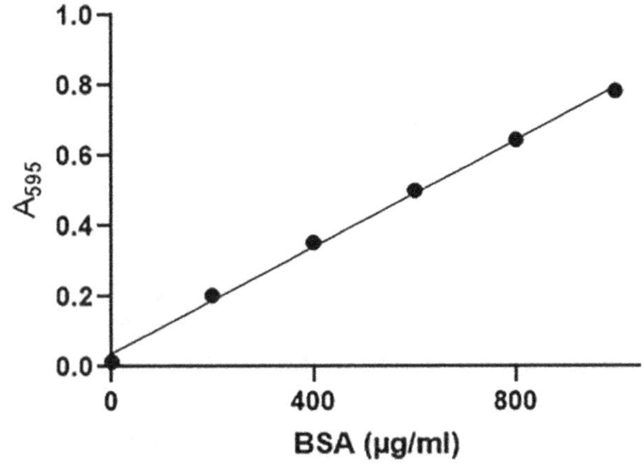

Fig. 2 Illustrative standard curve using BSA as standard protein (adapted from [1])

Problem	Possible causes	Action
HCl volume higher than 50 mL	High protein concentration	Decrease the sample weight or use a more concentrated standardized HCl solution

3.2 Spectrophotometric Methods

General Considerations for Spectrophotometric Methods

1. The calculation of protein concentration is based on a linear regression equation obtained by evaluating standard curves of bovine serum albumin. Figure 2 shows an illustrative standard curve. Remember that the absorbance of the sample must be within the standard curve range, ideally in its center. If the absorbance reading falls out of the curve, repeat the analysis, changing the sample concentration to fit the curve.
2. To perform the readings, the volume in the cuvette should cover at least 2/3 of its total capacity. If the method work volume does not cover 2/3 of the cuvette capacity, increase all the volumes proportionally to perform the readings.
3. Spectrophotometric methods are based on the absorption of a colored product or specific structure at a specific wavelength. The quantity of light absorbed by the substance is directly proportional to the concentration of the substance within a proper range. This linear range varies depending on each measured substrate. Above the absorbance value of 1.0, the absorbance measurement loses linearity; thus, the absorbance will not be directly proportional to the concentration of the substance. Ideally, the absorbance values for the standard curve

should be between 0.1 and 0.8. The samples should be diluted to reach absorbance values in this range, ideally in the center of the curve.

4. Samples evaluated by spectrophotometric measurement must be completely soluble. The solution must be clear and without turbidity to prevent interference in the results. The transmittance at 50—600 nm should be >90% for a good reading.
5. Since all the reagent proportions are maintained, the colorimetric assays can be carried out using a 96-well plate in a microplate reader instead of a spectrophotometer and cuvettes. Compare the results with the full method to validate them. Using multi-well plates, the reagents are added nearly simultaneously for all wells using a multichannel pipette, which can significantly decrease the variation between samples. However, the use of calibrated pipettes is of crucial importance. The smaller the volume, the higher the impact on the results due to deviation in pipetting. In the case of using a microplate reader for the BCA method, the equipment can be set to include the incubation step at 37 °C to speed up the analysis.
6. Remember, foam is composed of protein. Therefore, if the foam is formed during the procedures and sticks to glassware, it represents a loss of the analyte, and protein concentration might be underestimated.
7. In all spectrophotometric protocols, the absorbance is related to the concentration of the standard or sample. Therefore, if the sample shows a high absorbance value, it is necessary to dilute the sample and perform the reaction with the diluted sample. Remember to consider any sample dilution made previously to pipetting it on the reaction tubes.

Table 2
Set up the standard curve for the Bradford assay

	Distilled water (μl)	1.0 mg/mL BSA (standard solution) (μl)	BSA concentration (μg/ml)
Set 1	100	0	0
Set 2	80	20	200
Set 3	60	40	400
Set 4	40	60	600
Set 5	20	80	800
Set 6	0	100	1000

3.2.1 Based on Protein-Dye Interaction: Bradford Assay (See **Notes 6** and **7**)

Carry out all procedures at room temperature unless otherwise specified.

1. *Standard curve: One should perform a triplicate for each protein concentration*. Prepare a set of three tubes for each concentration of the standard solution. Add the appropriate volume of distilled water and standard solution according to Table 2—Bradford assay. Set 1 is the blank, with only water rather than a sample or standard. It can be used to zero the equipment. If purified, the protein of interest can be used at known concentrations to generate a more accurate standard curve.
2. If necessary, dilute the sample to reach an estimated concentration that fits the standard curve range. Consider this dilution factor in the final concentration calculation. Prepare sample tubes with 100 μL of the diluted sample (*see* **Note 8**).
3. Add 5 mL Bradford reagent to all test tubes. Vortex gently to prevent foaming, which will lead to poor reproducibility (*see* **Note 9**).
4. Incubate 5 min at room temperature (25 ± 2 °C).
5. Read the absorbance at 595 nm using disposable cuvettes (*see* **Note 10**).
6. Plot the absorbance at 595 nm versus the BSA protein standard concentrations (μg/mL). The standard curve (absorbance × concentration) must present an $R^2 > 0.98$.
7. Use the standard curve to determine the protein concentration of each unknown sample.

Troubleshooting: Bradford

Problem	Possible causes	Action
Absorbance values higher than expected	Sample turbidity	Some proteins do not solubilize in the acid Coomassie solution. If it is the case, select another method. Consider that most proteins tend to be more soluble in an alkaline environment; thus, assays such as Lowry, biuret, or BCA could be more appropriate
	Traces of dye bound to the cuvette	Remember to use disposable plastic cuvette.
Lower concentration than expected	Presence of peptides Low level of aromatic residues and/or arginine residues	If possible, check the composition of the tested protein

(continued)

Table 3
Set up the standard curve for the Biuret assay

	Distilled water (μl)	10 mg/mL BSA (Standard solution) (μl)	BSA concentration (mg/ml)
Set 1	1000	0	0
Set 2	800	200	2
Set 3	600	400	4
Set 4	400	600	6
Set 5	200	800	8
Set 6	0	1000	10

Problem	Possible causes	Action
	underestimate the protein measurement since the dye does not bind to the protein as efficiently	

For more troubleshooting, see General Troubleshooting: Spectrophotometric Methods.

3.2.2 Based on Copper-Protein Interactions

Biuret Method (*See* **Notes 11–17**)

Carry out all procedures at room temperature (25 ± 2 °C) unless otherwise specified.

1. *Standard curve: One should perform a triplicate for each protein concentration.* Prepare a set of three tubes for each concentration of the standard solution. Add distilled water and standard solution according to Table 3—Biuret assay. Set 1 is the blank, with only water rather than a sample or standard. It can be used to zero the equipment. If purified, the protein of interest can be used at known concentrations to generate a more accurate standard curve.
2. If necessary, dilute the sample to reach an estimated concentration that fits the standard curve range. Consider this dilution factor in the final concentration calculation. Prepare sample tubes with 1 mL of the diluted sample.
3. Add 4 mL of the Biuret reagent and mix well.
4. Incubate at room temperature for 30 min.
5. Read the absorbance at 540 nm.
6. Plot the absorbance at 540 nm versus the BSA protein standard concentrations (mg/mL). The standard curve (absorbance x concentration) must present an $R^2 > 0.98$.
7. Use the standard curve to determine the protein concentration of each unknown sample.

Table 4
Set up the standard curve for the Lowry assay

	Distilled water (μl)	300 μg/mL BSA (standard solution) (μl)	BSA concentration (μg/ml)
Set 1	500	0	0
Set 2	450	50	30
Set 3	400	100	60
Set 4	350	150	90
Set 5	300	200	120
Set 6	250	250	150
Set 7	200	300	180
Set 8	100	400	240

Troubleshooting—Biuret.
See General Troubleshooting: Spectrophotometric Methods.

Lowry Method (*See* **Notes 18–26**)

Carry out all procedures at room temperature.

1. *Standard curve: One should perform a triplicate for each protein concentration.* Prepare a set of three tubes for each concentration of the standard solution. Add the appropriate volume of distilled water and complete the volume to 500 μl with the standard solution (Table 4—Lowry assay). Set 1 is the blank, with only water rather than a sample or standard. It can be used to zero the equipment. If purified, the protein of interest can be used at known concentrations to generate a more accurate standard curve.
2. Dilute the sample, if necessary, to reach an estimated concentration that fits the standard curve range. Consider this dilution factor at the final concentration calculation. Ensure the sample is totally soluble and the solution is perfectly clear; otherwise, it can influence the absorbance reading. Consider that many insoluble proteins may dissolve readily in the alkaline copper reagent. A previous test on an alkaline solution can be carried out to check the protein solubility.
3. Add 500 μL of each sample to test tubes.
4. To obtain Lowry solution, mix reagents A and B in a 9:1 proportion. The final volume should consider that 5 mL is needed for each sample. This mixture must be freshly prepared.
5. Add 5 mL Lowry reagent to each test tube containing the sample.

6. Mix well and allow to stand for 10 min at room temperature (25 ° ± 2 °C) (*see* **Note 19**).
7. Dilute the Folin-Ciocalteu reagent to reach 1 eq/L concentration.
8. Add 500 μL of 1 eq/L Folin-Ciocalteu reagent and immediately vortex to mix it thoroughly. The complete mixing must be carried out just after the reagent addition. Thus, the reagent should be added, and the tube vortexed one at a time.
9. Incubate 30 min at room temperature (25 ° ± 2 °C).
10. Read the standards and samples on a spectrophotometer at 750 nm.
11. Plot the absorbance at 750 nm versus the BSA protein standard concentrations (μg/mL). The standard curve (absorbance x concentration) must present an $R^2 > 0.98$.
12. Use the standard curve to determine the protein concentration of the unknown sample.

Troubleshooting: Lowry

Problem	Possible causes	Action
Absorbance values higher than expected	Sample turbidity	Increase protein solubility by dispersing the sample in an alkaline solution prior to the analysis
	Aromatic compounds other than proteins	Interferents that can reduce the Folin-Ciocalteu reagent will increase the absorbance values. If it is not possible to remove those compounds, select another method to determine the protein concentration

For more troubleshooting, see General Troubleshooting: Spectrophotometric Methods.

Bicinchoninic Acid (BCA) Assay (*See* **Notes 27–34**)

Carry out all procedures at room temperature (25 ± 2 °C).

1. *Standard curve: One should perform a triplicate for each protein concentration*. Prepare a set of three tubes for each concentration of the standard solution. Add the appropriate volume of distilled water and standard solution according to Table 5—BCA assay. Set 1 is the blank, with only water rather than a sample or standard. It can be used to zero the equipment. If purified, the protein of interest can be used at known concentrations to generate a more accurate standard curve.

Table 5
Set up for the standard curve for the BCA assay

	Distilled water (μl)	1 mg/mL BSA (standard solution) (μl)	BSA concentration (μg/ml)
Set 1	100	0	0
Set 2	80	20	200
Set 3	60	40	400
Set 4	40	60	600
Set 5	20	80	800
Set 6	0	100	1000

2. If necessary, dilute the sample to reach an estimated concentration that fits the standard curve range. Consider this dilution factor in the final concentration calculation.
3. Prepare sample tubes with 100 μL pf sample (*see* **Note 29**).
4. Add 2 mL of BCA working solution to all tubes. Vortex gently for thorough mixing.
5. Incubate for 30 min at 37 °C (*see* **Notes 30** and **31**).
6. Allow the tubes to cool down to room temperature.
7. Read absorbance at 562 nm. All the tubes should be read within 10 min to avoid significant errors due to continuous color development.
8. Plot the absorbance at 562 nm versus the BSA protein standard concentrations (μg/mL). The standard curve (absorbance x concentration) must present an $R^2 > 0.98$.
9. Use the standard curve to determine the protein concentration of the unknown sample.

Troubleshooting: BCA

Problem	Possible causes	Action
Absorbance values higher than expected	High time incubation	Change temperature x time incubation. Higher temperatures such as 37 °C tend to increase color development. Incubation at room temperature might be more adequate in some cases

For more troubleshooting, see General Troubleshooting: Spectrophotometric Methods.

3.2.3 Based on Absorption at 280 nm (See Note 35)

This method can obtain rough estimates of protein in a solution and is more indicated for pure samples.

1. Prepare a solution with the sample with roughly 1 mg protein/mL. The A_{280} of 1 mg protein/mL should range from 0.8 to 2.2, depending on the protein composition. The sample concentration should be high enough to reach an absorbance reading of around 0.8. The sample must be completely soluble, and it is advised to filter the solution through a 0.22 μm filter or centrifuge to remove particles. Consider that particles may be protein aggregates, and their removal decreases the amount of total protein. To solubilize these aggregates, 1% SDS (sodium dodecyl sulfate-anionic detergent) can be used. Note that this procedure denatures proteins and is considered destructive. Consider this addition to the blank.
2. Fill a quartz cuvette (at least ¾) (*see* **Note 1**) with the chosen solvent and record the spectra from 250 to 350 nm to evaluate if compounds present in the solvent absorb at 280 nm and could interfere with the results. Save the results as blank for calculation.
3. Fill the quartz cuvette with the sample and record the results in the same way used for the solvent (blank).

Calculation

(2) If the ratio of the readings at 280 nm and 260 nm (A_{280}/A_{260}) is around 1.75, one can consider the following equation (Eq. 4):

$$[\text{protein (mg/ml)}] \sim A_{280} \tag{4}$$

(3) If the sample has nucleic acids, a rough protein estimation can be obtained by measuring the absorbance at 260 and 280 nm and using Eq. 5:

$$[\text{protein (mg/mL)}] = F \times A_{280} \tag{5}$$

where $F = A_{280}/A_{260}$.

(3) If working with a purified protein with known extinction coefficient ($\varepsilon_{1\text{mg/mL}}$), one can use Eq. 6, related to the Beer's law. In the case the sample is a pure protein, one can use a standard curve using such protein, following the serial dilution preparation as described in the previous methods in this Chapter.

$$A_{280} = \varepsilon_{1\text{mg/mL}} \times c \times l \quad (6)$$

where l = the cuvette pathlength (in general 1 cm).

c = protein concentration.

ε = extinction coefficient - usually expressed in 1% or 0.1%.

$\varepsilon_{1\text{mg / ml}}$ = 0.4–1.5 (values for most of the proteins).

Troubleshooting: UV

Problem	Possible causes	Action
Absorbance values higher than expected	Sample turbidity	Increase protein solubility by pretreatments (detergents, buffers, etc.). Take care not to use any compound that can absorb at 280 nm
	Glass/plastic cuvette	The cuvette must be made of quartz. Glass and plastic cuvettes absorb at 280 nm, and the result will be compromised
Absorbance values lower than expected	Protein lacks Phe, Tyr, or Trp residues	If known, check the protein composition. In case it lacks aromatic residues, choose a different method for total protein determination

3.2.4 General Troubleshooting: Spectrophotometric Methods

Problem	Probable causes	Action
Absorbance values higher than expected	Sample turbidity	Increase protein solubility by diluting the sample in an alkaline solution prior to the analysis
Absorbance values not directly proportional to the increase in sample concentration	High protein concentration means high absorbance values; thus, the increase in absorbance values will not be linear according to the Lambert-beer law	Dilute the sample before performing the assay. Do not dilute the solution from an already performed reaction. Absorbance values should ideally range from 0.1 to 0.8

Table 6
Useful detection range, interferents, difficulty level, reaction pH, and sample requirements for each of the main methods for total protein determination

Method	Useful detection range	Interferents	Difficulty level	Reaction pH	Sample requirements
Bradford	200 to 1400 μg/mL[a]	Low	Low	Acid	High protein solubility at acid pH
Biuret	500 to 80,000 μg/mL[a]	Moderate	Low	Alkaline (pH > 11)	High protein solubility at alkaline pH
Lowry	1 to 300 μg/mL[a]	High	Moderate	Alkaline (pH > 11)	High protein solubility at alkaline pH
BCA	200 to 1000 μg/mL[b]	Moderate	Low	Alkaline (pH > 11)	High protein solubility at alkaline pH
Abs 280 nm	Depends on the protein composition	Moderate	Low	Depends on the sample and solvent.	Soluble protein containing aromatic amino acid residues
Kjeldahl (semimicro)	1 mg nitrogen	Moderate	High	pH varies through the analysis, but it has no influence on the results	

Detection ranges from: [a] [1]; [b]Owusu-Apenten [6].

4 General Comments

1. Samples may contain substances that make them incompatible with any of the protein assay methods. In those cases, some pretreatment of the sample is necessary. If the sample has many interferent compounds, the best option is the Kjeldahl method. However, nonprotein nitrogen can overestimate the protein content.
2. Spectrophotometric methods require complete protein solubility under experimental conditions. The pH has a significant impact on protein solubility. All the copper-based methods are performed at alkaline pH, where the protein is more likely to be soluble. The Bradford method occurs at acid pH, which is not so favorable for protein solubility. The Kjeldahl method has no issues concerning the pH and protein solubility since the organic material is completely digested (Table 6).

The assay conditions, such as pH, concentration, reagents, and volume, might impact the result. It is of utmost importance to remember that the methods are based on different principles. Choosing the adequate assay for your sample and objectives must be considered, since each method for total protein determination does not measure protein directly but relies on different reactions or the presence of specific components. The method selection must consider speed, sample type, protein composition, presence of interferents, sensitivity required, and available apparatus.

5 Notes

1. To weigh sticky samples, use a piece of parchment paper (roughly 5 × 5 cm). Carefully put the sample on the paper, fold the edges over the sample, and transfer the packed sample to the tube. Run a blank assay with only a piece of parchment paper inside the tube.
2. Liquid samples can be analyzed by the Kjeldahl method. However, the total volume in each tube should not exceed 1/4 of the digestion tube's total volume. If higher volumes are needed, an evaporation step in a boiling water bath should be carried out until the sample reaches a volume of 1–2 mL. Then, add the catalyst and the concentrated H_2SO_4 (*see* **steps 2** and **3**) and proceed with the digestion (**step 4** onward).
3. High-fat samples may foam during the digestion step. Glass pearls are recommended to avoid overflow, which causes loss of the sample and damage to the heating block.
4. Remember that all nitrogen will be determined, including the nonprotein nitrogen from compounds, such as free amino acids, nucleic acids, nitrates, and urea, among others. If a product is adulterated with the addition of nitrogen compounds other than protein, this method cannot identify the adulteration.
5. One of the greatest advantages of this method is that there is no need to solubilize the protein; thus, matrices with a large range of characteristics can be analyzed.
6. This assay depends on the amino acid composition of proteins; thus, variation within different proteins should be expected.
7. The sensitivity of this method is 4–10 times higher than the Lowry assay.
8. The binding reaction between protein and Coomassie dye occurs when the peptide molecule has a minimum of nine amino acid residues. Thus, the Bradford assay is not suitable for quantifying compounds, such as protein hydrolysates and important bioactive peptides with a molecular mass smaller than 3000 Da.

9. The Bradford reagent should be added to standard and test samples sequentially, and the readings should be done in the same sequence due to the continuing color development of the protein-dye complex.
10. Plastic disposable cuvettes are highly recommended since the dye binds to the cuvette material and thereby interferes with the subsequent readings.
11. This method does not measure dipeptides and free amino acids because the reaction requires at least two peptide bonds to occur.
12. The reaction occurs under alkaline pH; thus, it is more likely to solubilize protein.
13. Anthocyanins and other flavonoids, lycopene, chlorophyll, lignins, and low-molecular-weight derivatives, DNA, glucose, starch, buffer components, TRIS, etc. (reactive) might interfere with the result.
14. The presence of interferents has much more impact on the result than the protein composition.
15. Copper-chelating compounds (such as EDTA) can complex the metal ion and underestimate the result.
16. If a sample shows turbidity under the analysis conditions, replace 200 μL of water with 200 μL 5% deoxycholate solution in 0.1 mol/L KOH. Deoxycholate is a strong detergent and can aid in clarifying solutions containing protein aggregates. If this strategy is used, remember to add the same amount of deoxycholate to the standard solutions and run a blank containing the detergent.
17. The Biuret assay can also be called the alkaline copper reagent (ACR) test.
18. To avoid mistakes in protein quantification, one should carefully consider the composition of the sample because some components can impair the results. The Folin reagent is a mixture of phosphomolybdate and phosphotungstate and can be reduced by any reducing compounds. It is used for determining total phenolic compounds. For instance, vegetal matrices generally present phenolic compounds, which can reduce the Folin reagent and yield an overestimated protein concentration.
19. The reaction with Cu^{2+} is nearly complete within 5–10 min at room temperature. However, longer times, up to 30 min, may increase the color development. The protocol can be optimized, but it is recommended that the time for each step be standardized.

20. Proteins with a high level of aromatic amino acids can yield higher absorbance values due to the reducing capacity of tyrosine and tryptophan residues.
21. The reaction depends on the complexation of Cu^{2+} to the protein chain to develop the measured color. Thus, chelating agents (such as EDTA) can complex the copper ions and compromise the reaction with protein chains. Samples should be free of any chelating agent.
22. This method is robust and highly sensitive. It is more sensitive than the Biuret assay and can be used for samples with lower protein concentrations.
23. The reaction depends on pH (10.0–11.5), but since the Folin-Ciocalteu reagent is unstable at alkaline pH, it must be freshly diluted just before the analysis.
24. The reaction partially depends on amino acid composition; thus, variation among different proteins might occur.
25. If the sample changes the pH of the reagent, the color might be decreased.
26. This method is very sensitive but susceptible to interferents.
27. Reducing sugars (e.g., glucose, fructose, lactose, etc.) can interfere with the results since they can also reduce Cu^{2+} to Cu^{+}. Make sure the sample does not contain reducing sugar; otherwise, another method should be selected.
28. Copper-chelating reagents such as EDTA can complex Cu ions and underestimate the result. Thus, samples containing chelating compounds are not recommended for analysis using Cu-protein-based methods.
29. Detergents such as SDS can be used to solubilize proteins. Unlike the Lowry method, the BCA method is less impacted by detergents in the samples.
30. The incubation time is not critical as long as the time reaction is the same for the standards and samples. Always run a blank assay.
31. For low protein concentrations, the incubation (*see* **step 5**) could be adapted to higher temperatures and/or longer incubation times for increased color development and, thus, increased sensitivity. If using higher temperatures, the possibility of sample loss through evaporation must be considered; therefore, remember to seal all the tubes properly and run a test assay to ensure no influence on the final volume. Also, the reaction tubes should be cooled down to room temperature before reading. Incubation at lower temperatures can slow down color development.

32. The increase of sensitivity by increasing the temperature has an advantage other than allowing the detection of low protein concentration: higher temperatures can increase the exposure of amino acids, which reduces the impact of variation due to amino acid composition within different protein samples.
33. Compared to the Bradford assay, the BCA assay is more straightforward since the peptide backbone also contributes to color formation.
34. There are commercial kits available to perform the BCA assay. In case of using one of these kits, follow the manufacturer's instructions.
35. It is essential to use a quality quartz cuvette in this method. One should check the cuvette quality by reading the absorbance of the cuvette with water at 280 nm to check if there is no absorption.

References

1. Copeland RA (1994) Methods for protein analysis. Chapman & Hall, New York, pp 62–71
2. Gornall AG, Bardawill CJ, David MM (1949) Determination of serum proteins by means of the biuret reaction. J Biol Chem 177(2): 751–766
3. Kruger NJ (2009) The Bradford method for protein quantitation. In: The protein protocols handbook. Springer, pp 17–24
4. Lowry OH, Rosebrough NJ, Farr AL, Randall RJ (1951) Protein measurement with the Folin phenol reagent. J Biol Chem 193:265–275
5. Mariotti F, Tomé D, Mirand PP (2008) Converting nitrogen into protein--beyond 6.25 and Jones' factors. Crit Rev Food Sci Nutr 48(2):177–184
6. Owusu-Apenten R (2002) Food protein analysis: quantitative effects on processing, vol 118. CRC Press
7. Peterson GL (1979) Review of the Folin phenol protein quantitation method of Lowry, Rosebrough, Farr and Randall. Analyt Biochem 100(2):201–220
8. Sáez-Plaza P, Navas MJ, Wybraniec S, Michałowski T, Asuero AG (2013) An overview of the Kjeldahl method of nitrogen determination. Part II. Sample preparation, working scale, instrumental finish, and quality control. Crit Rev Anal Chem 43(4):224–272
9. Smith PE, Krohn RI, Hermanson GT, Mallia AK, Gartner FH, Provenzano M et al (1985) Measurement of protein using bicinchoninic acid. Anal Biochem 150(1):76–85
10. Zaia DA, Zaia CTB, Lichtig J (1998) Determination of total protein by spectrophotometry: advantages and disadvantages of proposed methods. Química Nova 21(6):787–793

Chapter 8

Quantitation of Amine in Food Proteins and Hydrolyzed Proteins Using *Ortho*-Phthalaldehyde

Xiaohong Sun, Chibuike Udenigwe, and Apollinaire Tsopmo

Abstract

The importance of hydrolyzed food proteins in food science is on constant growth. Proteins are generally hydrolyzed to enhance their functional properties (e.g., food texture), to increase their intestinal absorption, or to release biologically active peptides. The property of the hydrolyzed proteins is dependent on several factors, including the physicochemical characteristics, free amine, nitrogen content, and the degree of hydrolysis. The quantification of proteins, peptides, and the determination of the free amino nitrogen content are important to assess the hydrolytic power of a protease and batch-to-batch reproducibility of protein hydrolysates. Procedures for the analysis of amine in proteins and mixtures of peptides, such as protein hydrolysates, have been developed. The measurement of free amine based on the derivatization with ortho-phthaldialdehyde (OPA) is advantageous as the detection can be done based on absorbance or fluorescence. This chapter discusses the basic principles, methodology, and calculations involved in amine quantitation in native and hydrolyzed food proteins using the OPA method.

Key words Protein hydrolysates, Degree of hydrolysis, Spectroscopy

1 Introduction

The analysis of peptides is important to assess the efficacy of a process designed to break down food proteins into small peptides commonly via enzymatic and microbial treatments. Hydrolyzed proteins or proteins hydrolysates can be used to improve nutritional needs for individuals who are unable to digest intact proteins, such infants with food hypersensitivity or adults with compromise health. Dietary hydrolyzed proteins rich in low-molecular-weight peptides and poor in free amino acids have high nutritional and therapeutic values [1]. Low-molecular weight peptides in protein hydrolysates have been associated their biological functions, such as antioxidant, anti-inflammatory, and antihypertensive activities [2–4]. In food products, the presence of larger polypeptides in protein hydrolysates has beneficial effects on the texture due to improved solubility, emulsifying, or oil binding properties ([5, 6]. Various

Lilian Mariutti and Frank Vriesekoop (eds.), *Food Chemistry: Basic Protocols*, Methods and Protocols in Food Science,
https://doi.org/10.1007/978-1-0716-4678-6_8,

procedures are available for the analysis of peptides in protein hydrolysates. The direct analysis will require the determination of the sequence of each peptide in the hydrolysate or their peptidic profile using tandem mass spectrometry [7]. However, there are few studies using this technique due to the limitation on the size of peptides (>500 Da) that can be detected and the complexity of the instrumentation. Most studies determine the amount of total nitrogen or the amount of α-amino groups present in the hydrolysate. The common methods for determining the total nitrogen content are Kjeldahl and Dumas [8]. There is, however, a lack of analytical selectivity for protein, as nitrogen can be quantified in both organic and inorganic molecules present in the food matrix. Other procedures, such as Bradford assay, are unable to detect peptides that are less than 3000 Da [9]. There are other methods that make use of the reactivity of the amine group with fluorescent and or ultraviolet light probes [1, 10]. Some are based on the complete hydrolysis of proteins or peptides into individual amino acids before their conjugation with the probe [11], while some, such as *ortho*-phthalaldehyde (OPA), can react directly with the primary amine ($—NH_2$) moieties of peptides or proteins with or without prior hydrolysis [12]. The OPA method was initially developed to measure available lysine in food proteins based on the reaction of ϵ-NH_2 groups with OPA to produce a fluorescent molecule in approximately 1 min [13]. Currently, the OPA is widely used as a peptide assay or pre-column derivatization agent for the analysis of amino acids by liquid chromatography. In fact, OPA has been recommended for routine analysis of food products such as milk and cereals [14]. The reaction of OPA with amine is shown in Fig. 1. The reaction can be performed in the presence of 2-mercaptoethanol or ethanethiol.

2 Material

The measured samples consisted of four proteins from different food sources and their hydrolysates, including whey protein isolate and hydrolysates (pepsin+pancreatin), fish collagen and hydrolysates (Alcalase), pigeon pea protein isolate and hydrolysates (papain), as well as non-cuticular insect protein and hydrolysates (papain).

o-Phthaldialdehyde (OPA, catalogue No. 40025454), sodium phosphate dibasic (catalogue No. SLBS1095V), sodium phosphate monobasic (catalogue No. BCBS2112V), sodium dodecyl sulfate (SDS, catalogue No. SLCD1580), β-mercaptoethanol (β-ME, catalogue No. SHBH7147), and L-serine (catalogue No. SLBN5053V) were purchased from Millipore Sigma (Burlington, MA, USA). Ethyl alcohol anhydrous (catalogue No. 025631) was bought from Commercial Alcohols Inc. (Brampton, ON, Canada). Distilled water was used to prepare all solutions in this study.

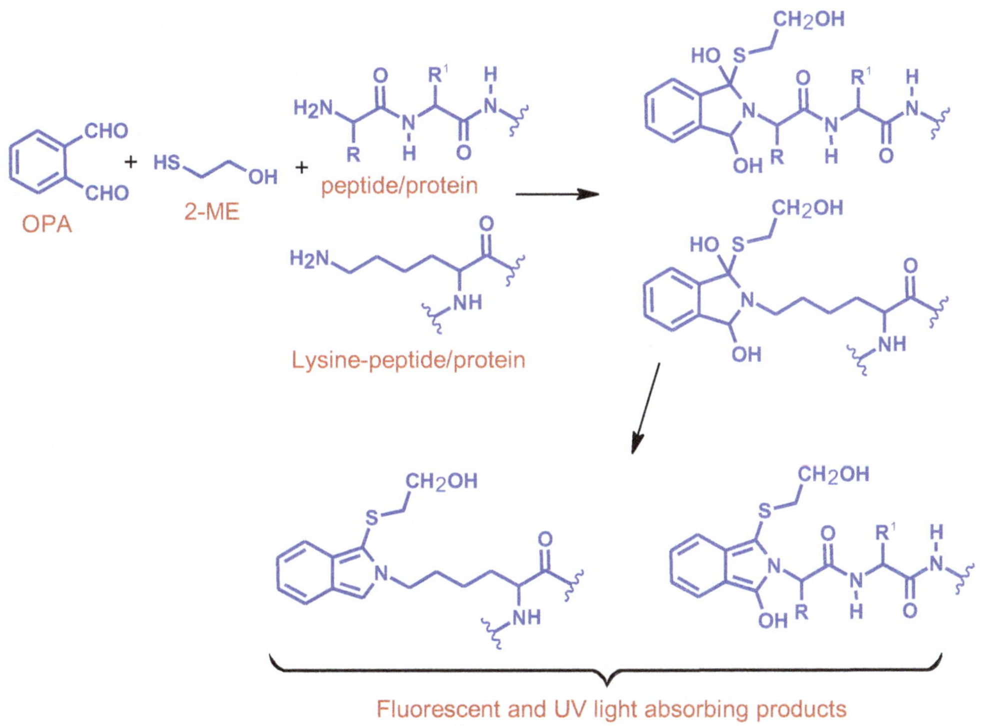

Fig. 1 Reaction of *ortho*-phthalaldehyde (OPA) with the primary amine of peptides or proteins to form fluorescent and ultraviolet light-absorbing derivatives in the presence of β-mercaptoethanol (β-ME)

3 Methods

3.1 Preparation of Phosphate-Buffered Saline (PBS, 100 mM, pH 6.9)

1. Weigh 1.590 g of sodium phosphate dibasic and 1.056 g of sodium phosphate monobasic, and transfer the powders to a glass beaker.
2. Add 200 mL of water to the beaker and stir until completely dissolved.
3. Check the pH of the PBS buffer using a pH meter.

3.2 Preparation of OPA Reagent

1. Weigh 50 mg of OPA in a 50-mL plastic centrifuge tube that is covered with aluminum foil.
2. Add 1 mL of absolute ethanol to the tube and mix on a vortex.
3. Add about 35 mL of PBS and 50 mg of SDS to the above mixture, and vortex mix until completely dissolved.
4. Add 100 μL of β-ME directly to the above solution in a fume hood. Avoid adding the reagent to the formed foams.
5. Make up the volume to 50 mL using PBS buffer.
6. OPA reagent is used within 2 h after preparation.

3.3 Preparation of Serine and Sample Solutions

1. Weigh about 1 mg of serine in a 2-mL microcentrifuge tube, and add PBS buffer to achieve a serine concentration of 9.52 mM (1000 μg/mL).
2. Perform a serial dilution of serine solution using PBS buffer to 7.61 mM (800 μg/mL), 5.71 mM (600 μg/mL), 2.85 mM (300 μg/mL), 1.43 mM (150 μg/mL), 0.71 mM (75 μg/mL), and 0.36 mM (37.5 μg/mL).
3. Based on sample weight, prepare solutions of the proteins and hydrolysates at a concentration of 1000 μg/mL using PBS.
4. Further dilute the samples with PBS buffer to 500 and 250 μg/mL.

3.4 Quantification of Peptide Concentrations Using OPA Method

1. Add 10 μL of blank (PBS buffer), standard (serine), or sample (proteins or hydrolysates) to a 96-well clear, flat bottom, polystyrene microplate (Corning, No. 3595).
2. Add 250 μL of OPA reagent to each well using a single channel or multichannel pipette and avoid making bubbles inside the wells.
3. After adding OPA reagent, quickly transfer the microplate to a multimode microplate reader (Tecan, Switzerland) or any other plate reader that can measure absorbance or fluorescence.
4. Allow the mixture to incubate for 2 min at room temperature with shaking at 1440 rpm.
5. Measure the absorbance at 340 nm, bandwidth 3.5, flashes 10, and settle time 50 ms.
6. Alternatively, measure the fluorescence intensity at the excitation wavelength of 360 nm (bandwidth 20.0), emission wavelength of 460 nm (bandwidth 20.0), bottom mode, flashes 30, gain optimal, Z-position 20,000 μm, and settle time 0 ms.
7. Plot the graph of absorbance at 340 nm (or fluorescence intensity) vs. serine concentrations as the standard curves.
8. Use the standard curves to determine the free α-amino nitrogen content of the protein and hydrolysate samples (expressed as mM serine equivalent). Considering the sample concentrations, the free α-amino nitrogen content was calculated using the following equation:

$$\text{Free } \alpha - \text{amino nitrogen content (mmol Ser Eq/g)} = \frac{\textit{Serine equivalent (mmol/L)}}{\textit{Sample concentration (g/L)}}$$

3.5 Absorption Spectral Scan

1. Obtain the absorption spectral scan of serine (9.52 mM), protein (1000 μg/mL), and protein hydrolysate (1000 μg/mL). For this chapter, whey protein and hydrolysate were used as examples.

2. Follow the method outlines in **steps 1–4** in Subheading 3.4.
3. Obtain the spectral scan in the wavelength range of 300 nm to 400 nm. The bandwidth and step size are set at 3.5 and 2, respectively.

4 Notes

1. The serine standard curves for the absorbance and fluorescence assays are shown in Fig. 2a, b. The respective regression

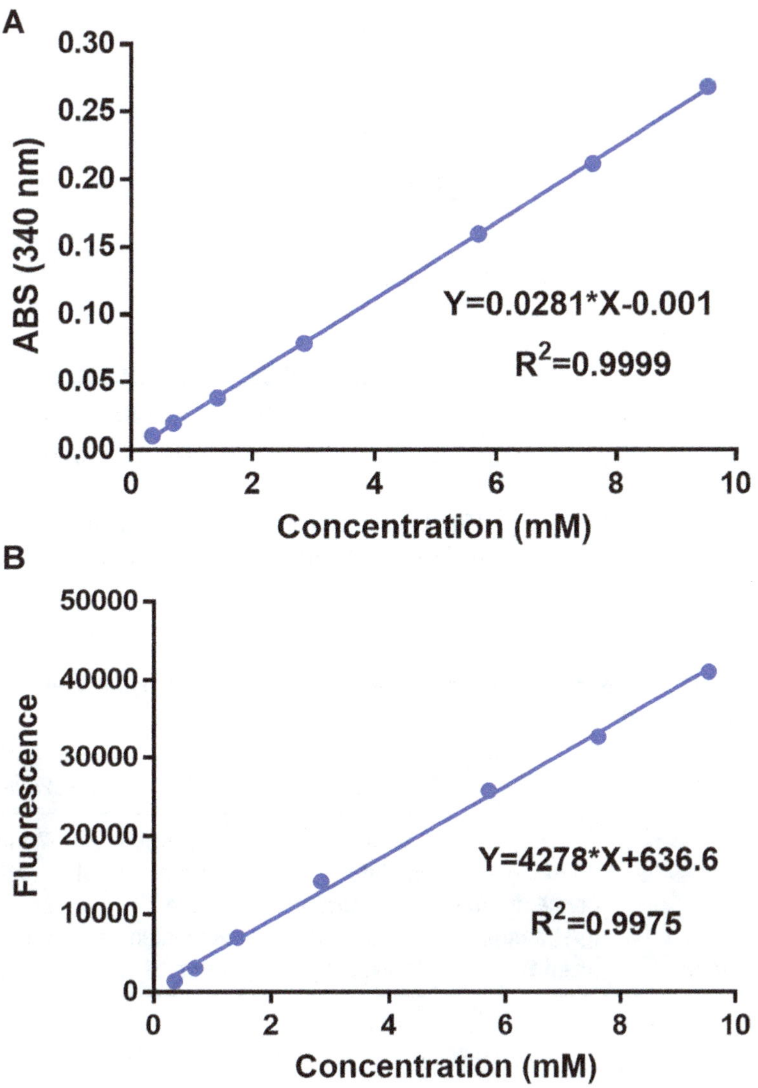

Fig. 2 (**a**) Serine standard curve (absorbance) in the o-phthaldialdehyde (OPA) assay. (**b**) Serine standard curve (fluorescence) in the o-phthaldialdehyde (OPA) assay

equations, $\Upsilon = 0.0281*X-0.001$ ($R^2 = 0.9999$) and $\Upsilon = 4278*X + 636.6$ ($R^2 = 0.9975$), were used to quantitate the amine content of the samples.

2. As shown in Fig. 3, the absorption spectral scan (from 300 nm to 400 nm) of serine (standard) and representative samples (whey protein and its hydrolysates) resulted in the selection of the wavelength of 340 nm for the OPA assay.
3. The measured amine concentrations of the protein and hydrolysate samples using the OPA method are listed in Table 1.

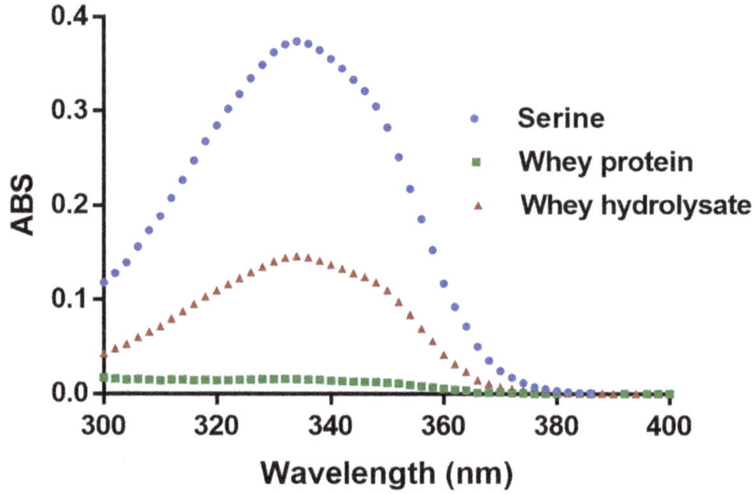

Fig. 3 Absorption spectral scan of serine (standard) and samples (whey protein and its hydrolysates) after mixing with o-phthaldialdehyde

Table 1
Quantification of peptide concentrations of proteins and hydrolysates using the o-phthaldialdehyde (OPA) method

		Absorbance		Fluorescence	
Sample	Concentration (g/L, sample weight)	Measured peptide concentration (mmol/L serine equivalent)[a]	Free amino content (mmol serine equivalent/g sample)	Measured peptide concentration (mmol/L serine equivalent)[a]	Free amino content (mmol serine equivalent/g sample)
Whey protein	1	0.45 ± 0.09	0.45	0.47 ± 0.04	0.47
Whey protein	0.5	0.21 ± 0.07	0.42	0.05 ± 0.03	0.10
Whey protein	0.25	0.02 ± 0.18	0.08	0	0

(continued)

Table 1
(continued)

Sample	Concentration (g/L, sample weight)	Absorbance		Fluorescence	
		Measured peptide concentration (mmol/L serine equivalent)[a]	Free amino content (mmol serine equivalent/g sample)	Measured peptide concentration (mmol/L serine equivalent)[a]	Free amino content (mmol serine equivalent/g sample)
Whey protein hydrolysate	1	4.63 ± 0.06	4.63	4.43 ± 0.28	4.43
Whey protein hydrolysate	0.5	2.31 ± 0.12	4.62	2.30 ± 0.20	4.60
Whey protein hydrolysate	0.25	1.07 ± 0.08	4.28	1.08 ± 0.02	4.32
Collagen	1	0.23 ± 0.06	0.23	0.35 ± 0.19	0.35
Collagen	0.5	0.10 ± 0.03	0.20	0.51 ± 0.12	1.02
Collagen	0.25	0.04 ± 0.01	0.16	0.02 ± 0.29	0.08
Collagen hydrolysate	1	3.71 ± 0.06	3.71	1.83 ± 0.12	1.83
Collagen hydrolysate	0.5	1.72 ± 0.02	3.44	1.00 ± 0.24	2.00
Collagen hydrolysate	0.25	0.88 ± 0.05	3.52	0.73 ± 0.18	2.92
Pigeon pea protein	1	0.49 ± 0.02	0.49	0.64 ± 0.16	0.64
Pigeon pea protein	0.5	0.16 ± 0.01	0.32	0.06 ± 0.15	0.12
Pigeon pea protein	0.25	0	0	0	0
Pigeon pea protein hydrolysate	1	1.97 ± 0.11	1.97	1.46 ± 0.13	1.46
Pigeon pea protein hydrolysate	0.5	0.91 ± 0.05	1.82	0.78 ± 0.29	1.56
Pigeon pea protein hydrolysate	0.25	0.44 ± 0.08	1.76	0.08 ± 0.01	0.32
Insect protein	1	1.60 ± 0.06	1.60	3.14 ± 0.18	3.14

(continued)

Table 1 (continued)

Sample	Concentration (g/L, sample weight)	Absorbance: Measured peptide concentration (mmol/L serine equivalent)[a]	Absorbance: Free amino content (mmol serine equivalent/g sample)	Fluorescence: Measured peptide concentration (mmol/L serine equivalent)[a]	Fluorescence: Free amino content (mmol serine equivalent/g sample)
Insect protein	0.5	0.75 ± 0.03	1.50	1.95 ± 0.18	3.90
Insect protein	0.25	0.35 ± 0.04	1.40	0.93 ± 0.12	3.72
Insect protein hydrolysate	1	3.09 ± 0.16	3.09	6.88 ± 0.17	6.88
Insect protein hydrolysate	0.5	1.51 ± 0.10	3.02	3.56 ± 0.21	7.12
Insect protein hydrolysate	0.25	0.67 ± 0.02	2.68	2.01 ± 0.40	8.04

[a]Calculated using the regression equations in Fig. 2a, b

5 Summary

1. Compared to native or parent protein samples, the corresponding hydrolysates showed significantly higher free amine contents (serine equivalent). This is because of the hydrolysate samples' higher concentration of peptides and amino acids.
2. The higher free amine contents demonstrate the hydrolytic power of the proteases. They can be used as a measure of the degree of protein hydrolysis and release of free amines in protein-containing food samples.
3. The fluorescence detection method did not produce uniform free amine content across different concentrations of each sample, except for whey protein hydrolysate. This may be because of the interference of fluorescence-active compounds in the samples. Background subtraction did not eliminate this matrix effect.

4. The absorbance detection method resulted in similar amine contents across the different concentrations analyzed. Therefore, the OPA absorbance method appears to be more accurate in quantitating the free amines than the fluorescence method.

References

1. Silvestre MPC (1997) Review of methods for the analysis of protein hydrolysates. Food Chem 60:263–271
2. Walters ME, Willmore WG, Tsopmo A (2020) Antioxidant, physicochemical, and cellular secretion of glucagon-like peptide-1 properties of oat bran protein hydrolysates. Antioxidants 9:557
3. Shittu A, Esfandi R, Tsopmo A (2020) Chromium and arsenic speciation analysis in meats by HPLC-ICP-MS in the presence of hydrolyzed oat proteins with radical scavenging activities. Heliyon 6:e03654
4. Udenigwe CC, Lin Y-S, Hou W-C, Aluko RE (2009) Kinetics of the inhibition of renin and angiotensin I-converting enzyme by flaxseed protein hydrolysate fractions. J Funct Foods 1:199–207
5. Ajibola CF, Fashakin JB, Fagbemi TN, Aluko RE (2011) Effect of peptide size on antioxidant properties of African yam bean seed (*Sphenostylis stenocarpa*) protein hydrolysate fractions. Int J Mol Sci 12:6685–6702
6. Nieto-Nieto TV, Wang YX, Ozimek L, Chen L (2014) Effects of partial hydrolysis on structure and gelling properties of oat globular proteins. Food Res Int 55:418–425
7. Agyei D, Tsopmo A, Udenigwe CC (2018) Bioinformatics and peptidomics approaches to the discovery and analysis of food-derived bioactive peptides. Anal Bioanal Chem 410:3463–3472
8. Moore JC, DeVries JW, Lipp M, Griffiths JC, Abernethy DR (2010) Total protein methods and their potential utility to reduce the risk of food protein adulteration. Compr Rev Food Sci Food Saf 9:330–357
9. Nwachukwu ID, Aluko RE (2019) A systematic evaluation of various methods for quantifying food protein hydrolysate peptides. Food Chem 270:25–31
10. Aalaei K, Rayner M, Sjöholm I (2019) Chemical methods and techniques to monitor early Maillard reaction in milk products; a review. Crit Rev Food Sci Nutr 59:1829–1839
11. Boogers I, Plugge W, Stokkermans YQ, Duchateau ALL (2008) Ultra-performance liquid chromatographic analysis of amino acids in protein hydrolysates using an automated pre-column derivatisation method. J Chromatogr A 1189:406–409
12. Panasiuk R, Amarowicz R, Kostyra H, Sijtsma L (1998) Determination of α-amino nitrogen in pea protein hydrolysates: a comparison of three analytical methods. Food Chem 62: 363–367
13. Goodno CC, Swaisgood HE, Catignani GL (1981) A fluorimetric assay for available lysine in proteins. Anal Biochem 115:203–211
14. Perez-Locas C, Yaylayan VA (2010) The Maillard reaction and food quality deterioration. In: Skibsted LH, Risbo J, Andersen MLBT-CD, P.I. of F. and B (eds) Woodhead Publishing Series in food science, technology and nutrition. Woodhead Publishing, pp 70–94. ISBN 978-1-84569-495-1

Chapter 9

Analysis of Free and Total Amino Acid and Biogenic Amines by Liquid Injection Gas Chromatography (LI-GC)

Barry Hucker and Frank Vriesekoop

Abstract

Amino acids and biogenic amines are essential molecules for sustaining cellular metabolism. While the use of high-performance liquid chromatography may be a preferred method for the analysis of amino acids, gas chromatography has its advantages with simple, fast methods with high resolution and accuracy. The following method describes a simple and effective method of analysis using gas chromatography coupled with flame ionization detection. Using propyl chloroformate as a derivatization agent, 17 amino acids and five biogenic amines can be successfully analyzed with a high accuracy (>90%) and precision (RSD <3.5).

Key words Amino acids, Biogenic amines, Gas chromatography, Propyl chloroformate, Flame ionization detection

1 Introduction

Amino acids are the building blocks for proteins, and as such, amino acids are an essential part of a healthy diet for growth, development, and maintaining life. There is a wide variety of amino acids, but they all share the same basic structure, comprising a central carbon flanked by an amino and a carboxyl group on one plane and a single hydrogen and a side chain (R) on another plane [1] (Fig. 1). It is the side chain that distinguishes one amino acid from another. There are 20 basic amino acids, where the side chain imposes different physical (size) and chemical properties on the individual amino acids but also on the peptides or proteins in which these amino acids are incorporated into [1]. Peptides are amino acids linked together via peptide bonds (Fig. 2), where the standard carboxyl group on one amino acid undergoes a condensation reaction with the standard amino group on another amino acid. While most peptides and proteins are encoded for on messenger RNA and their synthesis facilitated by ribosomes, there are a small number of non-ribosomal synthesized peptides. This means

Lilian Mariutti and Frank Vriesekoop (eds.), *Food Chemistry: Basic Protocols*, Methods and Protocols in Food Science,
https://doi.org/10.1007/978-1-0716-4678-6_9,

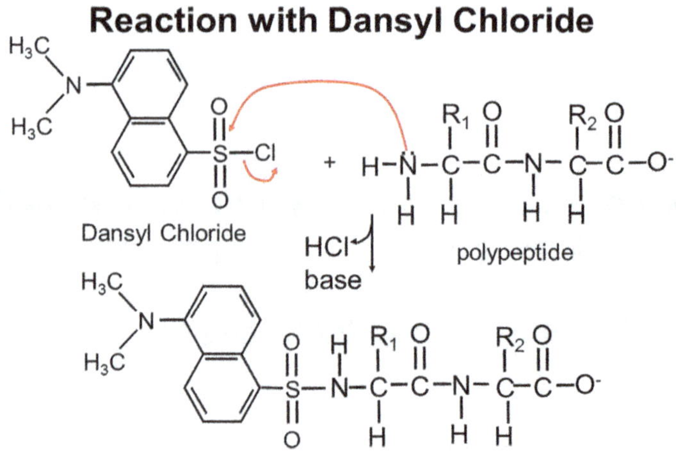

Fig. 1 Dansyl chloride derivatization of a polypeptide

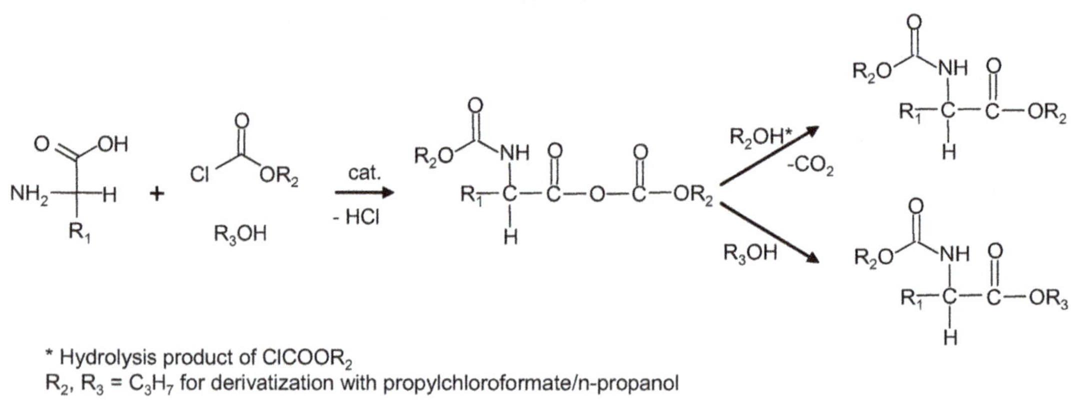

Fig. 2 Propyl chloroformate derivatization of amino acids

that ribosomes are the principal facilitators of peptide bond synthesis, while non-ribosomal peptides are typically synthesized by dedicated enzymes that might not always facilitate the linkages through peptide bonds [1].

Among the 20 common amino acids, there is variation that is facilitated by the chemistry of the side chains; these can be grouped based on their nonpolar, polar, neutral, and charged chemical side chains.

Biogenic amines (BA) are generally decarboxylated from amino acids that are found naturally in a variety of foods. The most important of these BA in food are histamine, putrescine, cadaverine, tyramine, tryptamine, 2-phenylethylamine, spermine, and spermidine. When present in high concentrations, they can have significant toxic effects, such as hypertension, headaches, nausea, and localized inflammation [2–5].

In general, the presence of BA in food is due to the presence of bacteria, such as *Bacillus, Citrobacter, Clostridium, Klebsiella, Lactobacillus*, and *Pediococcus*, which can break down a variety of free amino acids to BA [6]. In particular, histamine content is regulated in a number of countries for foods such as fish [7] and cheese due to the potential for a serious allergic response.

When undertaking amino acid/biogenic amine analysis, one of the crucial steps is the hydrolysis of the peptide bonds that hold the various amino acids together. Some peptide bonds between specific amino acids are stronger than others, meaning that not all peptide bonds are hydrolyzed at the same time. The nonenzymatic hydrolysis of the peptide bonds is best facilitated at an elevated temperature under either acidic or alkaline conditions. Both acidic and alkaline conditions can be detrimental to the stability of the side chain groups on some but not all amino acids. This means that a fine-tuned and targeted hydrolysis is required to favor the extraction of certain amino acids. For instance, most amino acids are recovered at 90 + % under acid conditions; however, tryptophan is very poorly recovered under acidic conditions [8]. Tryptophan, however, is quite stable during alkali protein hydrolysis, and if tryptophan determination is required a separate high-temperature, alkali hydrolysis tends to be facilitated [8, 9]. Enzymatic hydrolysis of proteins for amino acid analysis is seldom applied. Enzymatic hydrolysis has the advantage that it allows the quantification of otherwise sensitive amino acids, which are typically destroyed during chemical digestion; however, its widespread application has been hampered by the relative specificity of the proteases for certain residues, thus skewing quantitative results [8]. For complete enzymatic digestion of protein samples, several enzymes will be required, and the presence of multipole enzymes in a sample mix could artificially influence the amino acid profiles if the various proteolytic enzymes act on each other. In light of the varied consequences related to the protein hydrolytic methods, that choice will be steered by the sample matrix and the underpinning reasons for trying to establish the amino acid profile of the samples. This chapter will therefore only focus on the quantitative detection of free amino acids.

Because the chemical properties of amino acids are determined by their side chain, the first notion to consider is the separation of the amino acid mix into the individual amino acids. This is best achieved by some chromatographic method, whether that is via HPLC or GC. Amino acids can relatively easily be separated by means of HPLC, where the chemistry of the amino acid side chains is the principal distinguishing feature that allows the separation. However, visualization of amino acids following separation on a HPLC column is quite difficult since the response on a basic detector (UV or IR) is very poor. In order to achieve the detection of amino acids with a sensitive signal, the procedures for amino acid analyses have typically focused on the derivatization of the free amino acids with highly fluorescent tags. For instance, the dansyl chloride derivatization aims to create a sulfonic bond at each of the amino groups on amino acids, facilitating tagging of the amino acids by the fluorescent dansyl group (Fig. 1). A major drawback with dansylation reactions is the need to quench the reaction to prevent unwanted side product formation. This results in the quencher forming an unwanted peak in the chromatogram, which must be accounted for. This reaction method, while simple in nature, can result in artifacts in the chromatograms that may affect the final resolution of target anolytes.

The alternative to the method is to utilize gas chromatography. While amino acids are not volatile enough to be separated by GC, the derivatization of individual amino acids by chloroformates can result in highly specific separation of amino acids while avoiding large interferences. Depending on the matrix used, methyl, ethyl, and propyl chloroformates can be used to derivatize target amino acids [10–17]. The selection of the chloroformate will greatly influence the elution of the derivatized amino acids, and for fast run times, methyl or ethyl variants should be used, while propyl-chloroformates can be used to elute out many amino acids over longer run times with better efficiencies (Fig. 2) [10–17].

With all three formats, the same end result is achieved by the esterification of the carboxylic acid groups in the presence of a chloroformate-pyridine-HCl solution, resulting in the release of carbon dioxide and elimination of hydrogen from the amino group. This esterification allows for the chloroformate to react, producing volatile derivatives [10, 14]. This allows for rapid quantitative determination by GC-FID.

2 Materials

2.1 Analysis of Amino Acids and Biogenic Amines by Gas Chromatography

2.1.1 Basic Reagents

All reagents/standards listed below were sourced through Sigma-Aldrich (St. Louis, MO) and were of 90% purity or greater. Amino acids (AA)—glycine, valine, alanine, leucine, isoleucine, proline, threonine, serine, asparagine, aspartic acid, methionine, glutamic acid, phenylalanine, cysteine, histidine, tyrosine, tryptophan, and norvaline (internal standard 500 mg/L stock solution used at 50 mg/L); biogenic amines (BA)—putrescine, cadaverine, histamine, tyramine, 3-hydroxyltyramine (3-HTA). All amino acid/biogenic amine standards were made dissolved in 0.1 M HCl, which was prepared by diluting 32% HCl in 18 MΩ water, which was purified using a Sartorius Arium 611VF water purification system (Sartorius AG, Göttingen, Germany).

Derivatization mix—18 MΩ water, ethanol (HPLC grade), and pyridine (HPLC grade) were combined to create a derivatization mix at a ratio of 50:40:10, respectively. Propyl chloroform (PCF) was added to volatilize the AA/BAs. Derivatized AA/BAs were then extracted using a 1% v/v PCF in chloroform (HPLC grade).

2.1.2 Instrumentation

A Varian 3400 gas chromatograph with a Varian 8100 autosampler (10 μL syringe) was utilized for this research with the following settings: column head pressure, 10 psi; split flow, 10 ml/min; injector temperature, 240 °C; injection volume, 1 μL; initial column temperature of 120 °C (no hold) with a ramp rate of 8 °C/min to 280 °C, held for 8 minutes (total 28 minutes); detector, FID 280 °C; carrier gas, nitrogen; column utilized, Phenomenex Zebron Column ZB1701; 0.25 μm film thickness 30 m x 0.32 mm ID (Phenomenex, Torrance, California).

AA/BA elution order—glycine, valine, alanine, leucine, isoleucine, proline, threonine, serine, asparagine, aspartic acid, methionine, glutamic acid, phenylalanine, cysteine, putrescine, cadaverine, histamine, tyramine, histidine, tyrosine, tryptophan, 3-hydroxyltyramine (3-HTA).

2.1.3 Free Amino Acid and Biogenic Amine Preparation

Free AA and BA were extracted from samples by weighing 2g into a 50 ml centrifuge tube and 20 ml of 18 MΩ water was added. The sample was then placed in a mechanical roller (~12 rev/minutes; 20 x 6 cm arm holders; Industrial Equipment and Control, Thornbury, Australia) for 15 minutes. Samples were centrifuged at 2000 rcf in a Hermile Z323 centrifuge (Hermile Labortechnik, Wehingen, Germany). 5 mL of supernatant was then taken and filtered through a 0.45 μm RC syringe filter (Phenomenex, Torrance, California) and stored in a clean centrifuge tube at 4 °C for up to 24 hours prior to analysis. Samples were diluted 1:4 in 18 MΩ water prior to derivatization to dilute the AA/BAs to be within the calibration range of the equipment.

3 Methods

3.1 Total Amino Acid and Biogenic Amine Analysis

Total AAs and BAs were extracted from the samples by weighing 1 g of sample weighed into a 50 mL centrifuge tube and 10 mL of 6 N HCl was added. The sample was then autoclaved at 121 °C for 4 hours. The solution was allowed to room temperature, where it was neutralized with 6 N NaOH and made to 25 mL with 18 MΩ water. Samples were then centrifuged at 2000 rcf in a Hermile Z323 centrifuge (Hermile Labortechnik, Wehingen, Germany). 5 mL of supernatant was then taken and filtered through a 0.45 μm RC syringe filter (Phenomenex, Torrance, California) and stored in a clean centrifuge tube at 4 °C for up to 24 hours

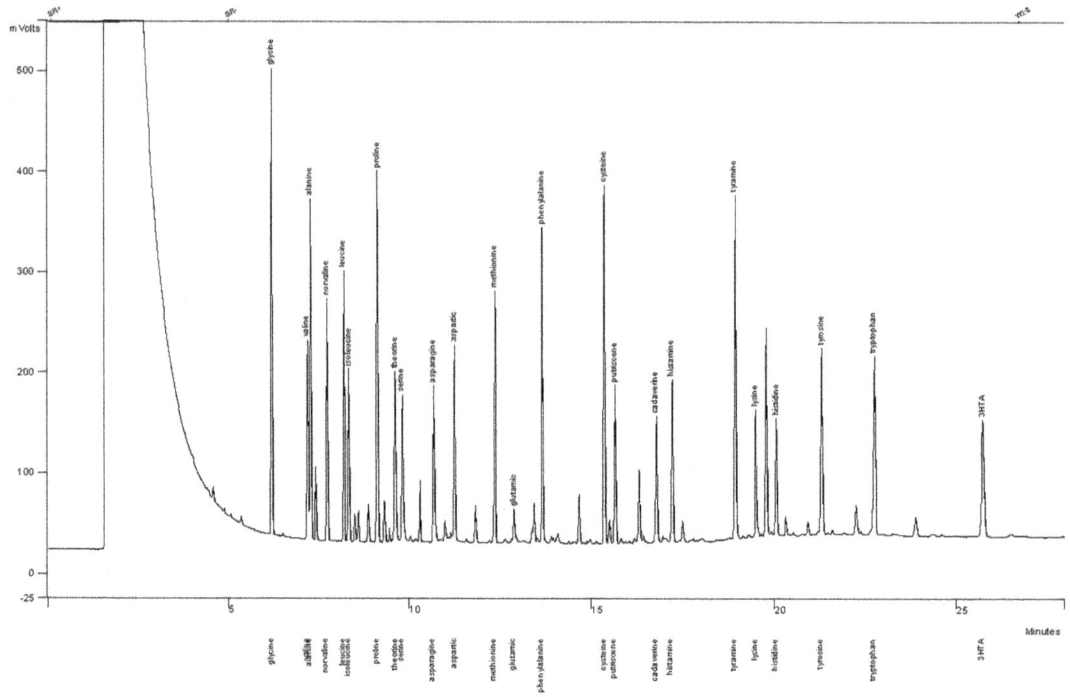

Fig. 3 Amino acid and biogenic amines via GC-FID

prior to analysis (Subheading 4.5). Samples were diluted 1:4 in 18 MΩ water prior to derivatization to dilute the AA/BAs to be within the calibration range of the equipment.

3.2 Derivatization of AAs and BAs

All standards and samples were derivatized by transferring 450 μL into a microfuge tube. 50 μl of norvaline (internal standard—50 mg/L) was added, followed by the derivatization mix (50:40:10 water:ethanol:pyridine) and 25 μL of propyl chloroformate (*see* **Note 1**). This solution was then vortexed for 30 seconds, and 300 μl of a 1% PCF in chloroform was added and vortexed for a further 30 seconds. This was then centrifuged at 130,000 rpm (microfuge) for 2 minutes. The aqueous (top) layer was aspirated and discarded and the organic layer removed and placed into a 300 μL limited vial insert. Samples were then analyzed within 6 hours to limit evaporation losses.

3.3 Separation and Analysis of AAs and BAs

Figure 3 depicts the elution order of the described AA and BA compounds. Separation using the described instrumentation set up means a simple and effective analysis can be observed. Table 1 shows the accuracy, precision LOD, and RSD of this analysis. This method also achieved ppm detection limits of the target analytes.

Table 1
Accuracy, precision, limit of quantitation and linearity of calibration of the free and total amino acid/ biogenic amine analysis

	Accuracy—free (av. % spike recovery)	Accuracy Total (av. % spike recovery)	Precision Free (% RSD)	Precision Total (% RSD)	LOQ/ LOD (μg/g)	Range of linearity (μg/g)	LOC (R^2 value)
Glycine (Gly)	96.3	94.2	2.4	3.5	5	5–106	0.9996
Valine (Val)	99.9	96.5	2.5	3.1	5	5–107	0.9997
Alanine (Ala)	99.6	96.2	2.3	3.6	4	4–100	0.9996
Leucine (Leu)	99.3	97.2	2.6	3.8	5	5–100	0.9995
Isoleucine (Isoleu)	94.5	90.6	3.2	3.2	4	4–96	0.9994
Proline (Pro)	95.8	93.2	2.9	2.5	5	5–111	0.9996
Threonine (The)	99.5	97.1	2.7	2.9	5	5–100	0.9997
Serine (Ser)	99.2	96.3	2.5	3.4	5	5–111	0.9991
Asparagine (Asp)	98.1	43.3	2.7	3.4	5	5–104	0.9993
Aspartic acid (Asp acid)	97.4	91.8	2.5	2.9	5	5–100	0.9996
Methionine (Met)	99.5	97.3	2.1	3.5	5	5–107	0.9999
Glutamic acid (Glu)	96.3	93.6	2.8	3.3	5	5–102	0.9998
Phenylalanine (Phen)	97.5	92.8	2.9	3.2	5	5–98	0.9999
Cysteine (Cys)	97.2	90.2	2.4	3.6	5	5–108	0.9995
Putrescine (Put)	97.3	92.1	2.6	3.1	5	5–102	0.9996
Cadaverine (Cad)	99.4	95.9	2.7	3.5	5	6–130	0.9993
Histamine (Hist-a)	96.5	91.3	2.5	2.8	5	5–117	0.9996
Tyramine (Tyra)	94.3	91.8	2.8	3.1	5	5–103	0.9994
Lysine	92.1	94.7	3.1	3.4	5	5–116	0.9989
Histidine (Histid)	96.4	92.7	2.5	3.7	5	5–113	0.9995
Tyrosine	98.4	95.6	2.4	3.4	5	5–99	0.9998

(continued)

Table 1 (continued)

	Accuracy—free (av. % spike recovery)	Accuracy Total (av. % spike recovery)	Precision Free (% RSD)	Precision Total (% RSD)	LOQ/ LOD (μg/g)	Range of linearity (μg/g)	LOC (R^2 value)
Tryptophan (Tryp)	99.2	90.1	2.6	3.4	5	5–101	0.9994
3-HTA	94.5	90.2	3.6	3.3	5	5–97	0.9997

4 Notes

1. Methyl- and ethyl-chloroformate can be substituted instead of the propyl form; however, elution time on the column can be quite fast and coelution can be observed in complex matrices; interfering compounds from the samples were not successfully separated out. If using the methyl and ethyl forms, adjustments to the column conditions will need to be made to stop coelution.

References

1. Damodaran S (1996) Amino acids, peptides and proteins. In: Fennema OR (ed) Food chemistry, 3rd edn. Marcel Dekker, New York
2. Suzzi G, Tooriani S (2015) Editorial: biogenic amines in foods. Front Micobiol 6:472
3. Tabanelli G (2020) Biogenic amines and food quality: emerging challenges and public health concerns. Food Secur 9(7):859
4. Ruiz-Capillas C, Herrero AM (2019) Impact of biogenic amines on food quality and safety. Food Secur 8(2):62
5. Shalaby A (1996) Significance of biogenic amines to food safety and human health. Food Res Int 29(7):675–690
6. Karocicova J, Kohajdova Z (2005) Biogenic amines in food. Chem Pap 59(1):70–79
7. Cicero A, Cammilleri G, Galluzzo G, Calabrese I, Pulvirenti A, Giangrosso G, Cicero N, Cumbo V, Vella A, Macaluso A, Ferrantelli V (2020) Histamine in fish products randomly collected in southern Italy: a 6-year study. J Food Prot 83(2):241–248
8. Fountoulakis M, Lahm HW (1998) Hydrolysis and amino acid composition analysis of proteins. J Chromatog A 826(2):109–134
9. Allenspach MD, Fuchs JA, Doriot N, Hiss JA, Schneider G, Steuer C (2018) Quantification of hydrolyzed peptides and proteins by amino acid fluorescence. J Pept Sci 24(8–9):3113
10. Husek P (1995) Simultaneous profile analysis of plasma amino and organic acids by capillary gas chromatography. J Chromatogr B 669: 352–357
11. Husek P, Farkas T (2001) Rapid body fluid amino acid profiling and sample workup. Am Biotechnol Lab 19(12):14–16
12. Husek P, Simek P (2001) Advances in amino acid analysis LCGC. Anal Bioanal Chem 19(9): 986–999
13. Husek P, Sweeley C (1991) Gas chromatographic separation of protein amino acids in four minutes. J High Resolut Chromatogr 14(11):751–753
14. Dettmer K, Stevens AP, Fagerer SR, Kaspar H, Oefner PJ (2012) Amino acid analysis in physiological samples by GC–MS with propyl chloroformate derivatization and iTRAQ–LC–MS/MS. In: Amino acid analysis. Humana Press, Totowa, pp 165–181
15. Matsumura S, Kataoka H, Makita M (1996) Determination of amino acids in human derum by capillary gas chromatography. J Chromatogr B 681:375–380
16. Nozal M, Bernal J, Toribio M, Diego J, Ruiz A (2004) Rapid and sensitive method for

determining free amino acids in honey by gas chromatography with flame ionization or mass spectrometric detection. J Chromatogr A 1047:137–146

17. Wang J, Huang Z-H, Gage D, Watson JT (2009) Analysis of amino acids by gas chromatography—flame ionization detection and gas chromatography—mass spectrometry: simultaneous derivatization of functional groups by an aqueous-phase chloroformate-mediated reaction. J Chromatogr A 663(1): 71–78

Chapter 10

Techno-Functional Properties of Proteins

Livia A. Santiago, Anne A. Sales, and Guilherme M. Tavares

Abstract

The presence of proteins in food products is highly relevant from a nutritional point of view. Regarding formulated food products, in addition to their nutritional relevance, proteins may also play an essential techno-functional role. Since proteins are complex macromolecules, their molecular and colloidal behaviors are dictated by their chemical nature. Proteins from different sources or subjected to different processes may display different techno-functional behaviors. In this context, characterizing proteins in terms of their techno-functionalities appears as an important strategy to reduce empiricism in their applications by the food industry during product development. This chapter will address different methods for characterizing the techno-functional properties of food proteins, grouping them into (i) solubility and swelling power, (ii) interfacial properties, and (iii) gelation.

Key words Food proteins, Colloidal properties, Techno-functionalities

1 Introduction

Proteins are macromolecules present in significant proportions in most food raw materials. They are of great importance from a nutritional point of view, particularly in terms of supplying essential amino acids. In addition to their nutritional importance, during food processing, changes in reaction conditions are often applied aiming to modify the protein structure of proteins and consequently alter some sensory aspects of the products [7]. The texture changes between milk and yogurt during the fermentation process are dictated by structural modifications of casein micelles induced by acidification, for instance. Regarding formulated foods, protein-rich ingredients are highly relevant as techno-functional agents [1, 23]. These ingredients are very often used for their ability to retain water or stabilize immiscible interfaces, as in the case of emulsions.

In this sense, this chapter compiles different methods for evaluating the techno-functional properties of proteins, grouped into (i) solubility and swelling power, (ii) interfacial properties, and (iii)

Lilian Mariutti and Frank Vriesekoop (eds.), *Food Chemistry: Basic Protocols*, Methods and Protocols in Food Science,
https://doi.org/10.1007/978-1-0716-4678-6_10,

gelation. Solubility in aqueous media is one of the primary techno-functional properties of proteins in protein-rich ingredients. This property is highly dependent on parameters of the reaction medium, such as pH, ionic strength, and temperature [19, 30]. Furthermore, the interfacial properties of proteins will be discussed in terms of foaming properties and emulsifying properties. In this section of the chapter, different methods for evaluating the foaming or emulsifying capacity, as well as to assess the morphology and stability of the obtained foams or emulsions, will be discussed [20, 25]. Regarding the gelling properties of proteins, visual and instrumental methods for evaluating the critical gelation concentration will be discussed, as well as methods for the evaluation of different characteristics of the obtained gels, such as their water retention capacities [23, 24, 27].

Since the techno-functional behavior of a protein is dependent on its chemical nature and its colloidal behavior, the characterization of the techno-functional performance of proteins and protein-rich ingredients is essential to guide the application of these materials by food industries.

2 Solubility and Swelling Power

2.1 Solubility in Aqueous Medium

The protein solubility in water is influenced by several factors such as pH, ionic strength, and the presence of other molecules, such as sodium dodecyl sulfate (SDS). The measured solubility in an aqueous medium displaying different composition may help predict the behavior of proteins under specific conditions. The most applied method to verify this techno-functionality is by centrifuging the protein dispersions and quantifying the protein content that remains in the supernatant [1, 7, 19, 23, 30].

Material

1. Sample: Prepare a protein dispersion at a fixed concentration, and adjust the pH using HCl, NaOH, or a buffer, as needed for your research (*see* **Note 1**).

Method

1. Mix the dispersions for 30 to 60 min, then centrifuge (*see* **Note 2**).
2. Determine the protein content of the supernatant and calculate the solubility (*see* **Note 3**).

2.2 Water and Oil Holding Capacity

The water and oil holding capacity are determined by the amount of liquid (water or oil, respectively) retained by a protein after centrifugation, applying pressure or filtering force. It is usually calculated as the weight of water or oil absorbed per sample mass [1, 30].

Material

1. Sample: Stir the protein powder with the same amount of water or vegetable oil (*see* **Note 4**).

Method

1. Keep the sample at room temperature for 30 min, then centrifuge (*see* **Note 2**).
2. Discard the supernatant and weigh the precipitate (*see* **Note 5**).

Notes (2—Solubility and Swelling Power)

1. The usual range of protein dispersion concentration is between 0.1% and 0.5% (w/v), and the pH values vary depending on the research objective.
2. Centrifugation usually ranges between 2,400 g and 8,000 g for 15 min to 30 min.
3. Different methods can be used to determine the protein content of the supernatant. The most used are the Bradford assay, the Biuret assay, the Lowry assay, the Dumas method, and the Kjeldahl method.

   The solubility is calculated as follows:

   $S (\%) = [(P_S/P_T) * 100]$, where P_S is the protein content in the supernatant and P_T is the total protein content in the sample.
4. The usual range of protein used is 0.1 g to 1 g suspended in 1.5 mL to 10 mL of water or vegetable oil, mixed for 1 min in a pre-weighed centrifuge tube.
5. The water and oil holding capacities are expressed in grams of water or oil held per gram of protein powder.

3 Interfacial Properties

3.1 Foaming Properties

The ability to form and stabilize foams is a useful techno-functionality of proteins in food processing, for example, in aerated desserts and beers. Foams are dispersions of gas in a continuous phase, usually an aqueous one. During whipping, proteins are able to adsorb at the air/water interface due to their amphiphilic nature, forming a flexible and cohesive film around the surface of the gas bubbles, which reduces the interfacial tension and makes the foam more stable [32, 35]. Four different methods applied to evaluate the foaming properties of proteins will be discussed, as described below.

3.1.1 Foam Analyzers

Foam analyzers are used to characterize foam properties (foamability and foam stability) by injecting some gas (Air, N_2, O_2, CO_2...) through a protein dispersion under controlled conditions, such as temperature and pressure. These devices provide measurements of the foam volume and the drained liquid fraction as a function of time. Furthermore, the size and distribution of the gas bubbles may be determined with some degree of automatization [5, 25, 28, 29].

Material

1. Sample: Prepare a protein dispersion at a fixed concentration, and adjust the pH using HCl, NaOH, or a buffer as needed for your research (*see* **Note 1**).
2. A foam analyzer.

Method

1. To determine foam capacity, foam expansion, liquid stability, and foam volume stability: Put a protein dispersion on the glass column, maintaining a fixed temperature, and pressurize some gas through the solution to achieve foaming. Evaluate subsequent changes in foam (*see* **Note 2**).
2. To determine the size and distribution of the gas bubbles, follow the same procedure described previously. The equipment will capture images of the foam (every 5 – 20 s), and from these data, it is possible to determine the number of bubbles and their diameters at different times using appropriate software (*see* **Notes 2** and **3**).

3.1.2 Determination of Foaming Properties Using "Dispersing Instruments"

A simple and widely applied technique to check the foaming properties is to mix the protein dispersions using high-stirring speed dispersing instruments. Foamability (FA) is often calculated by mathematical models that involve the dispersion volume before mixing and the foam volume formed immediately after it. Additionally, the foam stability (FS) may be determined as a function of the volume of foam formed immediately after mixing and the remaining volume of foam after a period of time [17].

Material

1. Sample: Prepare 50 mL of protein dispersion at a fixed concentration (the concentration of protein may vary according to the protein source but is usually in the range of 1 to 3% (w/v)) in a buffer (*see* **Note 4**) or Milli-Q water.

Method

1. Transfer the protein dispersion to a graduated cylinder (*see* **Note 5**).
2. Mix the dispersion using a high-stirring speed instrument for a fixed time (*see* **Note 6**).
3. Record the volume of foam formed immediately after stopping the stirring and after 5, 10, 15, 30, 45, 60 min, and every 1 h until no more foam (*see* **Notes** 7 and **8**).

3.1.3 Determination of Foaming Properties by Bubbling Technique

Another way to evaluate the FA of proteins and the stability of the formed foam is by the bubbling technique. Briefly, in this analysis, the foam is prepared in a graduated cylinder containing an air diffuser in the base coupled with an air pump. The protein dispersion is aerated for a specific time at constant airflow, and the volume of foam formed is measured, as shown in Fig. 1. The FA and FS may be calculated as aforementioned [24, 31].

Material

1. Sample: Prepare 50 mL of protein dispersion at a fixed concentration (the concentration of protein may vary according to the protein source but is usually in the range of 1 to 3% (w/v)) in a buffer (*see* **Note 4**) or Milli-Q water.

Method

1. Place an air diffuser coupled with the air pump on the base of a graduated cylinder (*see* **Note 5**).
2. Transfer the protein dispersion to the graduated cylinder.
3. Turn on the air pump and aerate the protein dispersion at a constant flow rate for 1 min.
4. Record the volume of foam formed immediately after stopping the flow of air and after 5, 10, 15, 30, 45, 60 min, and every 1 h until no more foam (*see* **Notes** 7 and **8**).

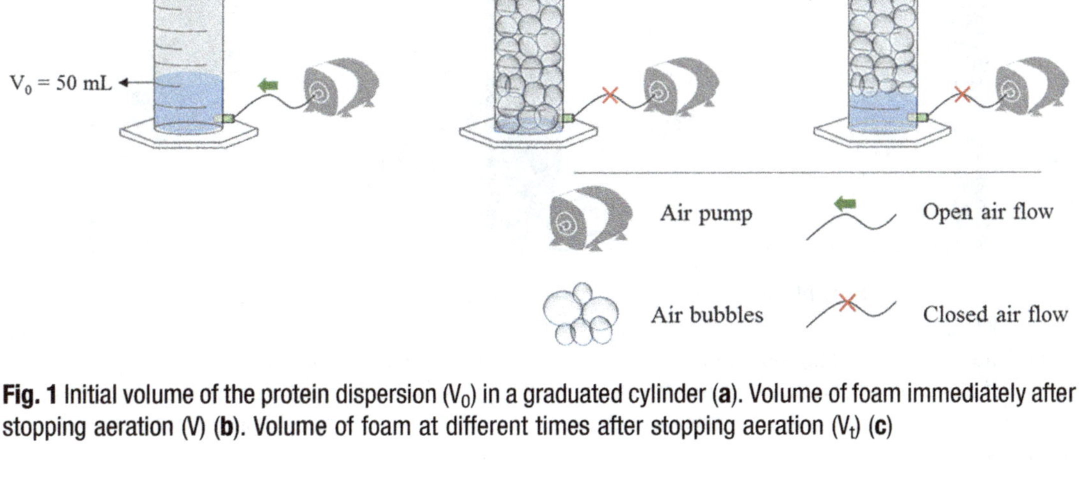

Fig. 1 Initial volume of the protein dispersion (V_0) in a graduated cylinder (**a**). Volume of foam immediately after stopping aeration (V) (**b**). Volume of foam at different times after stopping aeration (V_t) (**c**)

3.1.4 Determination of the Air Bubbles Morphology by Optical Microscopy

Optical microscopy may be useful in assessing the morphology, size, and distribution of air bubbles in foams. To perform this experiment, it is first necessary to produce the foam. Subsequently, for the observation at the microscope, a part of the foam at different incubation times is transferred to a concave glass slide and covered with a cover glass [2, 4, 16, 25, 27].

Material

1. Sample: With the dispersion of the protein in fixed concentration, prepare the foam by shaking or pressurizing some gas through it (*see* **Note 2**).

Method

1. Transfer a small portion of the foam to a concave glass slide and cover with a cover glass. It is important to control the room temperature (usual range 20–25 °C).
2. Examine the morphology of the air bubbles using an optical microscope.
3. Observe the change of air bubbles by a camera, as exemplified in Fig. 2 (*see* **Notes 9** and **10**).
4. Analyze the results using suitable software for image processing and analysis (*see* **Notes 11** and **12**).

Notes (3.1—Foaming Properties)

1. There is no standard for the concentration of protein dispersion or the pH used; this will depend a lot on your sample and

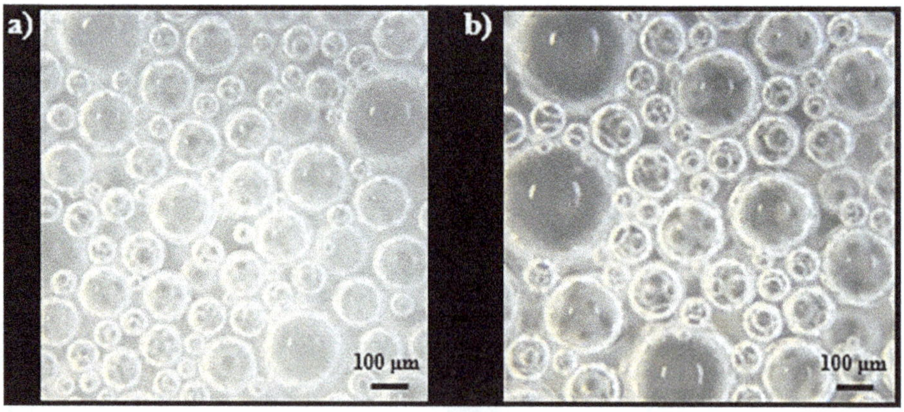

Fig. 2 Optical microscopy of foam at different incubation times. The foam was prepared by stirring a whey protein isolate dispersion at 2% (w/v), using a high-speed dispersion instrument (1,130 rpm for 5 min). Immediately after preparing the foam, the average diameter of the air bubbles was 69.3 μm (**a**), and after 20 min of incubation, the average diameter was 84.9 μm (**b**). The increase in the size of the air bubbles in this time interval may indicate the coalescence phenomenon. The diameter of air bubbles was measured by ImageJ software

the proposed application. If you need to use a buffer, check its compatibility with the protein. For example, phosphate buffer should not be used when casein micelles are in the system.

2. The technical specifications for gas flow, liquid volume, and gas can vary depending on the method applied. The usual range for gas flow is between 25 mL min^{-1} and 80 mL min^{-1}, and it is interrupted when it reaches at least twice the initial volume (usually, $v_i = 60$ mL).
3. Some foam analyzers do not determine the size and distribution of the air bubbles. Check the functions of the device you are using.
4. The buffer depends on the pH at which the analysis will be performed. But generally, this analysis is carried out at pH 7 using 10 mM phosphate buffer. To prepare this buffer, follow these steps: Add 80 mL of Milli-Q water in a 100 mL volumetric flask. Weigh 1.549 g of Na_2HPO_4-$7H_2O$ and 583 mg of $NaHPO_4H_2O$ and transfer to the cylinder. Adjust the solution to pH 7 with HCl or NaOH if necessary. Add Milli-Q water until the volume is 100 mL.
5. The graduated cylinder must have enough volume to hold all the foam that will be formed.
6. The high-stirring-speed instrument is used to aerate the dispersion and, consequently, form foam. The speed applied can vary depending on the function of the equipment used, generally between 7,000 and 16,000 rpm.
7. Do not move the cylinder or touch the foam for drainage to occur naturally.
8. The foamability is calculated by the formula:

   FA (%) $= [(V - V_0)/V_0] * 100$, where V is the volume of foam recorded immediately after stopping the aeration/mixing and V_0 is the initial volume (50 mL). The foam stability is calculated as follows: FS (%) $= [(V - V_t)/V] * 100$, where V_t is the volume of foam registered at different times. Using a semi-log scale during data processing can simplify the visualization of foam decomposition in the graph, especially in the first minutes of analysis.
9. The light of the microscope may heat the foam, inducing its decomposition. Therefore, the necessary time to collect the images is an important variable for this analysis. Depending on the characteristics of the samples and the room luminosity, the light of the microscope may not even be necessary.
10. The different times to analyze the variation in the size and distribution of the bubbles depend on the stability of your sample.

11. Some software programs that can be used to assess the size of the bubbles are CAS software and ImageJ software.
12. The reliability of the results is improved by increasing the number of images analyzed.

3.2 Emulsifying Properties

The emulsifying properties of proteins are also related to their amphiphilic character. This is due to their ability to decrease the interfacial tension between hydrophobic and hydrophilic components in food [32]. As discussed below, some different ways to evaluate the emulsifying properties of proteins will be addressed.

3.2.1 Determination of the Emulsifying Properties by Turbidity Measurements

To determine the emulsifying activity index (EAI) and the emulsion stability index (ESI) of emulsions stabilized by proteins, the emulsion needs to be produced by mixing protein dispersion and an appropriate amount of oil, using a high-speed dispersion instrument, a homogenizer, or microfluidic equipment. The emulsion sample is immediately taken and diluted in a buffer containing 0.1% SDS, and then the absorbance of the diluted emulsion is measured at 500 nm. EAI and ESI may be calculated by mathematical models correlating several variables such as protein concentration, oil fraction, and absorbance measured at different times [1, 15, 19, 20, 34].

Material

1. Sample: Prepare a protein dispersion at a fixed concentration, and adjust the pH using HCl, NaOH, or a buffer as needed for your research (*see* **Note 1**).
2. A vegetable oil.
3. A buffer supplemented with 0.1% sodium dodecyl sulfate (SDS).

Method

1. To determine EAI and ESI: Mix the protein dispersion and an appropriate amount of vegetable oil using a high-speed dispersion instrument, a homogenizer, or microfluidic equipment (*see* **Notes 2** and **3**).
2. Transfer an aliquot of the emulsion into a suitable container to dilute it in a buffer containing 0.1% (w/v) sodium dodecyl sulfate (SDS), then mix it using a vortex for 10 s (*see* **Note 4**).
3. Measure the absorbance of the diluted emulsion at 500 nm using a spectrophotometer immediately after mixing and also after 10 min. Use the absorbance values to calculate the EAI and ESI (*see* **Note 5**).

3.2.2 Determination of the Emulsifying Capacity by Measuring the Emulsion Conductivity

The emulsifying capacity (EC) may be assessed by the amount of oil needed to achieve the inversion point where the oil in water emulsion turns into the water in oil emulsion, which can be monitored by a sudden drop in emulsion conductivity [3, 10, 15, 21].

Material

1. Sample: Prepare a protein dispersion at a fixed concentration and adjust the pH using HCl, NaOH, or a buffer as needed for your research (*see* **Note 1**).
2. A vegetable oil.

Method

1. Use a high-speed dispersion instrument or another appropriate device to prepare the emulsion.
2. Progressively add the oil into the protein dispersion and follow the conductivity of the emulsion using an appropriate device.
3. Determine the amount of oil needed to promote a sudden drop in emulsion conductivity, which represents the amount of oil needed to convert the oil in water emulsion into the water in oil emulsion (*see* **Note 6**).

3.2.3 Determination of the Emulsifying Capacity by Measuring the Amount of Non-emulsified Oil

The EC of the protein can also be assessed by determining the volume of oil that a fixed amount of protein is capable of emulsifying. For this purpose, the emulsion prepared with a known excess of oil is centrifuged, and the amount of non-emulsified oil (free oil layer) is weighed [23].

Material

1. Sample: Prepare a protein dispersion at a fixed concentration, and adjust the pH using HCl, NaOH, or a buffer as needed for your research (*see* **Note 1**).
2. A vegetable oil.

Method

1. To prepare the emulsion, mix the protein dispersion and an appropriate amount of vegetable oil using a high-speed dispersion instrument or another appropriate device (*see* **Notes 3** and **7**).
2. Transfer the emulsion obtained to centrifuge tubes and then centrifuge (*see* **Note 8**).
3. After centrifugation, weigh the leftover oil not consumed in the emulsification process to calculate the emulsifying capacity (free oil layer, as shown in Fig. 3) (*see* **Note 9**).

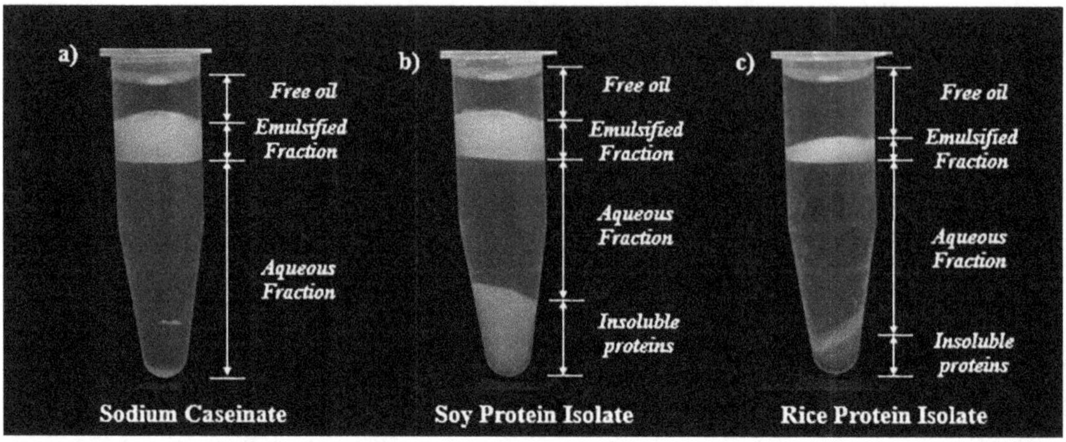

Fig. 3 Coarse oil-in-water emulsions stabilized by different proteins after centrifugation. The emulsions were prepared by mixing protein dispersions at 1% (w/v) of total protein content with sunflower oil in the proportion of 2:1 (v/v), using a high-speed dispersion instrument (7,000 rpm for 2 min). Emulsion (**a**) was prepared using commercial sodium caseinate, emulsion (**b**) using commercial soy protein isolate, and (**c**) using commercial rice protein isolate. The stratification of the different layers is highlighted: free oil layer, emulsified fraction, aqueous fraction, and insoluble proteins

3.2.4 Determination of the Emulsifying Properties Measuring the Height of the Emulsified Fraction

The emulsifying activity (EA) can be determined by measuring the height of the emulsified fraction after centrifugation. Briefly, the protein dispersion is mixed with vegetable oil using a high-stirring speed instrument or another appropriate device, followed by centrifuging. The EA is determined as a function of the height of the emulsified fraction observed after centrifugation and the total height of the mixture [12, 36].

Material

1. Sample: Prepare a protein dispersion at a fixed concentration, and adjust the pH using HCl, NaOH, or a buffer as needed for your research (*see* **Note 1**).
2. A vegetable oil.

Method

1. To prepare the emulsion, mix the protein dispersion with an appropriate amount of vegetable oil using a high-speed dispersion instrument or another appropriate device (*see* **Notes 2** and **3**).
2. Transfer immediately the emulsion obtained to centrifuge tubes and centrifuge (*see* **Note 8**).
3. Record the height of the emulsified layer, as shown in Fig. 3, and the total height of the mixture to calculate the emulsifying activity (*see* **Note 10**).

3.2.5 Determination of the Morphology, Size, and Stability of Emulsions by Optical Microscopy

Optical microscopy can be used to evaluate the morphology and droplet size of the emulsions. In addition, this method may provide information on emulsion stability. A small drop of the emulsion at different incubation times is transferred to a glass slide, covered with a cover slip, and observed with an optical microscope [13, 14, 22].

Material

1. Sample: Prepare a protein dispersion at a fixed concentration, and adjust the pH using HCl, NaOH, or a buffer as needed for your research (*see* **Note 1**).
2. A vegetable oil.

Method

1. To prepare the emulsion, mix the protein dispersion with an appropriate amount of vegetable oil using a high-speed dispersion instrument or another appropriate device (*see* **Note 3**).
2. If necessary, dilute the emulsion with vegetable oil or water (*see* **Note 11**).
3. Transfer immediately a drop of the sample to a glass slide and cover with a cover slip.
4. Using an optical microscope, observe the emulsion morphology and take pictures (Fig. 4).

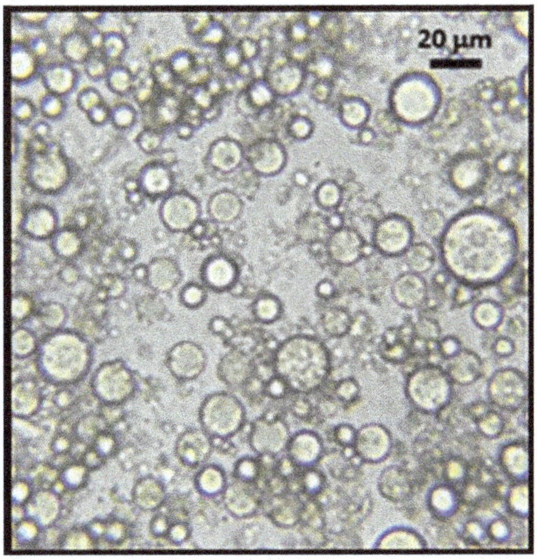

Fig. 4 Optical microscopy of an oil-in-water emulsion. The emulsion was prepared by mixing corn gluten meal dispersion at 2% (w/v) with corn oil (final oil concentration of 10% v/v), using a high-speed dispersion instrument (10,000 rpm for 2 min). The average diameter of the oil droplets was 10.5 μm, measured by ImageJ software

5. To assess the stability of the emulsion, repeat the procedures described in **items 3** and **4** at different incubation times.
6. Analyze the droplets' size in the emulsion's dispersed phase using suitable software for image processing and analysis (*see* **Notes 12** and **13**).

3.2.6 Determination of the Droplet Size of the Dispersed Phase of the Emulsions by Laser Diffraction

The droplet size of the dispersed phase influences many characteristics of the emulsion, such as its appearance, rheology, and stability. By determining the initial size of the droplets and their evolution during the aging of the emulsion, it is possible to follow and measure the emulsifying properties of the proteins. Commonly, a laser diffraction particle size analyzer or the electrical pulse counting technique are used for this [13–15, 22].

Material

1. Sample: Prepare a protein dispersion at a fixed concentration, and adjust the pH using HCl, NaOH, or a buffer as needed for your research (*see* **Note 1**).
2. A vegetable oil.

Method

1. To prepare the emulsion, mix the protein dispersion with an appropriate amount of vegetable oil using a high-speed dispersion instrument or another appropriate device (*see* **Note 3**).
2. Measure the emulsion droplet size distribution using a laser diffraction particle size analyzer equipped with a wet sample dispersion unit (*see* **Notes 14** and **15**).
3. To evaluate the stability of the emulsion, measure the droplet size distribution of the emulsion at different incubation times.

Notes (3.2—Emulsifying Properties)

1. There is no standard for the protein dispersion concentrations or the pH used; this will depend on the sample to be analyzed and the proposed application. The range commonly used for protein dispersion is between 0.1 and 2% (w/v), usually prepared at pH 7.
2. The total oil concentration used to prepare the emulsion is highly variable. The total oil concentration of 10% (v/v) is very often used.
3. When high-speed dispersion instruments are used to prepare an emulsion, the intensity of the applied treatment may be highly variable according to the objective of the experiment. Speeds between 7,000 and 20,000 rpm may be applied for 1 to 5 min. Ideally, the entire oil fraction should be added to the protein dispersion within the first minute of stirring or before.

4. Usually, the emulsion is diluted 1:100 (v/v) in the buffer supplemented with SDS.
5. The emulsifying activity index is calculated by the formula: EAI $(m^2/g) = [(2 * 2.303 * A_0 * N)/(c * \varphi * 10000)]$, where A_0 is the absorbance of the diluted emulsion immediately after homogenization, N is the dilution factor, c is the protein concentration (g/mL), and φ is the oil volume fraction of the emulsion. To calculate the emulsion stability index, the following formula is used: ESI (min) $= [(A_0/\Delta A) * t]$, where ΔA is the change in the absorbance between 0 and 10 min (A_0–A_{10}), and t is the time interval between initial and final measurements (10 min).
6. The result is expressed as a g of oil added just before phase inversion per g of protein.
7. The amount of oil used in this method needs to be higher than the emulsifying capacity of the protein.
8. The applied centrifugation needs to allow the stratification of the different fractions of the sample. Centrifugation at 1,000 to 1,500 g for 5 to 10 min may be applied.
9. The emulsifying capacity is calculated as the percentage of the total initial oil content actually emulsified by the proteins, as follows:

   EC (%) $= [1—(g_e * \rho_e / V_a)] * 100$, where EC (%) is the emulsifying capacity, g_e the weight of the non-emulsified oil (g) (free oil layer), ρ_e the density of the oil, and V_a the total volume of oil in the sample (mL) (initial amount).
10. The EA is calculated using the following formula:

    EA (%) $= (H_{EL}/H_T) * 100$, where H_{EL} is the height of the resulting emulsified fraction and H_T is the height of the total contents in the centrifugal tubes. The height measurement can be performed with a ruler.
11. The diluent used must be the original continuous phase of the emulsion. The dilution may be necessary to better observe the droplets.
12. Some of the software programs that can be used to assess the size of the droplets are Image-Pro PLUS software of Media Cybernetics, CAS software, and ImageJ software.
13. An appropriate number of pictures should be analyzed to improve the reliability of the results.
14. For instance, the equipment used in this analysis can be a Mastersizer (Malvern Instruments) or an LS 13320 (Beckman Coulter).
15. Some parameters used in this measurement are the refractive index of the diluent and the refractive index of the dispersed phase. The equipment software may calculate the droplet size distribution according to the intensity of the scattered light.

4 Gelation

Gelation is a protein techno-functionality essential to producing some food products, such as jellies, puddings, and meat products [35]. Gels are dispersed systems of at least two components in which a solid phase (proteins, in our case) forms a cohesive three-dimensional network in a liquid phase [32, 35].

4.1 Critical Gelling Concentration

A widely used technique for verifying a protein gelling ability is by assessing the least protein concentration required to form a cohesive gel. In general, the smaller the critical gelling concentration of the protein, the better its gelling property [24].

Material

1. Sample: Prepare dispersions with different concentrations of protein, and adjust the pH using HCl, NaOH, or a buffer, as needed for your research (*see* **Note 1**).

Methods

(a) *Heat-induced gels*

1. Transfer each protein dispersion to a test tube, then heat the tubes in a water bath (*see* **Note 2**).
2. After heating, immediately cool the tubes to room temperature in an ice bath.
3. Invert the tubes and observe which one of them the content does not fall or slip, as shown in Fig. 5 (*see* **Note 3**).

(b) *Acid-Induced Gels*

1. Transfer each protein dispersion to a test tube, add a fixed amount of glucono-δ-lactone (GDL), and stir quickly (*see* **Note 4**).
2. Keep the mixture at a given temperature for 24 h (usually room temperature) (*see* **Note 5**).
3. After 24 h, invert the tubes and observe which one of them the content does not fall or slip (*see* **Note 3**).

4.2 Water Holding Capacity (WHC) of Gels

The WHC provides information on the amount of water that remains within the protein gel structure. This property may be determined by centrifuging a fixed weight of the formed gel and then comparing it with its weight after centrifugation [27, 33].

Material

1. Sample: Prepare a protein dispersion at a fixed concentration, and adjust the pH using HCl, NaOH, or a buffer as needed for your research (*see* **Note 6**).

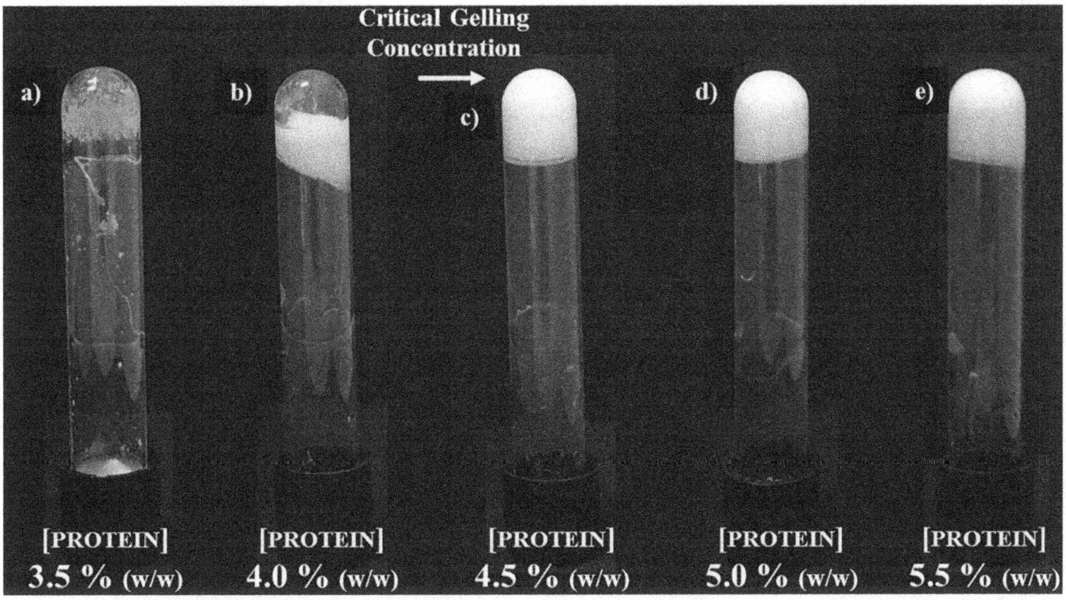

Fig. 5 Commercial Whey Protein Isolate dispersions at different total protein concentrations heat treated at 90 °C for 15 min. Total protein concentration: (**a**) 3.5% w/w; (**b**) 4.0% w/w; (**c**) 4.5% w/w; (**d**) 5.0% w/w, and (**e**) 5.5% w/w

Method

1. To prepare a heat-induced gel, transfer the protein dispersion to a suitable container, heat in a water bath, and cool the gel to room temperature or in an ice bath (*see* **Note 7**). Alternatively, you can prepare an acid-induced gel by adding GDL to the protein dispersion (*see* **Note 4**).
2. Take samples of the gel and weigh them (*see* **Note 8**).
3. Centrifuge the gel samples, dispose of the free water, and weigh again (*see* **Notes 9** and **10**).

4.3 Rheological Properties

Texture characteristics, strength, and viscoelasticity of the protein gels may be assessed through rheological analyses. Besides that, these analyses may also be applied to explore specific gelling conditions, such as the gelation temperature or pH [6, 8, 9, 11, 18, 23, 26].

Material

1. Sample: Prepare a protein dispersion at a fixed concentration, and adjust the pH using HCl, NaOH, or a buffer as needed for your research (*see* **Note 6**).

Methods

(a) *Texture Profile Analysis (TPA)*

1. To prepare a heat-induced gel, transfer the protein dispersion to a suitable container, heat it in a water bath, and cool the gel to room temperature in an ice bath (*see* **Note 7**). Alternatively, add GDL to the protein dispersion to prepare an acid-induced gel (*see* **Note 4**).
2. Take samples of the gel with a fixed height (*see* **Note 11**).
3. Using a texture analyzer equipped with a probe, perform a double compression of the gel sample up to a fixed strain value at a determined trigger force and probe speed. It is important to have intervals between compressions for material recovery (*see* **Note 12**).
4. Measure the textural parameters (*see* **Note 13**).

(b) *Rheological Measurements*

1. Use a rheometer (*see* **Note 14**) to determine the viscoelasticity of the sample (*see* **Note 15**) in terms of the storage (G ′) and loss (G″) moduli by frequency sweep, stress/strain sweep, time sweep, or temperature sweep, depending on the constant parameters used (*see* **Note 16**). The experiments must be conducted using an appropriate geometry with fixed diameter and gap and equilibrated at a particular starting temperature (*see* **Note 17**).
2. To determine the recovery of the structure after the shear treatment, thixotropic analysis is performed by submitting the sample at low and high strain percentage intervals at a determined time (*see* **Note 18**).
3. To evaluate the behavior of the material, the flow curves are obtained with an up-down-up program with the shear rate variation at a given temperature (*see* **Note 19**).

Notes (4—Gelation)

1. The concentrations to prepare the dispersions will vary depending on the protein source. To perform the preliminary tests, it is recommended to use an intermediary protein concentration (e.g., 5%). If gelation does not occur, increase protein concentration.
2. The temperature and time used will depend on the proposed application for the sample. In addition, to determine the minimum temperature and time for gelation, tests can be performed in different time and temperature binomials.
3. The critical gelling concentration is the least concentration of protein at which the tube content does not slip or fall after inverting it.

4. The amount of GDL used is fixed as the one necessary to reach a pH of 4 to 4.5 after 24 h at a given temperature (usually between 20 and 30 °C).
5. Usually, sodium azide 0.02% (w/w) is added to the protein dispersion to prevent bacterial growth.
6. The concentration will depend on the protein source, but it should be greater or equal to the critical gelling concentration (*see* Subheading 4.1).
7. The heat treatment is usually carried out between 70 to 90 °C for 10 to 30 min.
8. The gel samples used usually weigh 3 to 5 g.
9. The centrifugation is usually performed between 2,000 and 4,000 g for 15 to 30 min.
10. The water holding capacity of the gel is calculated as follows:

    WHC (%) = $(We_a/We_b) * 100$, where We_a and We_b are the gel weights after and before the withdrawal of free water (centrifugation), respectively.
11. There is no standard for the height of the gel samples. The sample geometry is usually a cube or a cylinder.
12. The type of probe depends on the equipment used and the sample, generally, a cylindrical probe is used. The strain value is usually 50% of the original height of the gel sample, and the probe speed is 1.0 mm/s or 2.0 mm/s. The usual range for the trigger force is 0.0045 to 5.0 N, and the interval between compressions is 3 to 5 s. It is important to control the temperature during this analysis, it is usually carried out at room temperature.
13. The main parameters obtained in the texture analysis are hardness, springiness, adhesiveness, cohesiveness, resilience, brittleness, chewiness, and gumminess. These parameters are obtained from the force-time graphic and calculated using software suitable for the texture analyzer.
14. Several types of rheometers exist, such as modular compacts equipped with a Peltier system, controlled stress, hood, and controlled rotational tension.
15. To prevent evaporation, the exposed edges of the samples can be covered with a thin layer of silicone oil.
16. The temperature sweep is one of the most used tests, and it is conducted at a frequency of 1 Hz and a strain between 0.1% and 1%, in which the temperature ramp is determined based on the transport temperature of the material using a rate of proper heating/cooling.
17. Depending on the rheometer and the sample, different geometries can be used, such as parallel plate, rough plate-plate,

and cone-and-plate, with diameters of plates varying from 25 to 50 mm. The gap of 1 mm between the plates is the most used.

18. Normally, the low percentage strain is 0.1%, and the high percentage strain is 100% at a frequency of 10 rad s^{-1}.
19. The variation of the shear rate will depend on the research and material used, which usually varies between 0 and 300 s^{-1} at room temperature.

References

1. Abbou A, Kadri N, Dahmoune F et al (2020) Optimising functional properties and chemical composition of Pinus halepensis Mill. Seeds protein concentrates. Food Hydrocoll 100: 105416. https://doi.org/10.1016/j.foodhyd.2019.105416
2. Alavi F, Emam-Djomeh Z, Momen S et al (2019) Effect of free radical-induced aggregation on physicochemical and interface-related functionality of egg white protein. Food Hydrocoll 87:734–746. https://doi.org/10.1016/j.foodhyd.2018.08.048
3. Alsalman FB, Tulbek M, Nickerson M, Ramaswamy HS (2020) Evaluation and optimization of functional and antinutritional properties of aquafaba. Legum Sci 2. https://doi.org/10.1002/leg3.30
4. Alves AC, Martha L, Casanova F, Tavares GM (2021) Structural and foaming properties of whey and soy protein isolates in mixed systems before and after heat treatment. Food Sci Technol Int 28:545
5. Amagliani L, O'Regan J, Schmitt C et al (2019) Characterisation of the physicochemical properties of intact and hydrolysed rice protein ingredients. J Cereal Sci 88:16–23. https://doi.org/10.1016/j.jcs.2019.04.002
6. Ashkar A, Laufer S, Rosen-Kligvasser J et al (2019) Impact of different oil gelators and oleogelation mechanisms on digestive lipolysis of canola oil oleogels. Food Hydrocoll 97: 105218. https://doi.org/10.1016/j.foodhyd.2019.105218
7. Boye JI, Aksay S, Roufik S et al (2010) Comparison of the functional properties of pea, chickpea and lentil protein concentrates processed using ultrafiltration and isoelectric precipitation techniques. Food Res Int 43:537–546. https://doi.org/10.1016/j.foodres.2009.07.021
8. Cai L, Feng J, Regenstein J et al (2017) Confectionery gels: effects of low calorie sweeteners on the rheological properties and microstructure of fish gelatin. Food Hydrocoll 67:157–165. https://doi.org/10.1016/j.foodhyd.2016.12.031
9. Casas-Forero N, Moreno-Osorio L, Orellana-Palma P, Petzold G (2021) Effects of cryoconcentrate blueberry juice incorporation on gelatin gel: a rheological, textural and bioactive properties study. LWT 138:110674. https://doi.org/10.1016/j.lwt.2020.110674
10. Cepeda E, Villarán MC, Aranguiz N (1998) Functional properties of faba bean (Vicia faba) protein flour dried by spray drying and freeze drying. J Food Eng 36:303–310. https://doi.org/10.1016/S0260-8774(98)00061-2
11. Chen H, Xie F, Chen L, Zheng B (2019) Effect of rheological properties of potato, rice and corn starches on their hot-extrusion 3D printing behaviors. J Food Eng 244:150–158. https://doi.org/10.1016/j.jfoodeng.2018.09.011
12. Flores I, Cabra V, Quirasco MC et al (2010) Emulsifying properties of chemically deamidated corn (Zea Mays) gluten meal. Food Sci Technol Int 16:241–250. https://doi.org/10.1177/1082013210366750
13. Gould J, Wolf B (2018) Interfacial and emulsifying properties of mealworm protein at the oil/water interface. Food Hydrocoll 77:57–65. https://doi.org/10.1016/j.foodhyd.2017.09.018
14. Hu YT, Ting Y, Hu JY, Hsieh SC (2017) Techniques and methods to study functional characteristics of emulsion systems. J Food Drug Anal 25:16–26
15. Karaca AC, Low N, Nickerson M (2011) Emulsifying properties of chickpea, faba bean, lentil and pea proteins produced by isoelectric precipitation and salt extraction. Food Res Int 44:2742–2750. https://doi.org/10.1016/j.foodres.2011.06.012
16. Li X, Murray BS, Yang Y, Sarkar A (2020) Egg white protein microgels as aqueous Pickering foam stabilizers: bubble stability and interfacial properties. Food Hydrocoll 98:105292.

https://doi.org/10.1016/j.foodhyd.2019.105292
17. Mishyna M, Martinez J-JI, Chen J, Benjamin O (2019a) Extraction, characterization and functional properties of soluble proteins from edible grasshopper (Schistocerca gregaria) and honey bee (Apis mellifera). Food Res Int 116: 697–706. https://doi.org/10.1016/j.foodres.2018.08.098
18. Mishyna M, Martinez JJI, Chen J et al (2019b) Heat-induced aggregation and gelation of proteins from edible honey bee brood (Apis mellifera) as a function of temperature and pH. Food Hydrocoll 91:117–126. https://doi.org/10.1016/j.foodhyd.2019.01.017
19. Mokni Ghribi A, Maklouf Gafsi I, Sila A et al (2015) Effects of enzymatic hydrolysis on conformational and functional properties of chickpea protein isolate. Food Chem 187:322–330. https://doi.org/10.1016/j.foodchem.2015.04.109
20. Mune MAM, Sogi DS (2016) Emulsifying and foaming properties of protein concentrates prepared from cowpea and Bambara bean using different drying methods. Int J Food Prop 19:371–384. https://doi.org/10.1080/10942912.2015.1023399
21. Neto VQ, Narain N, Silva JB, Bora PS (2001) Functional properties of raw and heat processed cashew nut (Anacardium occidentale, L.) kernel protein isolates. Nahrung—Food 45:258–262. https://doi.org/10.1002/1521-3803(20010801)45:4<258::AID-FOOD258>3.0.CO;2-3
22. Opedal NVDT, Sørland G, Sjöblom J (2009) Methods for droplet size distribution determination of water-in-oil emulsions using Low-field NMR. Diffusion-Fundamentalsorg 9:1–29
23. Pérez-Andrés JM, Álvarez C, Cullen PJ, Tiwari BK (2019) Effect of cold plasma on the techno-functional properties of animal protein food ingredients. Innov Food Sci Emerg Technol 58:102205. https://doi.org/10.1016/j.ifset.2019.102205
24. Santiago LA, Fadel OM, Tavares GM (2021) How does the thermal-aggregation behavior of black cricket protein isolate affect its foaming and gelling properties? Food Hydrocoll 110: 106169. https://doi.org/10.1016/j.foodhyd.2020.106169
25. Schmidt I, Novales B, Boué F, Axelos MAV (2010) Foaming properties of protein/pectin electrostatic complexes and foam structure at nanoscale. J Colloid Interface Sci 345:316–324. https://doi.org/10.1016/j.jcis.2010.01.016
26. Shang L, Wu C, Wang S et al (2021) The influence of amylose and amylopectin on water retention capacity and texture properties of frozen-thawed konjac glucomannan gel. Food Hydrocoll 113:106521. https://doi.org/10.1016/j.foodhyd.2020.106521
27. Shen H, Stephen Elmore J, Zhao M, Sun W (2020) Effect of oxidation on the gel properties of porcine myofibrillar proteins and their binding abilities with selected flavour compounds. Food Chem 329:127032. https://doi.org/10.1016/j.foodchem.2020.127032
28. Stubenrauch C, Hamann M, Preisig N et al (2017) On how hydrogen bonds affect foam stability. Adv Colloid Interf Sci 247:435–443. https://doi.org/10.1016/j.cis.2017.02.002
29. Teclis Scientific (2020) Foam Analyzers. https://www.teclis-scientific.com/products-foam-analyzers. Accessed 30 July 2020
30. Tontul İ, Kasimoglu Z, Asik S et al (2018) Functional properties of chickpea protein isolates dried by refractance window drying. Int J Biol Macromol 109:1253–1259. https://doi.org/10.1016/j.ijbiomac.2017.11.135
31. Vidotto DC, Tavares GM (2020) Impact of dry heating in an alkaline environment on the structure and foaming properties of whey proteins. Food Bioprocess Technol 13:1755–1764. https://doi.org/10.1007/s11947-020-02519-5
32. Weder JKP, Belitz H-D (2003) Functional properties. Encycl Food Sci Nutr:4835–4841
33. Xia X, Kong B, Xiong Y, Ren Y (2010) Decreased gelling and emulsifying properties of myofibrillar protein from repeatedly frozen-thawed porcine longissimus muscle are due to protein denaturation and susceptibility to aggregation. Meat Sci 85:481–486. https://doi.org/10.1016/j.meatsci.2010.02.019
34. Xu Y, Galanopoulos M, Sismour E et al (2020) Effect of enzymatic hydrolysis using endo- and exo-proteases on secondary structure, functional, and antioxidant properties of chickpea protein hydrolysates. J Food Meas Charact 14: 343–352. https://doi.org/10.1007/s11694-019-00296-0
35. Yada RY (2018) Proteins in food processing, 2nd edn. Woodhead Publishing, p 672
36. Yasumatsu K, Sawada K, Moritaka S et al (1972) Whipping and emulsifying properties of soybean products. Agric Biol Chem 36: 719–727. https://doi.org/10.1080/00021369.1972.10860321

Part IV

Carbohydrates

Chapter 11

Determination of Dietary Fiber Monosaccharide Composition

Lucimara Mach Côrtes Cordeiro

Abstract

This chapter will describe methods to determine neutral sugars by gas chromatography (GC) of alditol acetate derivatives and quantify uronic acids by a colorimetric assay using the m-hydroxydiphenyl method. Understanding the monosaccharide composition of polysaccharides is crucial for predicting their functionality, such as gelling, solubility, and rheological characteristics. It is also vital for comprehending the mechanisms associated with different physiological effects in the body.

Key words Dietary fiber (DF), Carbohydrates characterization, Monosaccharides, Indigestible carbohydrates

1 Introduction

The methods adopted for determining the monosaccharide composition of plant cell walls [1] or polysaccharides, in general, are also applied for the monosaccharide analysis of dietary fiber (DF). Monosaccharide composition analysis contains three steps: complete hydrolysis, separation, and detection.

The hydrolysis conditions are crucial and can vary depending on the nature of the DF (soluble or insoluble) and its component monosaccharides. The most common acids for hydrolysis are trifluoroacetic acid (TFA) and sulfuric acid [1, 2]. TFA is preferably used to hydrolyze soluble DFs, which, depending on the source, generally contain pectic polysaccharides, arabinoxylans (cereals), galactomannans, and glucomannans, among others. TFA is not adequate for the hydrolysis of insoluble DFs, mainly because of the recalcitrance of cellulose in its crystalline form. Sulfuric acid is a harsher acid better suited than TFA for complete hydrolysis of insoluble DFs, which generally contains cellulose and hemicellulosic polysaccharides [3]. While TFA is volatile and can be easily

Lilian Mariutti and Frank Vriesekoop (eds.), *Food Chemistry: Basic Protocols*, Methods and Protocols in Food Science, https://doi.org/10.1007/978-1-0716-4678-6_11,

removed from the sample by evaporation, sulfuric acid can be removed by precipitation with barium hydroxide.

Monosaccharides released from the hydrolysis can be identified and quantified by different techniques such as high-performance anion-exchange chromatography with pulsed amperometric detector (HPAEC-PAD), gas chromatography (GC), high-performance liquid chromatography (HPLC), and capillary electrophoresis [4–7].

GC with flame ionization detection (FID) and gas chromatography-mass spectrometry (GC-MS) has been extensively used for both quantitative and qualitative monosaccharide composition analysis of polysaccharides. These methods require derivatization to transform the monosaccharides into volatile compounds, such as trimethylsilyl (TMS) derivatives and alditol acetates. This latter is preferable since this procedure gives a single peak for each sugar in the analysis, and it is relatively more stable than TMS-derivatives [8]. Alditol acetates are formed under reduction ($NaBH_4$ or $NaBD_4$) and acetylation (acetic anhydride) of the monosaccharides released in the hydrolysis step. They are then analyzed by GC and identified by their retention time relative to an internal standard (typically *myo*-inositol but other compounds may also be used).

It should be emphasized that only neutral monosaccharides can be detected by GC analysis of their alditol acetate derivatives. This technique is not suitable for the acid sugar analysis since glycosidic linkages adjacent to uronic acids are difficult to hydrolyze efficiently or quantitatively [9, 10]. One strategy to overcome this is to perform the carboxyl reduction of the DF fraction before the hydrolysis step, transforming the uronic acids into their corresponding neutral sugar. The detailed protocol can be found in [11]. As mentioned above, GalA is present in pectins and glucuronic acids (GlcA) on pectins, type II arabinogalactan (proteins), and heteroxylans. Another strategy is to quantify the uronic acids by colorimetric assays. DF fractions are first hydrolyzed in sulfuric acid following the methods of [12, 13] and the uronic acid content estimated by the carbazole [14] or the *m*-hydroxydiphenyl method [15]. One disadvantage is that they only estimate total uronic acids and do not differentiate between GalA and GlcA. The Filisetti-Cozzi and Carpita [15] method is the colorimetric method of choice since it overcomes the interference caused by the presence of neutral sugars. The sulfamate suppresses the formation of brown pigments from the neutral sugars, and tetraborate increases the sensitivity of the reaction with uronic acids.

This chapter will describe the methodology to determine neutral sugars by GC quantification of alditol acetate derivatives and the quantification of uronic acids by colorimetric assay using the *m*-hydroxydiphenyl method [11].

2 Material

2.1 Reagents

Trifluoracetic acid (TFA).

Concentrated sulfuric acid.

Sugar standards: Rhamnose, Fucose, Arabinose, Xylose, Mannose, Galactose, Glucose, Galacturonic acid, and Glucuronic acid.

myo-Inositol,

NH_4OH (35%, v/v).

Distilled water.

Copper sulfate.

NaOH.

KOH.

Sulfamic acid.

Sodium tetraborate.

m-Hydroxydiphenyl,

Nitrogen gas (industrial grade).

Barium hydroxide.

$NaBH_4$.

Glacial acetic acid.

Acetic anhydride.

Pyridine.

Methanol.

Chloroform.

Acetone.

Anhydrous sodium acetate.

2.2 Equipment

Glass tube (4–7 mL, with a Teflon-lined screw cap).

Glass Pasteur pipettes.

Oven.

Heat-resistant test tube racks.

Analytical balance.

Benchtop centrifuge.

Vortex.

pH paper,

Micropipettes.

15 mL borosilicate glass tubes.

15 mL centrifuge tubes.

Volumetric flasks.

100 °C water bath.

Spectrophotometer.

Gas chromatograph equipped with a flame-ionization detector or GC-MS system.

GC columns (see below).

2.3 Solution Preparation

1. **TFA (2 M)**: Dilute 7.65 mL of TFA to 50.0 mL with distilled water. Store at 4 °C for up to 6 months.
2. **Sulfuric acid (72%, w/w)**: To prepare 100 mL of a 72% sulfuric acid solution, dilute 66.76 mL of 96% H_2SO_4 to a final volume with distilled water.
3. **Sugar mix**: Prepare a 20 mg/mL stock solution of each sugar (rhamnose, fucose, arabinose, xylose, mannose, galactose, glucose, and *myo*-inositol). Mix 100 μL of each solution. Store all solutions at −20 °C for up to 1 year.
4. **Copper sulfate (5%)**: Weigh 5 g of copper sulfate and dissolve it in 100 mL of distilled water.
5. **Galacturonic/Glucuronic acid standards**: Prepare a 20 mg/mL stock solution of galacturonic acid (or glucuronic acid) in water. Then, dilute this stock solution to 200 μg/mL by taking 100 μL of the stock solution and adding 9900 μL of water. This solution is then used to prepare the dilution series of the standards (5–40 μg/400 μL) according to Table 1. The stock solution can be frozen stored; however, prepare fresh dilutions from each experiment's standards.
6. **Sulfamic acid/potassium sulfamate solution (4 M, pH 1.6)**: Weigh 38.84 g of sulfamic acid and stir vigorously in 50 mL of water. Add saturated KOH solution dropwise until the sulfamic acid has dissolved. Allow the sulfamic acid solution to cool and carefully adjust the pH to 1.6 with saturated KOH

Table 1
Dilution of the uronic acid standard

Concentration of the uronic acid (μg/400 μL)	Volume of 200 μg/mL stock solution (μL)	Distilled water (μL)
5	125	1875
10	250	1750
15	375	1625
20	500	1500
30	750	1250
40	1000	1000

solution. Adjust the volume to 100 mL with water to give a final concentration of 4 M. Store at room temperature (24 °C).

7. **Sodium tetraborate solution (75 mM)**: Weigh 1.5 g of sodium tetraborate into a 100-mL volumetric flask, add approximately 80 mL of concentrated sulfuric acid, and stir until dissolved (it may require overnight stirring to completely dissolve the tetraborate). Then, adjust the final volume to 100 mL with sulfuric acid. Keep it at room temperature (24 °C).
8. ***m*-Hydroxydiphenyl solution**: Weigh 0.15 g of hydroxy diphenyl into a 100-mL volumetric flask, add approximately 80 mL of 0.5% (w/v) sodium hydroxide, and stir until dissolved. Then, adjust the final volume to 100 mL with 0.5% sodium hydroxide. Store in a refrigerator in darkness (using a dark bottle or wrapping the bottle in aluminum foil). The solution is stable for approximately 1 month.
9. **Sodium hydroxide, 0.5% (w/v)**: Weigh 0.5 g of NaOH into a 100-mL volumetric flask. Add 50 mL of water to dissolve the pellets, and then adjust the volume to 100 mL with water. It is recommended that the solution be prepared on the day of analysis.

Figure 1 shows the strategy for determining the monosaccharide composition (neutral and acid sugars) of soluble and insoluble dietary fibers described in the present chapter.

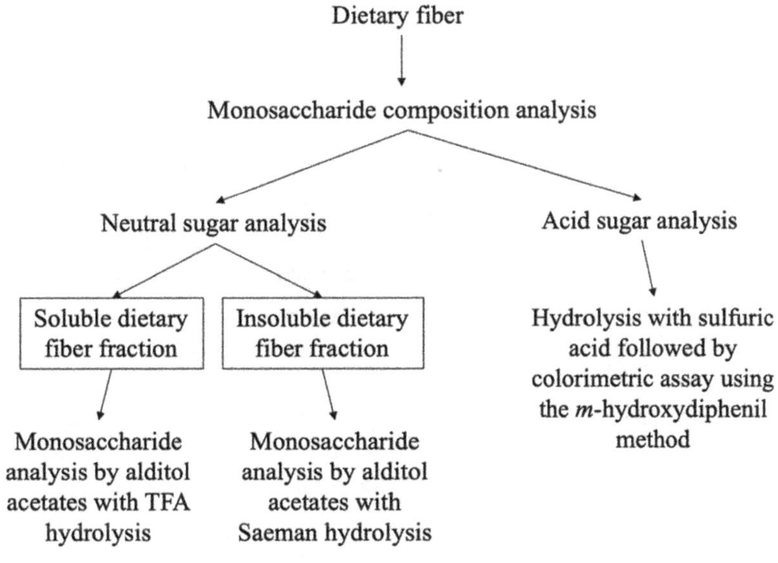

Fig. 1 Flow chart of the neutral and acid sugar analysis of soluble and insoluble dietary fibers

3 Methods

3.1 Determination of Neutral Sugars by Gas Chromatography

3.1.1 Hydrolysis

There are two options for hydrolysis conditions depending on the nature of the DF. TFA hydrolysis is used for soluble DFs, while Saeman hydrolysis is recommended for insoluble DFs containing cellulose. TFA does not fully hydrolyze cellulose. However, if only the amounts of noncellulosic neutral sugars are required, then the insoluble DF fraction can also be hydrolyzed with TFA like soluble DF.

TFA Hydrolysis

1. Weigh between 1 and 2 mg of freeze-dried sample into a glass tube (with a Teflon-lined screw cap) (*see* **Note 1**).
2. Add 500 μL of 2 M TFA to the sample and hydrolyze for 1 h at 121 °C (in an oven or heating block) (*see* **Note 2**).
3. Remove the samples from the oven and let them dry with air or N_2 stream.
4. Add the internal standard (*myo*-inositol, 25–50 μL of a 20 mg/mL solution) and dry with air or N_2 stream. Proceed to reduction.

Saeman Hydrolysis

1. Weigh between 2 and 5 mg of freeze-dried sample into a glass tube (with a Teflon-lined screw cap) (*see* **Note 1**).
2. Add 500 μL of 72% (w/w) sulfuric acid. Vortex intermittently for 1 h, keeping samples at room temperature (24 °C).
3. Dilute the acid to 8% by adding 4 mL of distilled water. Further, hydrolyze the sample by placing the tube in an oven at 100 °C for 3 h (*see* **Note 3**).
4. Cool and add barium hydroxide until the acid is neutralized (check the pH with pH paper). Vortex thoroughly (*see* **Note 4**).
5. Centrifuge and carefully transfer the upper aqueous phase with a Pasteur pipette to a clean tube. To increase the recovery of the hydrolyzed monosaccharides, add distilled water to the precipitated material, vortex thoroughly, and centrifuge again. Remove the upper aqueous phase with a Pasteur pipette, and combine it with the previously collected aqueous phase.
6. Add the internal standard (*myo*-inositol, 25–50 μL of a 20 mg/mL solution). Proceed to reduction.

3.1.2 Reduction

At this stage, it is recommended that a separate tube containing a standard sugar mix be prepared. This mix should be included to establish the response factors and GC retention times necessary for the experiment.

1. Dissolve the hydrolyzed sample in 2 M NH_4OH (500–1000 μL), add $NaBH_4$ (2 mg), and incubate it at 100 °C for 10 min, or at room temperature for 4 h, to reduce aldoses to alditols.
2. Excess $NaBH_4$ is eliminated by acetic acid addition (20 μL) (*see* **Note 5**).
3. Dry the sample with a stream of air or nitrogen and evaporate twice with methanol to remove the boric acid. Proceed to acetylation.

3.1.3 Acetylation

Several acetylation methods are available; the simplest one is described below, with acetic anhydride and pyridine as catalyst.

1. Add 500 μL of acetic anhydride and 500 μL of pyridine and incubate at 100 °C for 30 min (*see* **Note 6**).
2. Add 2 mL of water to destroy excess acetic anhydride. Mix and allow it to stand for 10 min.
3. Extract partially acetylated alditols with 1–2 mL of chloroform and transfer the chloroform phase to a clean tube with a cap.
4. Wash the chloroform phase with aliquots of 5% copper sulfate solution to remove the residual pyridine, aspirating the top aqueous phase using a Pasteur pipette.
5. After the last wash (*see* **Note 7**), remove as much of the aqueous phase from the chloroform phase as possible to prevent water from entering the GC-MS. Anhydrous sodium acetate can be added if necessary to remove small amounts of residual water.
6. Transfer the chloroform phase to another glass vial and let it dry at room temperature (*see* **Note 8**).
7. Dissolve the products in the required amount of acetone and analyze 1 μL aliquots by GC-FID or GC-MS.

3.1.4 GC-FID/GC-MS Analysis

Alditol acetates have generally been separated by GC-FID/GC-MS using packed columns, which typically utilize polar stationary phases (ECNSS-M, SP2330, SP2340, OV-275, OV-225, DB-225, DB-23, Silar 10C, or BPX70). However, low-polarity columns, such as CP-Sil 5CB, VF-5, or HP-1, have also been used.

For the quantitative analysis of the alditol acetates in our laboratory, we have used DB-225 or CP-Sil 5CB capillary columns (30 m × 0.25 mm i.d.). Helium is used as carrier gas at 1 mL/min, and 1 μL of sample is injected.

The alditol acetates can be identified by comparing the retention times relative to the samples and the sugar standard mix.

The quantification of each neutral sugar in the samples can be calculated relative to the internal standard *myo*-inositol by using response factors. The relative response of the detector for the individual alditol acetates can be calculated from the areas under the peaks for each alditol acetate relative to the area under the peak for *myo*-inositol acetate [9].

3.2 Determination of Uronic Acid Content of Dietary Fibers Using a Colorimetric Assay

The samples may need to be hydrolyzed in sulfuric acid before the colorimetric uronic acid determination to solubilize the insoluble dietary fiber fractions. Ahmed and Labavitch method [12] used concentrated acid, while Selvendran et al. [13] used 72% sulfuric acid in this hydrolysis step. The Ahmed and Labavitch method [12] is described below.

3.2.1 Hydrolysis

1. Weigh 5 mg (exact weight) of insoluble dietary fiber fraction into borosilicate glass tubes. Prepare four tubes.
2. Add 1 mL of concentrated sulfuric acid to all the tubes and cap. A reagent control tube containing only 1 mL of concentrated sulfuric acid should also be prepared. Place the tubes in a rack (inside a container filled with ice) and stir the tubes' contents gently for 5 min using magnetic bars.
3. Add another 1 mL of concentrated sulfuric acid to all the tubes and stir gently on ice for more than 5 min.
4. Add 0.5 mL water dropwise and stir gently for 5 min on ice.
5. Add another 0.5 mL water dropwise and stir until dissolution is complete.
6. Transfer the dissolved sample to a 10-mL volumetric flask and wash the tubes with water to fill it to its designated volume (*see* **Note 9**).

3.2.2 Colorimetric Assay

1. Take aliquots of 400 μL from each fraction and the reagent control tube and transfer them to 15-mL borosilicate glass tubes. A standard curve containing D-galacturonic acid or glucuronic acid should also be prepared for each batch of samples.
2. Add 40 μL of 4 M sulfamic acid/potassium sulfamate solution, pH 1.6, to all the tubes and vortex them.
3. Add 2.4 mL of 75 mM sodium tetraborate in sulfuric acid solution to all the tubes and vortex vigorously (*see* **Note 10**).
4. Place the tubes in a 100 °C water bath (boiling) for 20 min (*see* **Note 11**). After this incubation, cool the tubes by plunging them into an ice bath for 10 min.
5. Add 80 μL *m*-hydroxydiphenyl solution to 3 tubes of each sample and the reagent control and standard tubes. To the fourth tube of the sample (sample control), add 80 μL of 0.5% NaOH. Mix well all the tubes by vortexing.
6. Between 10 min and 1 h after the complete mixture (*see* **Note 12**), read the absorbance at 525 nm against the reagent

control tube. Subtract the values for the sample control from their corresponding sample absorbance.

7. Calculate the amount of uronic acids according to the prepared standard calibration curve.

4 Notes

1. Ensure that the tubes and Teflon seals on caps are intact to prevent the evaporation of the reagents during derivatization. TFA is highly volatile, and TFA will be lost if tubes are not properly sealed. Avoid the use of other caps that may contain plasticizers. They may contaminate the sample and complicate the interpretation of data by GC-MS.
2. Other conditions may also be applied, such as 2 M TFA for 8 h at 100 °C or 2.5 M TFA for 90 min at 121 °C. Information on the mechanism, hydrolysis rates, and kinetic parameters of acid-catalyzed hydrolysis of glycosides can be found in Adams [16], Aspinall [17], and Dutton [18]. The released monosaccharides are susceptible to degradation in the presence of hot concentrated acid; moreover, not all glycosidic linkages are cleaved at the same rate. Thus, care should be taken to balance these two needs; the hydrolysis must have sufficient strength and length to permit complete hydrolysis but not so long to lead to sample degradation. A hydrolysis curve may be performed for new or unknown dietary fibers by subjecting the sample to a chosen hydrolysis protocol and determining the quantity of each sugar at timed intervals [9, 10].
3. The time here may change according to the sample. Longer incubation time may be necessary to ensure complete hydrolysis.
4. Barium carbonate can also be used instead of barium hydroxide. Before adding the barium hydroxide/carbonate, transfer the hydrolyzed material to a centrifuge tube. Then, proceed to neutralization. Check the pH with a pH paper. If necessary, add more water to the tube to increase the recovery of the aqueous phase after the centrifugation step.
5. A fizz should be observed. Otherwise, the reduction was incomplete or the $NaBH_4$ was inactive.
6. The next steps should be performed in a fume hood due to the presence of pyridine (a volatile compound).
7. Copper sulfate complexes with pyridine and forms a dark blue solution. To completely remove the pyridine, repeat this washing step until the aqueous phase becomes light blue (the same color as the 5% copper sulfate solution).

8. Do not heat since it may cause the loss of the volatile derivatives. Samples can be stored until ready to run on GC-FID or GC-MS.
9. If a small amount of insoluble material remains, centrifuge the tubes to pellet any unhydrolyzed material. Thus, transfer the material from the volumetric flasks to centrifuge tubes and centrifuge them for 10 min at high speed at room temperature (24 °C).
10. Since the uronic acid assay involves using concentrated sulfuric acid, goggles, gloves, and protective clothing should be worn.
11. During the incubation in the boiling water bath, place glass marbles on top of the tubes (or alternatively cover them with aluminum foil) to prevent condensation from contaminating the samples.
12. A pink color develops within 5 to 10 min and is stable only for approximately 1 h. Take care to read the absorbance before this time since the color fades beyond after this. Ensure there are no bubbles in the cuvettes when reading the absorbances and no spillages down the sides of the cuvettes, as the sulfuric acid is very corrosive and may damage the equipment.

References

1. Albersheim P et al (1967) A method for the analysis of sugars in plant cell-wall polysaccharides by gas-liquid chromatography. Carbohydr Res 5(3):340–345
2. Saeman JF, Moore WE, Millet MA (1963) Sugar units present. In: Cellulose WRL (ed) Hydrolysis and quantitative paper chromatography. Academic, New York, pp 54–69
3. Mudgil D (2017) Chapter 3 – The interaction between insoluble and soluble fiber. In: Samaan RA (ed) Dietary fiber for the prevention of cardiovascular disease. Academic, pp 35–59
4. Blakeney AB et al (1983) A simple and rapid preparation of alditol acetates for monosaccharide analysis. Carbohydr Res 113(2):291–299
5. Corradini C et al (2004) High-performance anion-exchange chromatography coupled with pulsed amperometric detection and capillary zone electrophoresis with indirect ultra violet detection as powerful tools to evaluate prebiotic properties of fructooligosaccharides and inulin. J Chromatogr A 1054(1–2): 165–173
6. Honda S et al (1989) High-performance liquid chromatography of reducing carbohydrates as strongly ultraviolet-absorbing and electrochemically sensitive 1-phenyl-3-methyl-5-pyrazolone derivatives. Anal Biochem 180(2): 351–357
7. Sassaki GL et al (2008) Application of acetate derivatives for gas chromatography–mass spectrometry: novel approaches on carbohydrates, lipids and amino acids analysis. J Chromatogr A 1208(1):215–222
8. Melton LD, Smith BG (2001) Determination of neutral sugars by gas chromatography of their Alditol acetates. Curr Protoc Food Anal Chem 00(1):E3.2.1–E3.2.13
9. Brummer Y, Cui SW (2005) Understanding carbohydrate analysis. In: Cui SW (ed) Food carbohydrates: chemistry, physical properties, and applications. CRC Press, Boca Raton, pp 67–104
10. Sturgeon RJ (1990) Monosaccharides. In: Dey PM, Harbourne JB (eds) Methods in plant biochemistry. Academic, London, pp 1–25
11. Pettolino FA et al (2012) Determining the polysaccharide composition of plant cell walls. Nat Protoc 7(9):1590–1607
12. Ahmed AER, Labavitch JM (1978) A simplified method for accurate determination of cell wall uronide content. J Food Biochem 1(4): 361–365
13. Selvendran RR, March JF, Ring SG (1979) Determination of aldoses and uronic acid

content of vegetable fiber. Anal Biochem 96(2):282–292

14. Blumenkrantz N, Asboe-Hansen G (1973) New method for quantitative determination of uronic acids. Anal Biochem 54(2):484–489
15. Filisetti-Cozzi TM, Carpita NC (1991) Measurement of uronic acids without interference from neutral sugars. Anal Biochem 197(1): 157–162
16. Adams GA (1965) Complete acid hydrolysis. In: Whistler RL (ed) Methods in carbohydrate chemistry. Academic, New York, pp 269–276
17. Aspinall GO (1982) Chemical characterization and structure determination of polysaccharides. In: Aspinall GO (ed) The polysaccharides. Academic, New York, pp 35–131
18. Dutton GGS (1974) Applications of gas-liquid chromatography to carbohydrates: Part II. In: Tipson RS, Horton D (eds) Advances in carbohydrate chemistry and biochemistry. Academic, pp 9–110

Chapter 12

Determination of Dietary Fiber

Cinthia Baú Betim Cazarin

Abstract

This chapter will describe the enzymatic quantification of total dietary fiber content and its fractions method. The role of dietary fiber in preventing the development of noncommunicable diseases is well-established in the literature. Determining and quantifying this nutrient is essential to feed the food composition databases and help nutritionists plan dietary recommendations and elaborate public policies.

Key words Total dietary fiber, Insoluble dietary fiber, Soluble dietary fiber, Indigestible carbohydrates

1 Introduction

In recent years, with the advances in understanding the human intestinal microbiota and its relationship with the pathogenesis of systemic inflammatory diseases and immunity, the interest in clinical aspects of dietary fiber (DF) has increased [1, 2]. There is evidence that the gut-brain axis regulates cognitive functions [3], the development of obesity, type-2 diabetes, and cardiovascular diseases [4], for example. The human body lacks enzymes to hydrolyze the complex polysaccharides and oligosaccharides that constitute the dietary fibers; thus, they reach the distal intestine, where the microbiota can transform them into fermentable monosaccharides. Intestinal microorganisms transform them into short-chain fatty acids (SCFAs) and other metabolites that positively affect the host. SCFA formation can decrease the intestinal pH and improve mineral absorption [5], and butyrate, for example, can be used as a fuel by colonocytes, protecting this important epithelial barrier function [6–8].

According to the chemical structures, dietary fibers can regulate the type and ratio of the produced SCFAs and the gut microbiota composition by increasing beneficial intestinal microbes and reducing harmful microorganisms, thus improving human health [9, 10]. It is important to highlight that in addition to the prebiotic

Lilian Mariutti and Frank Vriesekoop (eds.), *Food Chemistry: Basic Protocols*, Methods and Protocols in Food Science,
https://doi.org/10.1007/978-1-0716-4678-6_12,

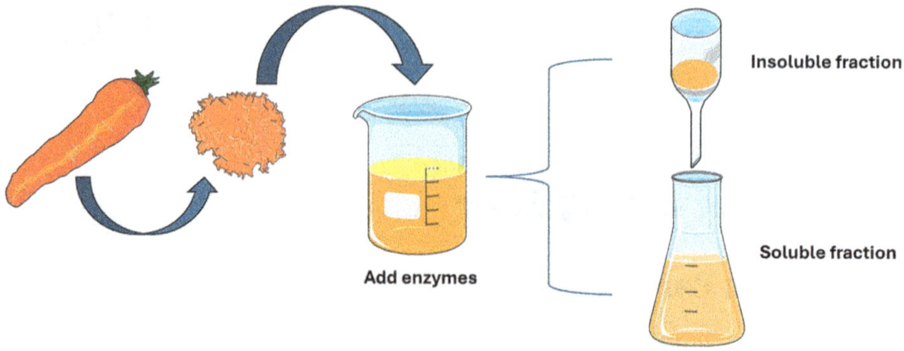

Fig. 1 Steps of food dietary fiber analysis. (This figure was created using Servier Medical Art templates, which are licensed under a Creative Commons Attribution 4.0 Unported License; https:/creativecommons.org/licenses/by/4.0/)

effect, modulating the microbiota, the soluble fraction, with its gelling activity, can regulate the digestion rate, absorption, and excretion. These and other physiological effects associated with dietary fiber make this food component essential for maintaining health. Therefore, information on food composition concerning dietary fiber content is extremely important to guide nutritional recommendations and educational actions.

Total dietary fiber and insoluble and soluble fractions can be determined by the enzymatic-gravimetric assay described by Prosky et al. [11] and McCleary et al. [12]. Many commercial kits are available; however, it is possible to analyze using different enzyme brands. It is only important to ensure enzyme activity. Determining the soluble fraction using this method has shown a high level of variability; thus, it is acceptable to predict the amount of soluble fraction by the difference between the total dietary fiber content and the insoluble fraction.

This chapter will describe the methodology to enzymatically quantify total dietary fiber and its fractions based on Prosky et al. [11] (Fig. 1).

2 Material

2.1 Reagents

- Enzymes (α-amylase, heat stable; protease; amyloglucosidase).
- 99% ethyl alcohol.
- Acetone.
- Disodium phosphate (Na_2HPO_4).
- Monosodium phosphate ($NaH_2PO_4.H_2O$).
- Sodium hydroxide.

- Hydrochloric acid.
- Distilled or deionized water.

2.2 Equipment

- Fritted crucibles (Gooch, porosity 40 to 60 microns, 50 mL)—*see* **Note 1**.
- Celite or glass wool.
- Beaker 250 and 400 mL.
- pH meter.
- Thermometer.
- Water bath, with temperature control and agitation.
- Vacuum pump.
- Oven, or air oven, set at 105 °C.
- Muffle.
- Desiccator.

3 Methods

3.1 Solution Preparation

1. 95% ethyl alcohol (v/v): To prepare 1000 mL of a 95% ethyl alcohol solution, dilute 960 mL of 99% ethyl alcohol to a final volume with distilled water.
2. 78% ethyl alcohol (v/v): Prepare 1000 mL of a 78% solution of ethyl alcohol, and dilute 788 mL of 99% ethyl alcohol to a final volume with distilled water.
3. 0.08 M phosphate buffer pH 6.0: Dissolve 1.4 g Na_2HPO_4 (1.753 g for $Na_2HPO_4{\cdot}2H_2O$) and 9.68 g $NaH_2PO_4.H_2O$ (it is possible to use 8.4 g NaH_2PO_4 anhydrous or 10.94 g $NaH_2PO_4{\cdot}2H_2O$ dihydrate) in 700 mL distilled or deionized water. Complete the volume to 1 liter with distilled or deionized water. Check the pH with a pH meter; if necessary, adjust the pH with NaOH 0.275 N or HCl 0.325 N.
4. 0.275 N sodium hydroxide: Dissolve 11 g NaOH (pellets) in 700 mL distilled or deionized water. Complete the volume to 1 liter with distilled or deionized water. This solution should be stored in a plastic bottle.
5. 0.325 N hydrochloric acid: Dilute 325 mL 1 N HCl in 700 mL distilled or deionized water. Complete the volume to 1 liter with distilled or deionized water.
6. 0.5 N hydrochloric acid: Dilute 500 mL 1 N HCl in 700 mL distilled or deionized water. Complete the volume to 1 liter with distilled or deionized water. This solution will be used to prepare the crucibles before the analysis.

3.2 Crucible Preparation

The crucibles should be prepared before starting the analysis.

- Wash the crucibles with distilled water and leave them for 24 hours in neutral pH liquid laboratory detergent (suggestion: Extran® 2%, v/v).
- Rinse the crucibles with distilled water using a vacuum pump (pass the water to both sides of the crucible to remove all detergent residues).
- Dry the crucible in an oven and then incinerate in a muffle at 525 °C for 3 hours.
- After cooling, wash the crucibles five times with 20 mL of 0.5 N HCl, before washing with distilled water until neutralization (check the water pH with an indicator paper). This step is important to remove all ash residue from the sintered glass plate.
- After neutralization, wash the crucible three times in the opposite direction with 20 mL of distilled water.

  Dry the crucibles in an oven at 105 °C for one night, cool them in a desiccator, and weigh them.
- Cover the bottom of the crucible with glass wool, dry them in an oven at 105 °C overnight, and weigh them.

$$\text{Mass of glass wood (g)} = \text{crucible} + \text{glass wood (g)} - \text{crucible (g)}$$

3.3 Sample Preparation

If the sample contains more than 10% lipids, proceed with the defatted step. The amount of lipids needs to be considered in the sample mass final calculus. The same should be performed for samples with high sugar content; they should be extracted using three times ten volumes of 85% methanol (v/v). The samples should be dried (overnight at 105 °C or freeze-dried) and ground in 0.3 to 0.5 mm mesh [11].

3.4 Total Dietary Fiber Quantification (TDF)

Consider running four beakers (2 for protein quantification +2 for ash) for the sample and *blank*. Run the *blank* without a sample to measure the reagents' contribution to the residue. The method described below was adapted from Prosky et al. [11, 13], Asp et al. [14, 15], and McCleary et al. [12].

- Weigh around 1.000 ± 0.020 g of the sample in a 250 mL beaker.
- Add 50 mL of phosphate buffer (pH 6.0) to each beaker.
- Add 100 μL of heat-stable alpha-amylase.
- Cover the beaker with aluminum foil and heat it in a boiling water bath for 30 minutes before the solution reaches 95–100 ° C (check the temperature of the beaker content with a thermometer) with continuous agitation.

- Remove the beaker from the bath and cool the samples to room temperature (24 °C).
- Adjust the pH to 7.5 by adding around 10 mL of 0.275 N NaOH (check the pH using a pH meter).
- Then, add 100 μL of protease (*see* **Note 2**).
- Cover the beaker again with aluminum foil and incubate at 60 °C for 30 minutes (*see* **Note 3**) with continuous agitation.
- Remove the beaker from the bath and cool the samples to room temperature (24 °C).
- Adjust the pH to 4.0–4.6 by adding around 10 mL of 0.325 N HCl (check the pH using a pH meter).
- Add 300 μL of amyloglucosidase, cover the beaker with aluminum foil, and incubate for 30 minutes at 60 °C (consider the incubation time after the internal solution reaches 60 °C; use a thermometer to check the temperature) with continuous agitation.
- Transfer the solution (hydrolyzate) quantitatively to a 500 mL beaker, and add 280 mL of 95% ethyl alcohol at 60 °C (preheated). Take care to remove all material stuck to the beaker walls, and ensure that everything is transferred to the new beaker. It is advised to use the ratio 4:1 to ethyl alcohol and the beaker volume.
- Cover the beaker with aluminum foil and leave the mixture (hydrolyzate +95% ethyl alcohol) at room temperature (24 °C) for 1 hour to precipitate all the soluble fiber.
- Wash the crucibles with 15 mL of 78% ethyl alcohol using a vacuum pump, and then filtrate the alcoholic solution containing the hydrolysate.
- Wash the residue in the crucible:
  - Three times with 20 mL of 78% ethyl alcohol.
  - Twice with 10 mL of 95% ethyl alcohol.
  - Twice with 10 mL of acetone.
- Dry the crucible with the residue at 105 °C overnight.
- Cool the crucibles in a desiccator and weigh them to record the weight (residue + crucible + wood glass).
- Use the content of two crucibles to determine the protein (micro-Kjeldahl method). Use total nitrogen × 6.25 as a conversion factor to quantify the amount of protein in the residue.
- Incinerate the contents of the other two crucibles at 525 °C for 5 hours to determine the ash content in the residue.
- Cool the crucibles in a desiccator and weigh them to record the weight (ash + crucible + wood glass).

$$\%\mathbf{TDF} = [\text{R-A-P-B/sample mass}] \times 100$$

R = mean of residue mass.

A = mean of ash mass.

P = mean of protein mass.

B = mean of blank mass.

3.5 Insoluble Dietary Fiber Quantification (IDF)

Consider running four beakers (2 for protein quantification +2 for ash) for the sample and *blank*. Run the *blank* without a sample to measure the reagents' contribution to the residue. The method described below was adapted from Prosky et al. [11, 13], Asp et al. [14, 15], and McCleary et al. [12].

- Weigh around 1.000 g of the sample in a 250 mL beaker.
- Add 50 mL of phosphate buffer (pH 6.0) to each beaker.
- Add 100 μL of heat-stable alpha-amylase.
- Cover the beaker with aluminum foil and heat it in a boiling water bath for 30 minutes before the solution reaches 95–100 ° C (check the temperature of the beaker content with a thermometer) with continuous agitation.
- Remove the beaker from the bath and cool the samples to room temperature (24 °C).
- Adjust the pH to 7.5 by adding around 10 mL of 0.275 N NaOH (check the pH using a pH meter).
- Then, add 100 μL of protease (Note: 50 mg of the enzyme should be dissolved in 1 mL of the phosphate buffer).
- Cover the beaker again with aluminum foil and incubate at 60 ° C for 30 minutes (consider the incubation time after the internal solution reaches 60 °C, use a thermometer to check the temperature) with continuous agitation.
- Remove the beaker from the bath and cool the samples to room temperature (24 °C).
- Adjust the pH to 4.0–4.6 by adding around 10 mL of 0.325 N HCl (check the pH using a pH meter).
- Add 300 μL of amyloglucosidase, cover the beaker with aluminum foil, and incubate for 30 minutes at 60 °C (consider the incubation time after the internal solution reaches 60 °C, use a thermometer to check the temperature) with continuous agitation.
- Filtrate the hydrolysate in the crucibles using a vacuum pump.
- Wash the residue two times with 10 mL of distilled water (preheated at 60 °C).

- Wash the residue twice with 10 mL of 95% ethyl alcohol and twice with 10 mL of acetone.
- Dry the crucible with the residue at 105 °C overnight.

  Cool the crucibles in a desiccator and weigh them to record the weight (residue + crucible + wood glass).
- Use the content of two crucibles to determine the protein (micro-Kjeldahl method). Use total nitrogen × 6.25 as a conversion factor to quantify the amount of protein in the residue.
- Incinerate the contents of the other two crucibles at 525 °C for 5 hours to determine the ash content in the residue.
- Cool the crucibles in a desiccator and weigh them to record the weight (ash + crucible + wood glass).

$$\%\mathbf{TDF} = [\text{R-A-P-B/sample mass}] \times 100$$

R = mean of residue mass.

A = mean of ash mass.

P = mean of protein mass.

B = mean of blank mass.

4 Notes

1. Pay attention to some crucible brands; the material should resist incineration at 525°C.
2. 50 mg of the enzyme should be dissolved in 1mL of the phosphate buffer.
3. Consider the incubation time after the internal solution reaches 60 °C; use a thermometer to check the temperature.

References

1. Cai Y et al (2020) Microbiota-dependent and -independent effects of dietary fibre on human health. Br J Pharmacol 177(6):1363–1381
2. Song Q et al (2021) Review of the relationships among polysaccharides, gut microbiota, and human health. Food Res Int 140:109858
3. Cussotto S, Aouizerate B, Capuron L (2021) From gut to brain: microbiota depletion in mice as a tool to explore causality. Brain Behav Immun 94:4
4. Richards P, Thornberry NA, Pinto S (2021) The gut–brain axis: identifying new therapeutic approaches for type 2 diabetes, obesity, and related disorders. Mol Metab 46:101175
5. Costa GT et al (2021) Systematic review of the ingestion of fructooligosaccharides on the absorption of minerals and trace elements versus control groups. Clin Nutrit ESPEN 41:68–76
6. Couto MR et al (2020) Microbiota-derived butyrate regulates intestinal inflammation: focus on inflammatory bowel disease. Pharmacol Res 159:104947
7. Liu J et al (2020) Beneficial effects of butyrate in intestinal injury. J Pediatr Surg 55(6): 1088–1093
8. Bertolo MRV, Bogusz Junior S. (2023). Quantification of Short-Chain Fatty Acids in Feces. In: Cazarin CBB (ed) Basic Protocols in Foods and Nutrition. Springer, Humana New York, NY, pp 73–92. https://doi.org/10.1007/978-1-0716-2345-9

9. Hamaker BR, Tuncil YE (2014) A perspective on the complexity of dietary fiber structures and their potential effect on the gut microbiota. J Mol Biol 426(23):3838–3850
10. Holscher HD (2017) Dietary fiber and prebiotics and the gastrointestinal microbiota. Gut Microbes 8(2):172–184
11. Prosky L et al (1992) Determination of insoluble and soluble dietary fiber in foods and food products: collaborative study. J AOAC Int 75(2):360–367
12. McCleary BV et al (2019) Determination of insoluble, soluble, and total dietary fiber (CODEX definition) by enzymatic-gravimetric method and liquid chromatography: collaborative study. J AOAC Int 95(3):824–844
13. Prosky L (1990) Collaborative study of a method for soluble and insoluble dietary fiber. In: Furda I, Brine CJ (eds) New developments in dietary fiber: physiological, physicochemical, and analytical aspects. Springer US, Boston, pp 193–203
14. Asp NG et al (1983) Rapid enzymic assay of insoluble and soluble dietary fiber. J Agric Food Chem 31(3):476–482
15. Asp N-G et al (1992) Dietary fibre analysis. In: Schweizer TF, Edwards CA (eds) Dietary fibre — a component of food: nutritional function in health and disease. Springer, London, pp 57–101

Chapter 13

Starch Method in Foods: Total Starch Determination

Mária Herminia Ferrari Felisberto, Ulliana Marques Sampaio, and Maria Teresa Pedrosa Silva Clerici

Abstract

Several methods are used to determine starch content in food; however, the enzymatic method is the most widely applied. The principle of the method involves the quantification of glucose molecules to make inferences about the starch content in the sample. Considering that foods are complex matrices and can present other carbohydrates with glucose in their structures, the knowledge of sample composition is essential for its correct preparation, and consequently, for the determination of the starch content.

Key words Carbohydrates, Color reaction, Digestion, Enzymes, Glucose, Starch

1 Introduction

Qualitative or quantitative methods can be used for the determination of starch in food samples. Qualitative methods range from the simplest ones, such as iodine-iodide, to near-infrared reflectance (NIR). The quantitative methods are gravimetry, redox titration, polarimetry, and chromatography, and the most modern ones are the enzymatic methods [1, 2].

The Association of Official Analytical Chemists (AOAC, 2006) [3] presents several official methods for determining starch, whereas, in the American Association of Cereal Chemists International (AACCI, 2010) [4], the methods are well established for cereals and derivatives. Eliasson (2004) [1] presents an overview of several methods; however, the enzymatic method has been the most used due to its specificity, and, for this reason, it will be emphasized in this protocol.

The first step to understand the quantitative analysis of starch by the enzymatic method is to remember that starch is in the group of carbohydrates, being a homopolysaccharide composed of glucose molecules, which can be connected by α-1,4 and α-1,6

Lilian Mariutti and Frank Vriesekoop (eds.), *Food Chemistry: Basic Protocols*, Methods and Protocols in Food Science, https://doi.org/10.1007/978-1-0716-4678-6_13,

linkages, according to the structure of amylose and amylopectin molecules.

Thus, in the sample, the presence of both free glucose (monosaccharide), as well as other disaccharides and oligosaccharides containing glucose in their composition, such as maltodextrins and dextrins, may overestimate the starch content. Complex food matrices, such as cakes, biscuits, and bread, which may contain a wide range of carbohydrates, require a carefully planned analysis for the correct preparation of the sample, especially about fat, protein, and moisture contents in the food product. Therefore, as a second step, it is essential to know the composition of the food sample and its correct preparation to eliminate any possibility of interference.

2 Material

The samples must be dry and contain low moisture (less than 15%), with a powder aspect and below a 32 mesh size. In the case of a sample with high protein content, *see* **Note 1**. Prepare all solutions using distilled water and analytical grade reagents, and specifically, sodium hydroxide solution should preferably be prepared inside a fume hood.

2.1 Enzymes

1. Thermostable α-amylase (*Bacillus* sp.): 3000 U/mL on Ceralpha reagent at pH 6.5 and 40 °C or 2500 U/mL on Ceralpha reagent at pH 5.0 and 40 °C.
2. Amyloglucosidase (*Aspergillus niger*): 3300 U/mL on soluble starch or 200 U/mL on *p*-nitrophenyl β-maltosidase, at pH 4.5 and 40 °C.

2.2 Reagents

1. D-Glucose standard solution: This solution can be obtained ready for use from Megazyme, with the following specifications: 5 mL, 1.0 mg/mL in 0.2% (w/v) benzoic acid.
2. Standardized regular maize starch: The standard maize starch can be obtained ready for use by Megazyme, with the following specification: 93% starch purity.
3. GOPOD reagent enzymes: This solution can be obtained ready for use by Megazyme, with the following specifications: Glucose oxidase plus peroxidase and 4-aminoantipyrine. Freeze-dried powder.
4. Sodium maleate buffer (100 mM, pH 6.9): Weigh 5.93 g of sodium tris-maleate in a beaker and add 200 mL of distilled water. Shake to dissolve and then transfer the entire volume to a 250 mL volumetric flask and make up to volume with distilled water. Leave the bottle to stand for about 24 hours, and then adjust the pH using a standard sodium hydroxide (NaOH)

solution (0.0375 M) or hydrochloric acid (HCl) with similar normality. Transfer the solution to a dark glass bottle, close tightly, identify, and store under refrigeration. This buffer solution can be stable for up to 1 year if stored at 4 °C. *See* **Note 2**.

5. Sodium acetate buffer (200 mM, pH 4.5): Add 11.6 mL of glacial acetic acid p.a. in 900 mL of distilled water and let it sit for about 24 hours, and only then adjust the pH to 4.5 using 4 M sodium hydroxide solution (NaOH). Add 0.2 g of sodium azide to prevent the growth of fungi and homogenize the mixture. Transfer the solution to a 1 L volumetric flask and make up to volume with distilled water. Then, transfer the content to a dark glass bottle, close it tightly, identify, and store it under refrigeration. This solution can be stable for 6 months if stored at 4 °C.
6. Ethyl alcohol 80%: Measure 804.02 mL of 99.5% ethyl alcohol and transfer to a 1 L volumetric flask, and make up to volume with distilled water (approximately 195.98 mL). Then, with the aid of an alcoholometer, check the concentration. Transfer the solution to a glass bottle, seal tightly, identify, and store at room temperature.
7. 4 M sodium hydroxide (NaOH) Solution: Weigh 160 g of sodium hydroxide (NaOH) in a plastic beaker, with a capacity of 1 L. Place this beaker in an ice bath and add approximately 700 mL of distilled water to the NaOH. Using a stick, carefully mix the solution until the base is completely dissolved. Allow the solution to cool to room temperature (25–30 °C), transfer to a 1 L volumetric flask, and make up to volume with distilled water. Transfer the solution to a plastic bottle, seal tightly, identify, and store at room temperature. *See* **Note 3**.
8. 0.275 M sodium hydroxide (NaOH) Solution (*see* **Note 4**): Weigh 11 g of sodium hydroxide (NaOH) in a plastic beaker, with a capacity of 1 L. Place this beaker in an ice bath and add approximately 700 mL of distilled water to the NaOH. Using a stick, carefully mix the solution until the base is completely dissolved. Allow the solution to cool to room temperature (25–30 °C), transfer to a 1 L volumetric flask, and make up to volume with distilled water. Transfer the solution to a plastic bottle, seal tightly, identify, and store at room temperature.

2.3 Glassware

1. Falcon 50 mL conical centrifuge tubes (Fig. 1a): 3 tubes will be used per sample.
2. Volumetric flasks (50 mL or 100 mL) (Fig. 1b): Volume of the volumetric flasks depends on the starch content of the samples (more concentrated samples, 100 mL flasks are recommended, and in those less concentrated samples, 50 mL flask is

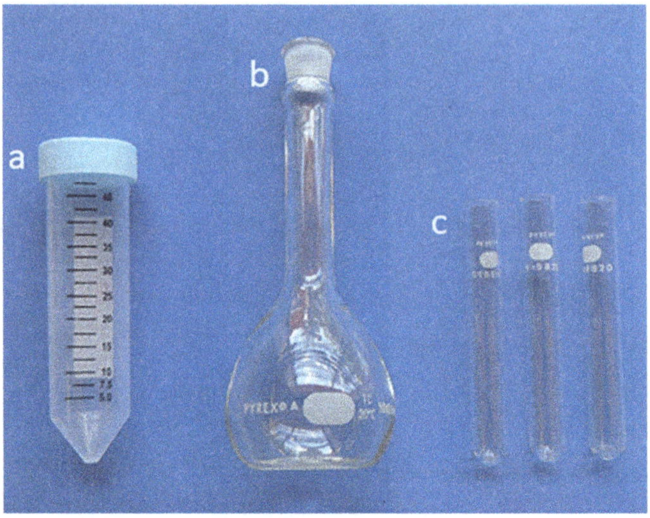

Fig. 1 Illustrative image of some glassware used in the analysis

recommended). Note that a total of 3 volumetric flasks will be used per sample.

3. Small glass test tubes without a cap (20 mL) (Fig. 1c): Nine tubes per sample, in addition to three tubes for the standard glucose solution, three tubes for the standard maize starch, and three more tubes for the control.
4. Spectrometer cuvettes (acrylic or polystyrene) (1.5 mL) (*see* **Note 5**): Nine cuvettes per sample, in addition to three cuvettes for standard glucose solution, three for the standard maize starch, and three more for the control.

2.4 Equipment

1. Analytical balance, 0.1 mg readability, accuracy, and precision.
2. Thermostatic water bath shaker, set at 50 °C, and 100 °C, and agitation.
3. Bench centrifuge, rating of approx. 3000 rpm (1500 g), and temperature control.
4. Adjustable micropipette/pipettor 2 μL, 30 μL, 50 μL, 100 μL, 200 μL, and 3 and 4 mL.
5. Spectrophotometer capable of operating at 510 nm.

3 Methods

The method described below is based on the procedures described in AOAC [3] Method 996.11 and AACC [4] Method 76–13.01, with adaptations of the general food analysis laboratory. It is important to know the moisture content of the sample and the standard starch, as it will be taken into account in the final calculation of the

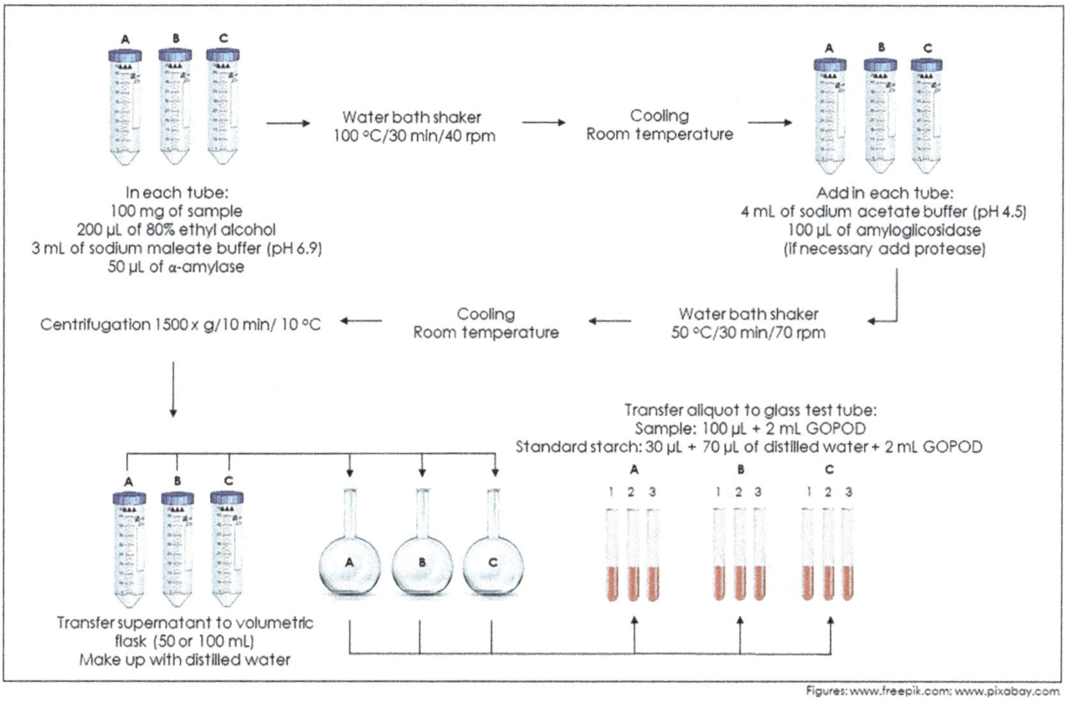

Fig. 2 Analytical scheme for the total starch determination procedure

starch content. At the end of the procedure, there is a schematic drawing to simplify each step (Fig. 2).

3.1 Sample Preparation

The sample must have its moisture content known and, if possible, should be dried and ground. If it contains lipids, it can be previously extracted with n-hexane or petroleum ether [5]. The defatted residue obtained can be dried and washed with 80% ethyl alcohol to extract soluble sugars. Thus, to separate the soluble sugars, the material must be centrifuged, where the supernatant with the soluble sugars will be discarded, and the precipitated material will be collected, dried, and ground. The material must be packed in airtight and well-sealed packaging and stored in a dry, ventilated place, and away from moisture until the moment of analysis.

3.2 Starch Hydrolysis

1. Standardized regular maize starch control (93% starch): Accurately weigh 100 mg ± 5 mg (SM) of regular maize starch, in triplicate, into Falcon 50 mL conical centrifuge tubes previously identified. Record the exact weight in Table 1.
2. Samples: Accurately weigh 100 mg ± 5 mg (S) of a test sample, in triplicate, into Falcon 50 mL conical centrifuge tubes, previously identified. Record the exact weight in Table 1.
3. In each Falcon conical centrifuge tube containing sample or not (regular maize starch), add 200 µL 80% ethyl alcohol and

Table 1
Spreadsheet for data treatment

Samples	Replicate	Id.	Sample weight wb (g) (SW wb)	Moisture (M)	Sample weight db[a] (g) (SW db)	Volumetric flasks (V)[b]	Abs (A) 510 nm	Glucose content (GC)	Glucose content sample (GCS) (mg of gluc./mg of sample)	Starch content (SC) (g/100 g)	Average starch content (ASC) (g/100 g)	Standard deviation (SD)
Sample 1	1	S 1	0.1007	7.5635	0.0931	100	0.5630	0.3219	0.3458	31.1255	31.84	1.49
						100	0.6070	0.3471	0.3729	33.5580		
						100	0.5580	0.3191	0.3428	30.8490		
	2	S 2	0.1002		0.0926	100	0.5950	0.3402	0.3673	33.0587	33.34	0.53
						100	0.6110	0.3494	0.3772	33.9477		
						100	0.5940	0.3396	0.3667	33.0032		
	3	S 3	0.1008		0.0932	100	0.5850	0.3345	0.3590	32.3097	33.19	0.94
						100	0.5990	0.3425	0.3676	33.0829		
						100	0.6190	0.3539	0.3799	34.1875		
Standardized regular maize starch	1	SM 1	0.1003	12.40	0.0879	100	0.4750	0.9053	1.0304	92.7358	91.43	1.93
						100	0.4730	0.9015	1.0261	92.3454		
						100	0.4570	0.8710	0.9914	89.2216		
	2	SM 2	0.1005		0.0880	100	0.4170	0.7948	1.0283	92.5513	90.54	1.87
						100	0.4560	0.8691	0.9872	88.8492		
						100	0.4930	0.9396	1.0024	90.2132		
	3	SM 3	0.1008		0.0883	100	0.4770	0.9091	1.0347	93.1263	89.13	3.84
						100	0.4570	0.8710	0.9864	88.7791		
						100	0.4400	0.8386	0.9497	85.4766		
Standard glucose	1	SG 1				100	0.5220	Average glucose absorbance (AGA)				
						100	0.5200					
						100	0.5320	0.5247				

[a]Correction for the sample weight on a dry basis (db)

[b]If any sample was diluted in the 50 mL flask, while the glucose was in the 100 mL flask, the obtained value should be divided by 2, taking into account the reduction of the glucose

3 mL of the sodium maleate buffer (pH 6.9). Then, add 50 μL of the thermostable α-amylase enzyme.

4. Transfer the centrifuge tubes to a water bath and incubate at 100 °C for 30 minutes, with constant agitation of 40 rpm, to homogenize the material (*see* **Notes 6** and 7).
5. Afterward, cool the material, and in the meantime, adjust the water bath temperature for the next incubation temperature. Only after the samples have cooled (you can use an ice bath to speed up this step) can you proceed with the analysis.
6. In case the sample requires the protease, its use will be made at this point. Otherwise, proceed to the next step, which is the addition of a sodium acetate buffer. Before adding the protease, the pH of the samples must be adjusted to 7.5 ± 0.1 using NaOH 0.275 M, and then, add 100 μL of protease. Transfer the centrifuge tubes to a water bath and incubate at 60 °C for 30 minutes, with constant agitation of 70 rpm, to homogenize the material.
7. Afterward, cool the material, and in the meantime, adjust the water bath temperature for the next incubation temperature. Only after the samples have cooled (you can use an ice bath to speed up this step), add 4 mL of the sodium acetate buffer (200 mM and pH 4.5) in each tube, and then 100 μL of amyloglucosidase.
8. Afterward, cool the material and centrifuge the tubes at 1500 × g (~ 3000 rpm) at 10 °C for 10 min. Transfer the supernatant to the volumetric flask (V = 50 or 100 mL) (note in Table 1) (Fig. 3), previously identified, and make up to the volume with distilled water. In this case, the flask volume will depend on the starch content of the sample so that the hydrolyzed glucose content is within the range (for starch extracted from young bamboo culm, for example, we use the 100 mL flask). At this point, we obtain the sample with all starch hydrolyzed into glucose molecules. The quantified glucose content will be correlated with the starch content of the evaluated samples.

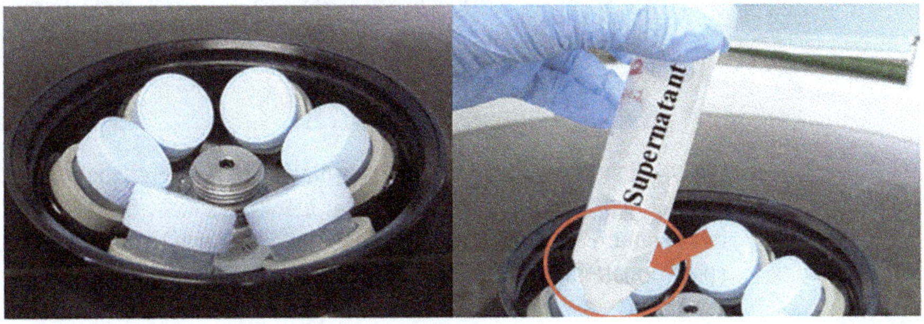

Fig. 3 Illustrative image of the tubes after centrifugation, with emphasis on the supernatant

3.3 Quantification of Glucose Content

1. Prepare the samples to perform the reading on a spectrophotometer, a triplicate aliquot must be removed for the small test tubes previously identified, as follows below.
2. Standardized regular maize starch control: Accurately transfer triplicate 30 μL aliquots to the glass test tubes, plus 70 μL distilled water.
3. Samples: Accurately transfer triplicate 100 μL aliquots to the glass test tubes of samples already hydrolyzed.
4. D-Glucose standard solution: Accurately transfer triplicate 30 μL aliquots to the glass test tubes, plus 70 μL distilled water.
5. Blank: Accurately transfer 100 μL aliquot to one glass test tube of distilled water.
6. In each test tube prepared above, add 2 mL of GOPOD (volume considering a ready-made solution, cold reaction), starting with the blank. Let it react for 20 minutes (timed), the time required for maximum color formation with 100 μg of glucose standard (Fig. 4a–d), and then read the absorbance in a spectrophotometer at 510 nm (Fig. 4e). A1.5 mL cuvettes (acrylic or polystyrene) should be used, and note the absorbance readings (A) in Table 1. Do not forget that the reading must start with the blank (100 μL of distilled water).

3.4 Calculation of the Total Starch Content

After filling in all the data obtained so far in Table 1, some adjustments need to be made due to the moisture content of the sample. The first calculation to be made is the conversion of the sample mass used from a wet basis (SW wb) to a dry basis (SW db) (Eq. 1),

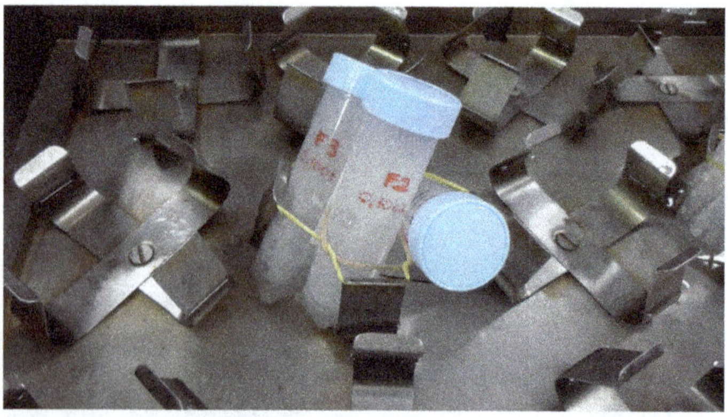

Fig. 4 Illustrative image of color formation from the reaction of samples with GOPOD and a spectrophotometer. *a: blank test tube, b: standardized regular maize starch test tube, c: sample test tube, d: standard glucose test tube, and e: spectrophotometer

Table 2
Equations to determine total starch content

N	Equation
1	$SW\ db = SW\ wb \times \left(\frac{100-M}{100}\right)$
2	$GC = \left(\frac{A}{AGA}\right) \times 0.3$
3	$GC\ standardized\ regular\ maize\ starch = \left(\frac{A}{AGA}\right)$
4	$GCS = \left(\frac{GC}{SW\ db}\right) \times 0.1$
5	$\%SC = GCS \times 0.9 \times 100$

if necessary. Therefore, it is important to know the moisture content (M) of the samples and the standard starch used. It is recommended to manipulate Table 1 in Excel as it facilitates data treatment and obtaining the results.

Volumetric flask ($V = 50$ or 100 mL) is a parameter that was previously determined, and absorbance (A) in 510 nm was obtained from the sample. Absorbance will be used to calculate the glucose content (GC), multiplying "A" by the average glucose absorbance (AGA) and dividing the result by 0.3 (Eq. 2). Except for standardized regular maize starch, since this sample used 30 μL of hydrolyzed starch to prepare the solution in the flasks (item 3.2) (Eq. 3).

Then, the value obtained must be converted to mg of glucose/ mg of sample (GCS), dividing the result obtained by the weight of the sample (pay attention to the choice of the correct weight of the sample, considering on a dry basis—db), and multiply the result by 0.1, which is the sample mass recommended for the test (considering that accurately 100 mg ± 5 mg of the sample had been used) (Eq. 4).

Finally, multiply the value obtained by 0.9 (factor of conversion of glucose molecules to starch) and by 100 for conversion to percentage, and obtain the percentage of starch content (% SC), dry basis (db) (Eq. 5). There, calculate the average of starch content obtained, dry basis (ASC) for each sample, and the standard deviation (SD). All equations can be seen in Table 2.

4 Notes

1. Purified protease may be necessary if the sample has a protein content that needs prior digestion, such as pulses, beans, processed foods, and others. Example: purified protease (10 mL, 50 mg/mL; ~ 350 tyrosine U/mL, with an optimum pH of 7.5).

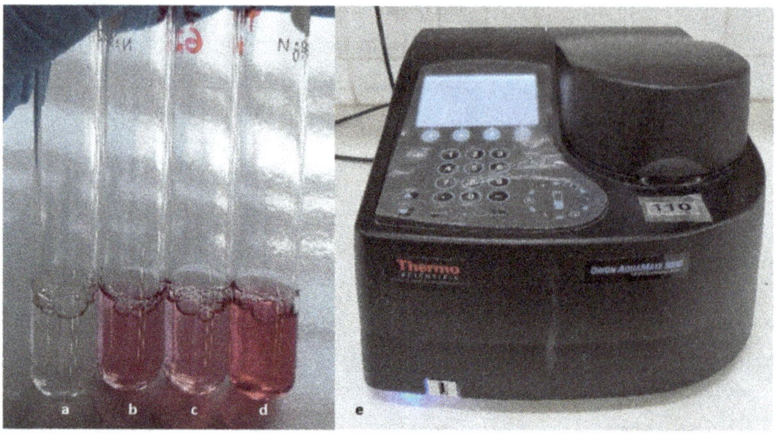

Fig. 5 Illustrative image of how to use the rubber bands to keep the centrifuge tubes submerged in the water bath

2. Check the amount of the sodium maleate buffer solution to be prepared, considering that each sample (triplicate) uses about 9 mL, in addition to the triplicate of the standard starch, which consumes an additional 9 mL.
3. The preparation of sodium hydroxide solution should preferably be carried out inside a fume hood.
4. The 0.275 M sodium hydroxide (NaOH) solution will be necessary only if protease is used in the sample hydrolysis process.
5. It is indicated to use only 1 type of cuvette, acrylic or polystyrene. It is not recommended to use cuvettes of different materials to carry out the analysis in the same batch.
6. The water bath must be turned on at 100 °C (or boiling, according to the atmospheric pressure of the place where the analysis is being carried out), about one hour before use, to stabilize the temperature.
7. It may be necessary to fix the tubes immersed in the water bath with a nylon rubber band (Fig. 5).

Acknowledgments

The authors are grateful to the Coordination for the Improvement of Higher Education Personnel—Brazil (CAPES)—Finance Code 001 and scholarship by U. M. Sampaio, and Conselho Nacional de Desenvolvimento Científico e Tecnológico (CNPq), for the scholarship awarded to M. T. P. S. Clerici (312660/2023-5).

References

1. Eliasson A-C (ed) (2004) Starch in food: structure, function and application. Woodhead Publishing, Cambridge, UK
2. Collins CH, Bonato PS, Braga GL (eds) (2006) FUNDAMENTOS de cromatografia. Editora da UNICAMP, Campinas
3. AOAC. Association of Official Analytical Chemists (2006) Official methods of analysis of the AOAC International, Arlington
4. AACCI (2010) International American Association of Cereal Chemistry. Approved methods of analysis, 11th edn. Saint Paul, Minnesota
5. Heloisa Mascia Cecchi (2003) Fundamentos teóricos e práticos em análise de alimentos, 2nd edn. Editora da UNICAMP, Campinas

Chapter 14

Analysis of Advanced Glycation End Products (AGEs)

Mustafa Yaman, Halime Uğur, Esra Yıldırım Servi,
and Ayşe Binnur Karataş

Abstract

Advanced glycation end products (AGEs) are formed both exogenously in foods and endogenously through metabolic processes. Their key precursors, glyoxal (GO), methylglyoxal (MGO), and 3-deoxyglucosone (3-DG), are reactive dicarbonyl compounds. This study outlines the analytical procedures for quantifying GO, MGO, CML, CEL, and furosine using high-performance liquid chromatography (HPLC) and liquid chromatography-tandem mass spectrometry (LC-MS/MS). Sample preparation includes derivatization, extraction, and purification steps specific to each analyte. The described methods enable precise detection of AGEs and their precursors in complex matrices.

Key words AGEs, Glyoxal, Methylglyoxal, CEL, CML, Furosine, HPLC, LC-MS/MS

1 Introduction

Advanced glycation end products (AGEs) can be formed exogenously in foods and endogenously in the human body. Glyoxal (GO), methylglyoxal (MGO), and 3-deoxyglucosone (3-DG) are precursors for AGE production and are known as dicarbonyls. N-ε-carboxymethyl lysine (CML), Nε-carboxyethyl-lysine (CEL), hydroimidazolone (MG-H1), and furosine are well-known AGEs. Here, we describe the analysis of precursors of AGEs and AGEs by HPLC or LC-MS/MS.

The Maillard reaction expresses a complex reaction chain, beginning with the covalent bond between amine groups and carbonyl compounds. As a result of this reaction, toxic substances often occur with mutagenic, carcinogenic, and cytotoxic effects [1]. With the continuation of the Maillard reaction, advanced glycation end products (AGEs) appear, which play an important role in the pathogenesis of chronic diseases such as diabetes, ath-

Lilian Mariutti and Frank Vriesekoop (eds.), *Food Chemistry: Basic Protocols*, Methods and Protocols in Food Science,
https://doi.org/10.1007/978-1-0716-4678-6_14,

erosclerosis, kidney failure, and Alzheimer's disease [2, 3]. Advanced glycation end products (AGEs) can be formed exogenously in foods and endogenously in the human body [4, 5].

The formation of AGEs can be divided into three stages: initial, intermediate, and final. All these stages are interrelated, can occur simultaneously, and can be affected by the reaction conditions [1]. In the first stage, the aldehyde group in the structure of reducing sugars such as glucose, fructose, ribose, and the N-terminal end of the free amino groups of proteins, fats, or nucleic acids (mostly lysine, arginine, sulfur-containing amino acids) react, and a Schiff base is formed [6, 7]. The Schiff base is chemically rearranged within days and forms Amadori products, also known as early glycation products. The reactions occurring at this stage are reversible [8, 9]. The products of the Amadori reaction, which takes place in the second stage, mainly depend on pH values and temperature. Further chemical regulations such as oxidation, reduction, and hydration occur in the presence of transition metal ions and at high temperatures and pH above 7 [10]. In this way, reactive carbonyl intermediates are formed, consisting of glyoxal (GO), methylglyoxal (MGO), and 3-deoxyglucosone (3-DG), which are precursors for AGE production and are known as dicarbonyls. In the final stage, these intermediates form the final AGEs by reacting with amino protein groups. These processes occur within weeks or months and are irreversible [4, 5, 11].

Pentosidine, one of the final AGEs, is produced by reacting pentose with lysine and arginine residues of proteins. On the other hand, GO, MGO, and 3-DG react with two lysine residues to form glyoxal-derived lysine dimer (GOLD), methylglyoxal-derived lysine dimer (MOLD), and 3-deoxyglucosone-derived lysine dimer (DOLD), respectively, while they react with a lysine and an arginine residue to form glyoxal derived imidazolium cross-link (GODIC), methylglyoxal dimer imidazolone cross-link (MODIC), and 3-deoxyglucosone-derived imidazolium cross-link (DODIC), respectively. Nε-carboxymethyl lysine (CML), N-ε-carboxyethyl-lysine (CEL), and pyralin are other common AGEs, and each occurs by the reaction of a lysine residue and carbonyl compounds (GO, MGO, 3-DG, respectively). In addition, the less common AGEs methylglyoxal hydroimidazolone (MG-H1), tetrahydropyrimidine, and argpyrimidine are formed as a result of the reaction of the arginine residue with MG and G-hydroimidazolone as a result of the reaction of the arginine residue with GO. The structures of AGEs in foods are complex and versatile due to the wide variety of amino acids and carbonyl compounds [10, 12]. Besides the many AGEs that can be formed, CML, CEL, and pentosidine are the more frequently occurring AGEs [4, 5, 13].

2 Material

2.1 Glyoxal and Methylglyoxal

Glyoxal (40%), methylglyoxal (40%), methanol, acetonitrile, sodium acetate anhydrous, 4-nitro-1,2-phenylenediamine, ortho-phosphoric acid, plastic falcon tube, 0.45 μm cellulose acetate filter.

2.1.1 Sodium Acetate Buffer (0.1 M, pH: 3)

1. Prepare 100 mL of sodium acetate buffer.
2. Weigh 8.2 g sodium acetate anhydrous into a 100 ml volumetric flask and add 50 mL deionized water.
3. Dissolve it using a magnetic stir bar or an ultrasonic water bath. Make up to 100 mL with deionized water.
4. Adjust the pH 3 using 85% ortho-phosphoric acid and store at room temperature.

2.1.2 Derivatization Solution

1. Weigh 0.1 g 4-nitro-1,2-phenylenediamine into a 100 ml volumetric flask.
2. Dissolve with 25 mL of methanol and make up to 100 mL with methanol.

2.1.3 Stock Standard Solution (4000 Ppm)

1. Weigh 1 mL of glyoxal and methylglyoxal into a 100 mL volumetric flask and dissolve with methanol.
2. Use the stock standard solution to prepare a working standard solution of between 1 ppm and 10 ppm.

2.2 Advanced Glycation End Products (CEL, CML) [14–16]

Nε-carboxymethyl lysine (CML) and Nε-carboxyethyl-lysine (CEL), sodium borate, sodium borohydride, hydrochloric acid (HCl), formic acid, methanol, acetonitrile, chloroform, MCX cartridge.

2.3 Furosine [16]

Furosine, ammonium acetate, methanol, trifluoroacetic acid, and cellulose acetate filter.

3 Methods

3.1 Analysis of Glyoxal and Methylglyoxal [17]

3.1.1 Sample Extraction

1. Homogenize the sample with a blender knife or coffee grinder.
2. Weigh 2–5 g of homogenized sample into a 50 mL plastic Falcon tube and add 20 mL of methanol.
3. Extract the sample with an ultra-turrax homogenizer for 2 min and centrifuge at 8000 rpm for 5 min.
4. Take 0.5 mL of supernatant using an automatic pipette into a 10 mL glass tube, and add 1 mL sodium acetate buffer.
5. Add 0.5 mL of derivatization solution and incubate at 70 °C for 20 min. Filter the samples using a 0.45 μm cellulose acetate filter, and inject it into the HPLC.

3.1.2 HPLC Conditions

1. Use a high-performance liquid chromatography (HPLC) system with a pump and a UV/vis detector (*see* **Notes 1** and **2**).
2. Prepare the mobile phase with water, methanol, and acetonitrile (42:56:2, v/v/v).
3. Use a UV-vis or photodiode array (PDA) detector.
4. Set the wavelength to 254 nm and separate the GO and MGO with an Inertsil ODS-3, 250 × 4.6 mm, 5 μm column with a 1 mL/min flow rate.
5. Set the column oven temperature to 30 °C.

3.2 Analysis of Advanced Glycation End Products (CEL, CML) [14–16]

3.2.1 Sample Preparation

1. Weigh 0.2000 g of homogenized sample in a glass tube.
2. Add 2 mL borate buffer (0.2 M, pH 9.2) and 0.4 mL sodium borohydride (2 M in 0.1 N NaOH), and incubate at 4 °C for 8 h.
3. Add 4 mL of methanol-chloroform (1:2, v:v), and centrifuge at 5000 rpm for 10 min to defat and precipitate protein.
4. Hydrolyze the precipitated protein with 4 mL of 6 M hydrochloric acid (HCl) at 110 °C for 24 h.
5. Dilute it with water to 20 mL and filter it with filter paper.
6. Eluate through an MCX cartridge (Waters Oasis MCX 3 cc/ 60 mg). First, pre-activate the MCX cartridge with 3 mL methanol and 3 mL water in sequence. Load the solution onto the MCX cartridge. Rinse with 3 mL water and 3 mL methanol, respectively. Eluate target compounds with 6 mL of ammonia: methanol (95:5, v/v) (*see* **Note 3**).
7. Dry the eluate to dryness under a stream of nitrogen gas at 50 °C.
8. Reconstitute the residue with 2 mL methanol:water (80: 20, v:v).
9. Vortex for 1 min and filter through a 0.22-μm membrane filter before injecting into the LC–MS/MS.

3.2.2 UHPLC Tandem MS Analysis

1. Analyze the derivatized CML and CEL by LC–MS/MS system using positive electrospray ionization.
2. Use the Waters XSelect HSS T3 column (4.6 mm × 150 mm, 5 μm) for separation.
3. Prepare mobile phase A (0.1% formic acid in water) and mobile phase B (0.1% formic acid in acetonitrile). Start the mobile phase composition with 20% mobile phase B. The elution composition linearly increases 0–6.5 min, from 20% to 90% B; 6.5–8 min, 90% B; linearly decreases 8–9 min, from 90% to 20% B; 9–13 min, 20% B. Set the flow rate at 0.4 mL/min. Total acquisition is completed in 13 min.

4. Set the nebulizing gas flow to 3 L/min, drying gas flow 15 L/min, desolvation line (DL) temperature at 250 °C, and heat block temperature at 400 °C.
5. Set the column oven temperature at 40 °C.
6. Inject 5 μL of sample solution.
7. The multiple reaction monitoring transitions for CML, CEL, and MG-H1 are given in the table below.
8. Calibration curve is made from standards ranging from 50 to 1000 ppb.

	Precursor and product m/z	Collision energy	Q1 voltage (V)	Q3 voltage (V)
CML	205.0 > 130	13	30	22
	205.0 > 84.1	23	30	30
	205.0 > 56.0	41	30	22
CEL	219 > 84.1	22	30	29
	219 > 130.1	13	30	12
	219 > 156	15	30	30

3.3 Furosine Analysis [16]

3.3.1 Sample Preparation

1. Weigh one gram of homogenized sample into a glass tube and mix with 8 mL 10.6 mol/L HCl.
2. Saturate it with nitrogen for 1 minute, close the lid tightly, and filter it with filter paper.
3. Transfer into a 10 mL volumetric flask and complete 10 mL with deionized water. Pipette 1 mL of the solution into a glass tube, and add 5 mL of ammonium acetate (6 mg/mL) solution.
4. Vortex for 1 min.
5. Filter through a 0.22-μm CA filter before HPLC analysis.

3.3.2 HPLC Conditions

1. Use HPLC-UV detector for furosine for identification and quantification.
2. Set the Zorbax C8 column (4.6 mm × 250 mm, 5 μm) for the separation (*see* **Note 4**).
3. Prepare mobile phase A (methanol) and mobile phase B (0.1% trifluoroacetic acid).
4. The elution gradient programs as mobile phase A linearly increases from 0% to 17% for 0–8 min, 17% to 0% for 8–10 min, and 0% for 10–15 min.
5. Flow 1 mL/min.
6. Injection volume 10 μL.

7. Set the wavelength at 280 nm.
8. Set the column temperature at 32 °C.

4 Notes

1. When performing GO and MGO analysis, the analysis process must be completed on the HPLC device within 24 hours after the samples are assigned for analysis.
2. HPLC column cleaning must be done thoroughly before starting the analysis, otherwise, contamination may cause problems.
3. To obtain appropriate peaks in LC/MS/MS in CEL and CML analysis, an MCX cartridge or similar quality should be used. In addition, proper washing should be done.
4. C8 column must be used in the determination of Furosine determination. Otherwise, proper peaks may not be obtained.

References

1. de Oliveira FC, Coimbra JSDR, de Oliveira EB, Zuñiga ADG, Rojas EEG (2016) Food protein-polysaccharide conjugates obtained via the Maillard reaction: a review. Crit Rev Food Sci Nutr 56(7):1108–1125
2. Corzo-Martínez M, Moreno FJ, Olano A, Villamiel M (2010) Role of pyridoxamine in the formation of the Amadori/Heyns compounds and aggregates during the glycation of β-lactoglobulin with galactose and tagatose. J Agric Food Chem 58(1):500–506
3. Echavarría AP, Pagán J, Ibarz A (2012) Melanoidins formed by Maillard reaction in food and their biological activity. Food Eng Rev 4(4):203–223
4. Gaens KH, Stehouwer CD, Schalkwijk CG (2013) Advanced glycation endproducts and its receptor for advanced glycation endproducts in obesity. Curr Opin Lipidol 24(1):4–11
5. Wetzels S, Wouters K, Schalkwijk CG, Vanmierlo T, Hendriks JJ (2017) Methylglyoxal-derived advanced glycation endproducts in multiple sclerosis. Int J Mol Sci 18(2):421
6. Vistoli G, De Maddis D, Cipak A, Zarkovic N, Carini M, Aldini G (2013) Advanced glycoxidation and lipoxidation end products (AGEs and ALEs): an overview of their mechanisms of formation. Free Radic Res 47(sup1):3–27
7. Aragno M, Mastrocola R (2017) Dietary sugars and endogenous formation of advanced glycation endproducts: emerging mechanisms of disease. Nutrients 9(4):385
8. Snelson M, Coughlan MT (2019) Dietary advanced glycation end products: digestion, metabolism and modulation of gut microbial ecology. Nutrients 11(2):215
9. Bonnefont-Rousselot D (2002) Glucose and reactive oxygen species. Curr Opin Clin Nutrit Metab Care 5(5):561–568
10. Wei Q, Liu T, Sun DW (2018) Advanced glycation end-products (AGEs) in foods and their detecting techniques and methods: a review. Trends Food Sci Technol 82:32–45
11. Ravichandran G, Lakshmanan DK, Raju K, Elangovan A, Nambirajan G, Devanesan AA, Thilagar S (2019) Food advanced glycation end products as potential endocrine disruptors: an emerging threat to contemporary and future generation. Environ Int 123:486–500
12. Serin Y, Akbulut G, Uğur H, Yaman M (2021) Recent developments in in-vitro assessment of advanced glycation end products. Curr Opin Food Sci 40:136
13. Luevano-Contreras C, Chapman-Novakofski K (2010) Dietary advanced glycation end products and aging. Nutrients 2(12):1247–1265
14. Scheijen JL, Clevers E, Engelen L, Dagnelie PC, Brouns F, Stehouwer CD, Schalkwijk CG (2016) Analysis of advanced glycation endproducts in selected food items by ultra-performance liquid chromatography tandem mass spectrometry: presentation of a dietary AGE database. Food Chem 190:1145–1150
15. Sun X, Tang J, Wang J, Rasco BA, Lai K, Huang Y (2015) Formation of advanced

glycation endproducts in ground beef under pasteurisation conditions. Food Chem 172: 802–807

16. Li Y, Wu Y, Quan W, Jia X, He Z, Wang Z et al (2021) Quantitation of furosine, furfurals, and advanced glycation end products in milk treated with pasteurization and sterilization methods applicable in China. Food Res Int 140: 110088

17. Cengiz S, Kişmiroğlu C, Cebi N, Catak J, Yaman M (2020) Determination of the most potent precursors of advanced glycation end products (AGEs) in chips, crackers, and breakfast cereals by high performance liquid chromatography (HPLC) using precolumn derivatization with 4-nitro-1, 2-phenlenediamine. Microchem J 158:105170

Chapter 15

Carbohydrate Analysis by High-Performance Anion-Exchange Chromatography with Pulsed Amperometric Detector (HPAEC-PAD)

Henrique Silvano Arruda and Gustavo Araujo Pereira

Abstract

Carbohydrates are the most abundant class of biomolecules on earth, performing numerous biological essential roles (e.g., energy source, energy storage, and structural components). The recent application of advanced analytical techniques, especially high-performance anion-exchange chromatography coupled with a pulsed amperometric detector (HPAEC-PAD), to analyze carbohydrates has enabled researchers to characterize and quantify carbohydrates in different matrices (e.g., foods/beverages, products from bioprocesses, and body fluids). HPAEC-PAD is a versatile, relatively simple, fast, sensitive, and selective technique to analyze and characterize different carbohydrates. We describe herein HPAEC-PAD methods that have been used successfully to investigate the profile and content of the most common carbohydrates. Specifically, we used food matrices as a model for carbohydrate analysis by HPAEC-PAD. We aim to provide the reader with guidelines on applying the HPAEC-PAD technique to identify and quantify different carbohydrates and highlight this approach's advantages and limitations.

Key words Monosaccharides, Oligosaccharides, Polysaccharides, Polyols, Sugars, Functional sugars, Prebiotic, Food analysis, Ion chromatography, Electrochemical detection

1 Introduction

Carbohydrates, the most abundant class of biomolecules on earth, are important constituents of all living organisms and perform numerous essential roles, such as energy source, storage, and structural components [1, 2]. Furthermore, recent research has shown that dietary carbohydrates are involved in both the prevention (e.g., fibers and prebiotics) and onset/triggering (e.g., excessive intake of sugars and starch) of various diseases [3, 4]. The biological importance of carbohydrates has highlighted the need to characterize and measure these compounds in body fluids, products from biotechnological processes, and especially in food matrices. This trend is leading to the development of several analytical methods for the

Lilian Mariutti and Frank Vriesekoop (eds.), *Food Chemistry: Basic Protocols*, Methods and Protocols in Food Science,
https://doi.org/10.1007/978-1-0716-4678-6_15,

separation, quantification, and structural analysis of carbohydrates in food ingredients and matrices.

Carbohydrates are chemically defined as polyhydroxy aldehydes or polyhydroxy ketones with the general chemical formula $C_n(H_2O)_n$. Monosaccharides are the simplest carbohydrate structures, which can be joined together by glycosidic bonds to give rise to larger structures (disaccharides, oligosaccharides, and polysaccharides) [3]. The multiple chemical functional sites of monosaccharides (polyhydroxylated nature of chain) permit multiple points of linkage and branching of carbohydrate chain that, when associated with their anomeric carbon forms (α or β) and chirality centers (stereoisomerism), contribute to a great diversity and complexity of carbohydrate structures found in nature. In addition, the absence of an analytically valuable chromophore in most carbohydrates makes carbohydrate analysis notoriously difficult and complex [3, 5, 6]. However, these analytical barriers have been successfully overcome with more advanced analytical techniques and sophisticated instruments.

Currently, there are a large number of analytical methods available to analyze carbohydrates in food matrices, such as gravimetry (e.g., AOAC 985.29 method); spectrophotometry (e.g., 3,5-dinitrosalicylic acid (DNS) method, phenol-sulfuric acid method, glucose oxidase/peroxidase method, among others); chromatography (e.g., high-performance liquid chromatography (HPLC) and gas chromatography (GC)); mass spectrometry (MS) (e.g., matrix-assisted desorption/ionization-quadrupole time-of-flight mass spectrometry (MALDI-QTOF-MS)), capillary electrophoresis (CE), nuclear magnetic resonance (NMR); and their combinations (e.g., HPLC and GC coupled with mass spectrometry) [7–9]. Although there are several analytical techniques available for analyzing carbohydrates, the choice of the appropriate analytical tool will depend on many factors, such as analytical purpose (detection, identification, structural elucidation, or quantification), amount of target compounds in the sample, availability of commercial standards, amount of sample available, previous knowledge about the sample and its composition, and financial resources [6].

Among the analytical techniques available for carbohydrate analysis in food matrices, high-performance anion-exchange chromatography coupled with a pulsed amperometric detector (HPAEC-PAD) has been the most used [10]. The combination of anion-exchange chromatography at alkaline pH and pulsed electrochemical detection in HPAEC-PAD analysis allows direct quantification of a wide variety of nonderivatized carbohydrates, including monosaccharides, sialic acids, sugar phosphates, and oligosaccharides, with minimal sample preparation and clean-up, in addition to providing selective separation and sensitive detection (low-picomole levels) of a mixture of compounds with high

resolution in a single run [11, 12]. The high selectivity and specificity of this method for carbohydrates are due to the capacity of (1) pulsed amperometry at a gold working electrode to detect only those compounds with oxidable functional groups at the employed detection voltage (in the case of carbohydrates, -COH groups); and (2) anion-exchange columns to elute neutral or cationic sample components in, or close to, the void volume of the column, avoiding interference of noncarbohydrate components even if such species are oxidizable. At the same time, the method sensitivity is enhanced due to the alkaline conditions (pH > 12) from the chromatographic run at the gold working electrode to be able to improve the electrocatalytic oxidation of carbohydrates [13–15].

Carbohydrate analysis by HPAEC-PAD is based on the capacity of anion-exchange columns in separating anionic carbohydrates that contain sialic acid, carboxylate, sulfate, or phosphate groups. Then, under optimized electrochemical conditions, these carbohydrates are oxidized at a gold working electrode, where aldehyde, ketone, and alcohol moieties are readily oxidizable at 0.10 V versus an Ag/AgCl reference electrode, thereby allowing their reliable identification. Neutral carbohydrates can also be analyzed by HPAEC-PAD since they are weak acids (pK_a values ranging from 12 to 14; *see* Table 1 for some examples) and, consequently, at high pH, their hydroxyl groups are at least partially transformed into oxyanions, thereby allowing their separation without the need to use additives as borates [14, 16, 17]. Thus, alkaline solutions are necessary for carbohydrate separation in HPAEC-PAD. The mobile phases most commonly used for carbohydrate separation via HPAEC-PAD are sodium hydroxide and sodium acetate. Mono-, di-, tri-, and tetrasaccharides and their sugar alcohols are typically

Table 1
Dissociation constants of some common neutral carbohydrates in the water at 25 °C

Carbohydrate	pK_a
Fructose	12.03
Mannose	12.08
Xylose	12.15
Glucose	12.28
Galactose	12.39
Dulcitol	13.43
Sorbitol	13.60
α-Methyl glucoside	13.71

Source: Bruggink [17]

eluted using only sodium hydroxide (10–20 mM) under isocratic conditions. On the other hand, combinations of sodium hydroxide and sodium acetate are employed to separate strongly retained species (e.g., larger carbohydrates and carbohydrates that possess a negative charge at neutral pH, such as sialic acids, phosphorylated sugars, and sulfated sugars). Larger and charged carbohydrates are usually separated by using sodium hydroxide (100 mM) and a gradient of sodium acetate (up to 500 mM) [5, 11].

Retention of monosaccharides on the HPAEC-PAD column can be directly predictable by extrapolating the hierarchy of monosaccharide hydroxyl acidities (pK_a values). The most acidic monosaccharide (with the lowest pK_a value) will bind more strongly to the column and elute last. This implies that the lower the pK_a value, the longer the retention time. However, in addition to pK_a values, monosaccharide separation in HPAEC-PAD is influenced by the number of ionizable hydroxyl groups, presence of negatively charged groups (e.g., sialic acid, sulfate, or phosphate groups), anomerism, positional isomerism, and accessibility of oxyanions to the functional groups of the anion-exchange column. It is known that the greater the negative charge of a carbohydrate, the longer the retention time. Unlike monosaccharides, it is not so simple to predict the elution order of oligosaccharides. In addition to the aspects mentioned above for monosaccharides, oligosaccharide retention is directly affected by other factors such as the degree of polymerization, monosaccharide composition, degree of branching, linkage position, and type. However, for members of a linear homologous series of oligosaccharides (e.g., inulin, maltodextrins, and xylooligosaccharides), the retention time becomes longer as the degree of polymerization increases. Meanwhile, for branched oligosaccharides, the retention time increases as the degree of branching increases [14, 17].

After the separation in anion-exchange columns, the carbohydrates are directed toward the pulsed amperometric detector. The coupling of PAD to HPAEC has some advantages, such as compatibility with gradient elution and enhanced sensitivity under alkaline conditions ($pH > 12$) from the chromatographic run. Moreover, PAD shows high sensitivity (permits detection of carbohydrates with excellent signal-to-noise ratios down to approximately 10 picomoles without requiring derivatization) and good selectivity (allows only the detection of compounds that are oxidable at a gold working electrode under selected potentials) [14, 15]. The carbohydrate detection in a pulsed amperometric detector is based on the measurement of electrical current generated by the carbohydrate oxidation in highly alkaline conditions at the surface of a gold working electrode by application of a positive potential (carbohydrates are oxidizable at 0.10 V when combined with an Ag/AgCl reference electrode). However, products from carbohydrate oxidation gradually poison the electrode surface, causing a loss of analyte signal. Thus, it is necessary to clean the detector between measurements. PAD employs a repeating sequence of three potentials to overcome this drawback. The first potential (detection potential) oxidizes the carbohydrates at the surface of the gold working electrode, generating the electrical current used for carbohydrate detection. The second potential (cleaning potential) oxidizes the gold working electrode, causing desorption of the carbohydrate oxidation products and, consequently, its cleaning. The third potential (restoring potential) reduces the gold oxide on the electrode surface back to gold, restoring the gold working electrode and permitting the detection during the next cycles. In addition to being an extremely sensitive and selective tool for carbohydrate detection, PAD can be used to quantify carbohydrates accurately since the electrical current generated is proportional to the carbohydrate concentration [15, 18].

This chapter is intended to introduce and guide carbohydrate analysis via HPAEC-PAD, especially in food matrices. The conditions described in this chapter have been found to give reliable separation and quantification of the common classes of carbohydrates found in food matrices by using HPAEC-PAD. Samples and their matrices vary; therefore, these conditions are intended to be used as guidelines only.

2 Material

2.1 Sample Preparation

Carbohydrates can be easily extracted from solid samples using ultrapure water at 1:10 (w/w) ratio. The equipment orbital shaker, ultra-homogenizer, and ultrasound bath or probe have been employed to extract different carbohydrates. We have used ultra-homogenizers and ultrapure water as extraction techniques and solvents, respectively [19–22]. This technique is effective, easy to operate, control, clean, and associated with low time consumption. Hydroethanolic solvents can also be employed, but the ethanol must be removed before sample injection into the chromatographic system. Ethanol, even at a very low concentration, can affect the column quality and, consequently, decrease the accuracy of analysis (e.g., may affect analyte retention and cause diminished electrode response) and column lifetime. Nonetheless, we have noted that using ethanol weakly increases the recovery content of target carbohydrates [22, 23]. Therefore, we recommend using ultrapure water to extract carbohydrates that will be analyzed into the HPAEC-PAD system. If necessary, an exhaustive extraction protocol should be carried out.

The extracts and liquid samples should be visually analyzed to identify particulate matter and colored compounds. These compounds at high content can affect the column quality, interfere with the chromatographic separation, and suppress the detector signal.

The particulate matter can be removed by filtering the sample. It is recommended that all samples be passed through a 0.45 μm (column with particle size >3 μm) nylon syringe filter before injection into the column to remove particulates. Polyvinylidene fluoride (PVDF) and polytetrafluoroethylene (PTFE) hydrophilic materials can filter aqueous samples [24]. If the chromatography column contains particle size <3 μm, a filter of 0.22 μm should be used [15, 22, 23].

Unlike particulate matter, colored compounds and proteins generally are not removed by passing the sample through a syringe filter. If the analyst verifies that the sample or diluted sample shows high-intensity color, proteins, etc., the compounds responsible for that should be removed. This can be done by using specific filter cartridges; for example, polyvinylpyrrolidone (PVP) filter cartridges can be used to remove phenolic compounds, while hydrophobic filter cartridges should be used to remove proteins from the sample. Certainly, sample preparation depends on sample complexity and the presence of interferents that can affect the carbohydrate analysis by ion chromatography, such as proteins, phenolic compounds, and metals [15].

The sample appropriately prepared can be diluted and injected into the chromatography column. The electrochemical detector is very sensitive, and the sample should be initially 1000-5000-fold diluted before injection. That depends on the target carbohydrate amount in the unknown samples and the range of the analytical curve. For example, glucose syrup should be diluted 10,000-fold before injection, while a filtered aliquot from the end of the glucose fermentation process can be directly injected (glucose was almost entirely consumed by microorganisms). Therefore, the dilution factor should be carefully selected to avoid clogging up the column, separation problems, peak asymmetry, detector poisoning, among other drawbacks. The analyst can initially use a high dilution factor (≥ 1000-fold) and calculate an ideal dilution factor from the data obtained.

The aqueous extracts and samples can be stored at freezing temperatures, and freeze/thaw cycles should be avoided. Nevertheless, the storage system should be evaluated according to sample characteristics.

2.2 Analytical Curves

Analytical mono- and oligosaccharide standards with purity ≥96% can be purchased from (but not limited to) Sigma-Aldrich (St. Louis, USA) and Megazyme (Wicklow, Ireland). Stock solutions of each carbohydrate standard (1 mg/mL) should be prepared in ultrapure water and stored in a freezer (−18 °C). The standard stock solution should be diluted in ultrapure water according to the analytical curve range. We have constructed analytical curves for mono- and oligosaccharide analysis using the concentration range of 0.25–12.50 μg/mL (at least 5 points) [19].

The mono- and oligosaccharides of an unknown sample can be identified by comparing the retention time of the standard and the sample. If necessary, the identity of the individual peaks can be confirmed by co-injection of samples spiked with commercial standards (*see* **Note 1**) [19, 23].

2.3 Mobile Phases

To prepare mobile phases, the analyst must use ultrapure deionized water (18 MΩ cm^{-1} at 25 °C) and analytical grade reagents, 50% (w/w) sodium hydroxide solution, and anhydrous sodium acetate. We recommended using a 50% (w/w) sodium hydroxide solution (Sigma-Aldrich, cat. no. 415413) and anhydrous sodium acetate (Sigma-Aldrich, cat. no. 32319) as the sources of sodium hydroxide and sodium acetate, respectively. However, the analyst can purchase these reagents from other suppliers if their reagents meet the quality criteria for HPAEC-PAD analysis. The water should be free from ionized impurities, organics, microorganisms, and particulate matter (*see* **Note 2**). The mobile phases (sodium hydroxide and sodium acetate/sodium hydroxide) should be stored at room temperature in plastic reservoirs, as they can leach borate and silicate out of glass. Blanketing the mobile phases with helium or nitrogen is recommended to create an inert gas headspace and avoid carbon dioxide absorption (it can produce sodium carbonate).

Sodium hydroxide pellets should never be used because they are covered with a thin layer of sodium carbonate. This substance binds strongly to the column, decreasing the column's selectivity, resolution, and efficiency. Furthermore, the sodium carbonate can precipitate at the bottom of the eluent reservoir. The 50% (w/w) sodium hydroxide solution shows a very low amount of sodium carbonate; consequently, it is the recommended reagent for HPAEC-PAD (*see* **Note 3**) [15]. It is important to consistently prepare the mobile phases using the same protocol to ensure reproducible chromatography.

Three eluents have been used to carry out HPAEC-PAD: ultrapure water, sodium hydroxide, and sodium acetate containing sodium hydroxide. In this system, isocratic and gradient programs can separate mono- and oligosaccharides. The eluting power of the mobile phases is as follows: sodium acetate/sodium hydroxide > sodium hydroxide > water. The concentration of sodium hydroxide and sodium acetate/sodium hydroxide eluents depends on the method and the target carbohydrates that will be analyzed [19–34]. If the system has a ternary and quaternary pump model, ultrapure water may be used as a diluting eluent of the alkaline eluents. This is very useful because the analyst can standardize the preparation/concentration of sodium hydroxide and sodium acetate/sodium hydroxide eluents (e.g., 0.2 M sodium hydroxide and 1 M sodium acetate containing 0.2 M sodium hydroxide) and adjust the concentration of them with ultrapure water just operating the chromatography system.

2.4 Chromatographic Columns

Mono-, di-, and some oligosaccharides have been separated by using the CarboPac PA1 analytical column (250 × 4 mm i.d.) coupled with a guard column (50 × 4 mm i.d.), while the CarboPac PA100 analytical column (250 × 4 mm i.d.) coupled with a guard column (50 × 4 mm i.d.) has been used to analyze oligosaccharides and linear polymers (*see* **Note 4**) [19, 21–23]. These columns show different particle sizes and resin compositions as follows: (1) CarboPac PA1 is packed with a pellicular resin 10 μm diameter, polystyrene/divinylbenzene substrate agglomerated with 580 nm MicroBead quaternary ammonium functionalized latex; (2) CarboPac PA100 is packed with a pellicular resin 8.5 μm diameter, ethylvinylbenzene/divinylbenzene substrate agglomerated with 275 nm MicroBead quaternary ammonium functionalized latex [15]. These pellicular anion-exchange resins can effectively separate mono- and oligosaccharides. We recommend reading the column manual to verify the operating parameters (such as pH range, temperature, flow rate, injection volume, pressure limits, and organic solvent compatibility) and storage conditions.

2.5 Anion-Exchange Chromatography System

We have been using HPAEC to analyze food and products from processes containing carbohydrates, especially mono-, di-, and oligosaccharides [19–34]. The equipment comprises a quaternary pump, an autosampler with a 25 μL injection loop volume, a column module, and an ED50 electrochemical detector (gold working electrode and Ag/AgCl reference electrode). The potential waveform of the electrochemical detector was set up according to the equipment pre-setting to carbohydrate analysis, as shown in Table 2 (*see* **Note 5**). The HPAEC-PAD system, model Dionex ICS-5000 (Thermo Fisher Scientific, Waltham, MA, USA), was employed to conduct the analyses described herein. The

Table 2
Detector configuration (quadrupole potential waveform) recommended for carbohydrate analysis by pulsed amperometric detection

Time (s)	Potential (V) vs. Ag/AgCl	Gain region	Integration
0.00	+0.1	Off	Off
0.20	+0.1	On	On
0.40	+0.1	Off	Off
0.41	−2.0	Off	Off
0.42	−2.0	Off	Off
0.43	+0.6	Off	Off
0.44	−0.1	Off	Off
0.50	−0.1	Off	Off

chromatographic system was operated by Dionex Chromeleon software version 7.0. We recommend reading the HPAEC-PAD system manual before operating it.

3 Methods

Prepare all solutions using ultrapure water (resistivity ≥18 MΩ cm^{-1} at 25 °C) obtained from a Milli-Q water purification system (Millipore, Bedford, USA), especially mobile phases (*see* **Note 2**) and chromatographic or analytical grade reagents. Prepare and store all solutions and reagents at room temperature (unless indicated otherwise). Carry out all procedures at room temperature unless otherwise specified. Diligently follow all waste disposal regulations when disposing of waste materials.

3.1 Sample Preparation

In this section, we presented a general protocol by HPAEC-PAD to prepare samples for analyzing carbohydrates. Several techniques are available for extracting carbohydrates, as discussed in Subheading 2.1. Here, we will present an easy, simple, and fast technique (homogenizer-assisted extraction) to extract carbohydrates safely for routine analysis. We point out that sample preparation depends on sample matrix complexity and target carbohydrate classes, and thus, the following recommendations should be considered as guidelines only.

1. Prepare the extract by weighing 1 g of sample into a 50 mL Falcon plastic tube and then add 20 mL of ultrapure water (*see* **Note 6**). Extract the sample with an ultra-homogenizer at 11,000 rpm for 30–60 s at room temperature. Centrifuge the mixture at 4000 *g* for 11 min at 5 °C. Collect and store the supernatant at −18 °C until analysis (*see* **Note 7**).
2. Properly dilute the sample in ultrapure water (tips on dilution factors for different samples can be found in Subheading 2.1). For example, to prepare a sample 1000-fold diluted, pipette 100 μL of extract into a 100 mL volumetric flask and bring it to the final volume with ultrapure water (*see* **Note 8**). Stir gently the mixture manually (*see* **Note 9**).
3. Filter the samples through a 0.22 μm nylon syringe filter to remove particulate matter (*see* **Note 10**).
4. Transfer an aliquot of filtered sample to a 1.5 mL injection glass vial.
5. Cap the vials and inject immediately or store under refrigeration for just a few hours until analysis.

3.2 Analytical Curve Preparation

This section presents a general protocol to prepare analytical curves for quantifying carbohydrates. We emphasize that the number and type of carbohydrates employed to prepare calibration solutions will depend on the target carbohydrates intended in each analysis and the number of available analytical standards in the laboratory. The analytical curves must be constructed with at least 5 points according to the recommendations of the EURACHEM guidelines [35]. We have constructed analytical curves for mono- and oligosaccharide analysis using the concentration range of 0.25–12.50 μg/mL (7 points). The results obtained by our group showed analytical curves with very good linearity and low limits of detection and quantification for all compounds in the concentration range analyzed (*see* **Note 11**) [19].

1. Prepare stock solutions at 1 mg/mL for each target carbohydrate by weighing 5 mg of each carbohydrate standard into a 5 mL volumetric flask and bring to the final volume with ultrapure water. Stir until completely dissolved (*see* **Note 12**).
2. Prepare a working standard solution at 15 μg/mL containing all the carbohydrate standards of interest by pipetting 75 μL of stock solution at 1 mg/mL of each target carbohydrate into a 5 mL volumetric flask. Adjust the volume to the required level (5 mL) with ultrapure water, then briefly stir this solution (*see* **Note 13**).
3. Prepare 7 different calibration levels at 0.25, 1.25, 2.50, 5.00, 7.50, 10.00, and 12.50 μg/mL by pipetting 15, 75, 150, 300, 450, 600, and 750 μL of the working standard solution at 15 μg/mL into 1.5 mL injection glass vials and then complete to the required volume (900 μL) with ultrapure water (885, 825, 750, 600, 450, 300, and 150 μL of ultrapure water, respectively) (*see* **Note 8**). Cap the vials and mix the solutions with a vortex.
4. Inject immediately or store under refrigeration for just a few hours until analysis.

3.3 Mobile Phase Preparation

In this section, we present a general protocol to prepare the eluents most often used in HPAEC-PAD (sodium hydroxide and sodium acetate containing sodium hydroxide) and some tips. We emphasize that the concentration and type of eluents used in the analysis will depend on several aspects related to the target carbohydrates (e.g., carbohydrate class, number of carbohydrates to be separated, degree of polymerization, and presence of charged groups) (*see* Figs. 1 and 2). Preparing the eluents following the same protocol is very important daily, which increases the reproducibility of analysis. Furthermore, it is recommended that eluents (e.g., A, B, and C) are always transferred to their respective plastic bottle on the HPAEC-PAD system (e.g., A, B, and C) to avoid cross-contamination between eluents.

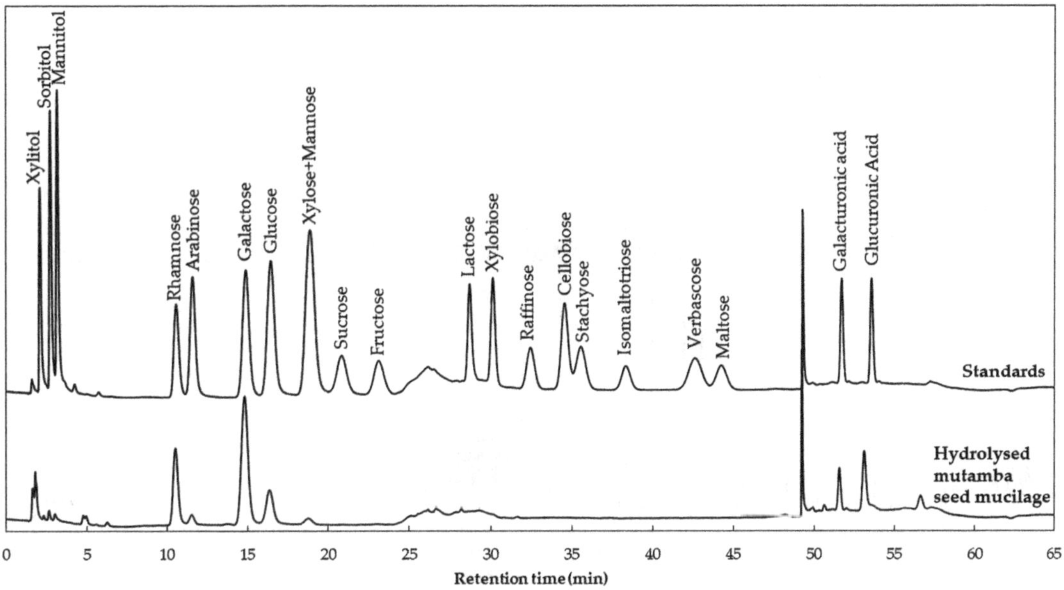

Fig. 1 HPAEC-PAD chromatogram showing the monosaccharides profile of a hydrolyzed mutamba seed mucilage. Dionex ICS-5000 system (Thermo Fisher Scientific, Waltham, MA, USA) was used to perform this analysis (*see* Subheading 2.5). Standard mix concentration was equal to 12.5 μg/mL. The sample (hydrolyzed mutamba seed mucilage, *see* Pereira et al. [27]) was diluted 60-fold and it was filtered through a 0.22 μm PTFE hydrophilic syringe filter. An aliquot of the diluted sample (25 μL) was injected into the chromatography column at 30 °C (CarboPac PA1, 250 × 4 mm i.d., particle size 10 μm, Thermo Fisher Scientific, Waltham, MA, USA). The mobile phases (constant flow rate of 1.0 mL/min) consisted of 0.2 M sodium hydroxide (eluent A), ultrapure water (eluent B), and 1 M sodium acetate containing 0.2 M sodium hydroxide (eluent C). The gradient was performed as follows: 0–22 min, 4% A and 96% B; 22–24 min, 4–50% A and 96–50% B; 24–45 min, 50% A and 50% B; 45–55 min, 84% B and 16% C; 55–60 min, 100% A; and 60–65 min, 4% A and 96% B. The gradient was developed by Pereira et al. [27] to separate different carbohydrates, such as polyols (xylitol, sorbitol, and mannitol), mono- and oligosaccharides (e.g., glucose, fructose, galactose, sucrose, and verbascose) and uronic acids (galacturonic and glucuronic acids). This method can be applied to analyze sugars in fruits, the monosaccharides composition of polymeric carbohydrates, sugars in milk derivates and raffinose family oligosaccharides (RFOs) from commonly consumed grains. Regarding the method, it is important to note that uronic acids elute only with the sodium acetate/sodium hydroxide eluent because they interact strongly with the column. Unlike uronic acids, the polyols eluted first, which demonstrated that these compounds interact weakly with the column. After 22 min run, it was necessary to carefully increase the sodium hydroxide concentration (eluent A) to eluate the other compounds. After that, sodium acetate/sodium hydroxide (eluent C) was employed to separate uronic acids (a ghost peak appears due to the changes in the mobile phase, 45–50 min). After eluting all compounds, the column was regenerated with 0.2 M sodium hydroxide (55–60 min) and re-equilibrated with 8 mM sodium hydroxide (60–65 min). The regeneration step is critical and must not be omitted, as it ensures the removal of contaminants, such as peptides and amino acids, that remain strongly bound to the chromatographic column. It is very important to keep the washing and re-equilibration times consistent from run to run

1. Degas the ultrapure water using one of the following methods: (1) Vacuum degas water by placing eluent bottles containing ultrapure water in an ultrasound bath. Vacuum degas the filled water bottles for 5–10 min with a vacuum pump while sonicating. (2) Sparging and blanketing the filled bottle with helium

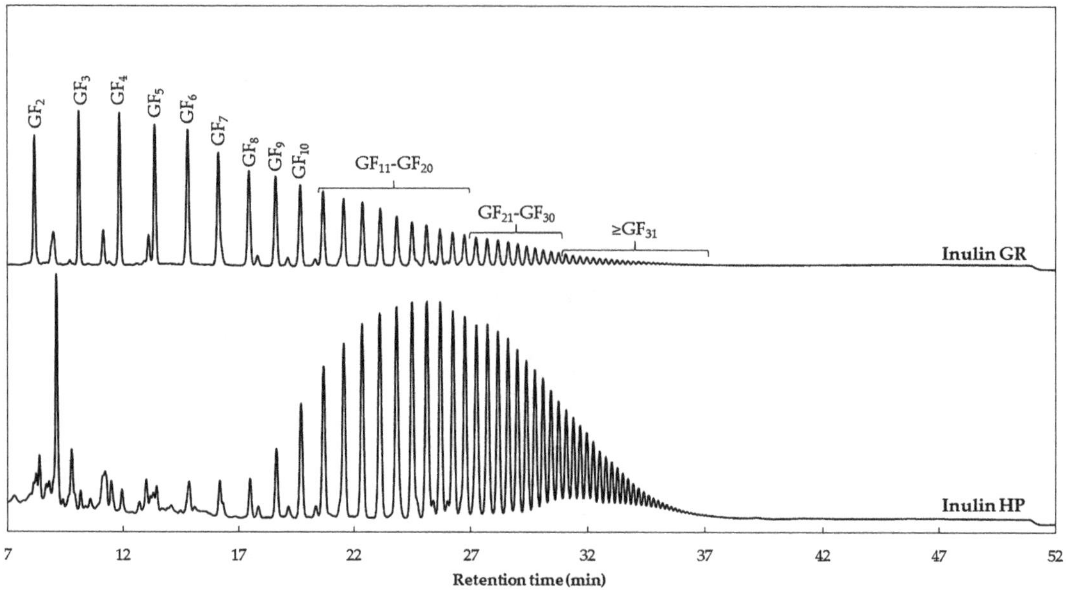

Fig. 2 HPAEC-PAD chromatogram showing the fructooligosaccharides profile of soursop whey beverages enriched with two commercially available inulin of different degrees of polymerization (DP) extracted from chicory root. Inulin GR (BENEO-Orafti®, average DP = 10). Inulin HP (BENEO-Orafti®, average DP = 23). Dionex ICS-5000 system (Thermo Fisher Scientific, Waltham, USA) was used to perform this analysis (*see* Subheading 2.5). The soursop whey beverages consisted of 34.95 g/100 g of distilled water, 30 g/100 g of whole pasteurized and homogenized milk, 15 g/100 g of frozen soursop pulp, 8 g/100 g of crystal sugar, 6 g/100 g of whey powder, 6 g/100 g of inulin powder (GR or HP), and 0.05 g/100 g gellan gum [24]. The soursop whey beverages containing inulins GR and HP were diluted 300- and 25-fold, respectively. Both diluted samples were filtered through a 0.22 μm PTFE hydrophilic syringe filter. An aliquot of the diluted sample (25 μL) was injected into the chromatography column at 30 °C (CarboPac PA100, 250 × 4 mm i.d., particle size 8.5 μm, Thermo Fisher Scientific, Waltham, MA, USA). The mobile phases (constant flow rate of 1.0 mL/min) consisted of 0.2 M sodium hydroxide (eluent A), 1 M sodium acetate containing 0.2 M sodium hydroxide (eluent B), and ultrapure water (eluent C). The gradient was performed as follows: 0–2 min, 48.5% A and 1.5% B; 2–44 min, 48.5–0% A and 1.5–50% B; 44–49 min, 50% B; and 49–54 min, 48.5% A and 1.5% B (the eluent C was maintained at 50% along the chromatographic run). Sodium hydroxide concentration was kept constant during the sodium acetate gradient to maintain baseline stability as acetate has no buffering capacity at high pH. Moreover, change in sodium hydroxide concentration over the run is not recommended as it alters the solution pH, affecting the detector response and, consequently, the baseline stability. The gradient was developed by Silva et al. [34] to separate homologous series of oligosaccharides, such as fructooligosaccharides, maltooligosaccharides, and xylooligosaccharides. This method can be applied to analyze homologous series of oligosaccharides extracted from natural sources, added to foods and beverages, and produced from biotechnological processes. Regarding the method, it is important to note that unlike monosaccharides and very short-chain oligosaccharides, medium- and long-chain oligosaccharides elute only with the sodium acetate eluent because they interact strongly with the column. Moreover, it was necessary to progressively increase the sodium acetate/sodium hydroxide concentration (eluent B) to eluate longer oligosaccharides. In a homologous series of oligosaccharides, as in this example (homologous series of fructooligosaccharides that make up the inulin structure), the retention time of the carbohydrates increases as the degree of polymerization increases. Another important factor to be noted in this analysis is the variability of the detector response with increasing retention time, since the detector response decreases for longer chains. In this way, peak area in HPAEC-PAD analysis does not directly reflect the correspondent carbohydrate amount and, therefore, the analytical curve of a compound in a homologous series of carbohydrates cannot be used to quantify all other

or nitrogen should be done for 10–30 min, depending on the water volume in the bottle.

2. Use the degassed water to prepare the sodium hydroxide and sodium acetate/sodium hydroxide eluents. If necessary, the degassed water can be directly used as a mobile phase (*see* Subheading 2.3).

3. Prepare the sodium hydroxide eluent as follows: Degas the required volume of ultrapure water (as previously described). Add the appropriate weight of 50% (w/w) sodium hydroxide into a plastic volumetric flask containing the degassed ultrapure water to give the required concentration (e.g., to prepare 1 L of 0.2 M sodium hydroxide eluent, either weigh 16 g or pipette 10.4 mL of 50% (w/w) sodium hydroxide solution into a plastic 1 L volumetric flask containing approximately 800 mL degassed deionized water) (*see* **Note 14**). Avoid bubbling air into the eluent when placing the reagent. Stir gently with a stirring bar or manually (15–30 s). Adjust the volume to the required level (1 L) with ultrapure water. Immediately transfer this solution to the respective plastic eluent bottle on the HPAEC-PAD system.

4. Prepare the sodium acetate containing sodium hydroxide eluent as follows: Degas the required volume of ultrapure water (as previously described) and transfer part to a plastic volumetric flask (e.g., 800 mL of 1000 mL). Add the required weight of anhydrous sodium acetate to the flask while it is gently stirring (use a stir bar) (e.g., to prepare 1 L of 1 M sodium acetate containing 0.2 M sodium hydroxide, dilute 82.04 g of anhydrous sodium acetate reagent into a plastic 1 L volumetric flask containing approximately 800 mL degassed deionized water). After the salt dissolves, remove the stir bar with a magnetic retriever. Using a plastic pipette, add the appropriate weight/volume of 50% (w/w) sodium hydroxide to the sodium acetate solution (e.g., prepare 1 L of 1 M sodium acetate containing 0.2 M sodium hydroxide, either weigh 16 g or pipette 10.4 mL of 50% (w/w) sodium hydroxide solution). Stir gently with a stirring bar or manually (15–30 s). Adjust the volume to the required level (1 L) with ultrapure water. Vacuum filter through a 0.2 μm nylon filter (*see*

Fig. 2 (continued) compounds in this series. On this basis, samples rich in long-chain carbohydrates (e.g., inulin HP) should be injected less diluted than samples containing high levels of short-chain carbohydrates (e.g., inulin GR) to obtain better results. After eluting all compounds, the column was regenerated with 0.5 M sodium acetate containing 0.1 M sodium hydroxide (44–49 min) and re-equilibrated with 97 mM sodium hydroxide and 15 mM sodium acetate containing 3 mM sodium hydroxide (49–54 min)

Note 15). Immediately transfer this solution to the respective plastic eluent bottle on the HPAEC-PAD system.

5. Once eluents have been prepared, immediately connect all bottles containing them to the HPAEC-PAD system and keep them blanketed under nitrogen at all times (*see* **Note 16**).

3.4 Anion-Exchange Chromatography System Preparation

This section presents a general protocol for operating the HPAEC-PAD system and some tips. General protocols for analyzing mono- and oligosaccharides are also discussed (*see* Figs. 1 and 2). We recommend always reading the equipment manual.

1. Connect the eluent bottles to the HPAEC-PAD system.
2. If necessary, blanket the bottles with the inert gas employed in the system (e.g., nitrogen).
3. Remove bubbles from the system by purging the filled bottles from the weakest to the strongest eluent (ultrapure water, sodium hydroxide, and sodium acetate/sodium hydroxide). Before starting the bottle purge, loosen the pump valve to avoid overpressure. Close the pump valve when purging all bottles is complete.
4. Remove bubbles from the injection system by purging the bottles (*see* **Note 17**).
5. Set the electrochemical detector parameters for the required analysis as shown in Table 2 (*see* **Note 5**).
6. Set the oven temperature (30 °C) and flow rate (1.0 mL/min) (*see* **Note 18**).
7. Clean the HPAEC-PAD system with ultrapure water (*see* **Note 19**). The baseline stability indicates the appropriate cleaning time.
8. Connect the column to the chromatography system.
9. Clean the column with the appropriate solvent for at least 20 min (*see* **Note 20**). The baseline stability indicates the appropriate cleaning time.
10. After cleaning, the column should be re-equilibrated with the mobile phase, which will start the elution program for approximately 10 min or until the baseline stabilizes.
11. Set the equipment with the desired elution program (*see* **Note 21**).
12. Inject 25 μL of blank (ultrapure water) to check the baseline.
13. Inject 25 μL of a standard mix (containing the target compounds) to check the carbohydrate separation. If necessary, the analyst could adapt an established elution program or develop one to obtain a good peak separation and resolution.

14. Inject 25 μL of each calibration point of the analytical curve in triplicate (*see* Subheading 3.2) (start from the lowest concentration to the highest one). It is recommended to inject a blank every 3–5 runs to check baseline and column stability.
15. Integrate the peaks and construct the analytical curves, plotting the peak area against standard concentrations.
16. Inject 25 μL of the diluted samples as previously discussed in Subheading 3.1.
17. Adjust the dilution factor of the sample to the range of the analytical curve and inject the samples appropriately diluted, at least in triplicate.
18. Integrate the peaks adequately and attempt to identify if peak displacement occurred (*see* **Note 1**).
19. Calculate and express the data according to the study goal (mg/g, mg/mL, g/L, etc.).
20. After finishing the analysis, clean the column with the appropriate eluent (*see* **Note 20**) and wash it with the storage eluent (*see* **Note 22**).
21. Remove the column from the system to avoid the formation of precipitate in the column inlet/outlet. These precipitates can clog up the column and decrease its lifetime.
22. If the equipment is not used, clean the system with ultrapure water and turn it off.
23. The eluent bottles can remain attached to the system if they remain closed and blanketed with inert gas (nitrogen or helium) (*see* **Note 15**).
24. Dispose of waste materials according to the waste disposal regulations (*see* **Note 23**).

4 Notes

1. The displacement of chromatogram peaks may occur because of the matrix effect (food and products from bioprocesses show a diversity of components, which can affect the chromatography analysis). The matrix effect may occur even though the interferents are removed (e.g., proteins, phenolic compounds, and metals). To overcome this limitation, the target sample can be spiked with analytical standards and injected into the column to prove the peak identity.
2. Mobile phases must be prepared with ultrapure deionized water. The use of distilled water can significantly affect the equipment performance and, consequently, the quality of the analysis, especially the column lifetime, the electrochemical

detector signal intensity, and the eluting power of the solvents. For example, carbon dioxide dissolved in water becomes carbonate in alkaline conditions (pH > 12) from all HPAEC-PAD eluents. This divalent anion is a stronger eluent than hydroxide; thus, excessive amounts can cause a loss of carbohydrate retention and, consequently, reduce resolution. Another problem related to low-purity water is the presence of metallic impurities that, at alkaline pH, can precipitate in the system in the form of metal hydroxides, compromising the quality of the analysis or even clogging up the column inlet. Therefore, ultrapure deionized water is the recommended solvent for HPAEC-PAD. We recommend the use of ultrapure deionized water obtained from a Milli-Q water purification system (Millipore, Bedford, USA) since ultrapure water obtained from this system meets the minimum purity requirements for analysis by HPAEC-PAD (resistivity ≥18 MΩ cm^{-1} at 25 °C). We recommend reading Thermo Fisher Scientific Dionex Technical Notes 20 and 71 [15, 36].

3. The 50% (w/w) sodium hydroxide solution has a low sodium carbonate content. Still, this solution should be carefully handled to avoid that sodium carbonate is formed and contaminate the chromatography system, especially the column. Jeff Rohrer showed the following tips in the Thermo Fisher Scientific Dionex Technical Note 20 [15]: (1) Store the 50% (w/w) sodium hydroxide close to where the eluent will be prepared to avoid mixing the sodium carbonate precipitate into the solution. Do not shake or stir this solution. (2) Never pour the solution from the bottle. (3) Pipette sodium hydroxide from the center of the solution, not from the edges or bottom. Do not allow the pipette to stir the solution at the bottom. (4) Use only plastic pipettes, as sodium hydroxide leaches borate and silicate from glass. Borate will be complex with carbohydrates and will thus alter its chromatographic behavior. (5) Never return unused liquid to the bottle. (6) Close the bottle immediately after each use and leave it open for the shortest time possible to avoid carbon dioxide absorption. (7) Discard the 50% (w/w) sodium hydroxide bottle when 2 to 3 cm or less of the solution remains.

4. Carbohydrate chromatographic separations can be performed using only the analytical columns. However, we strongly recommend coupling a guard column in front of the analytical column to prevent sample contaminants from eluting onto the analytical column and prolonging its lifetime. Also, it is less expensive to clean or replace a guard column than an analytical column.

5. Different detector configurations are available for carbohydrate analysis by PAD [18]. However, Waveform A (parameters

shown in Table 2) is the most recommended since it provides the best analysis reproducibility and minimizes wear electrode. For other detector configurations for carbohydrate analysis via HPAEC-PAD, please consult Thermo Fisher Scientific Dionex Technical Note 21 [18].

6. We recommend that carbohydrate extraction be carried out with ultrapure water instead of in combination with organic solvents (e.g., hydroethanolic solutions). Carbohydrate extraction with solutions containing organic solvents leads to more steps in sample preparation (e.g., removing the organic solvent is mandatory before injection into the HPAEC-PAD system; *see* Subheading 2.1) while not significantly increasing extraction yields [22, 23]. Moreover, if target carbohydrates are long-chain oligosaccharides or polysaccharides, they can be partially precipitated in the extraction solution containing organic solvents, interfering in the analysis (particularly in the compound quantification).
7. We recommend this sample preparation procedure, especially for solid, semi-solid, and viscous liquid samples. Liquid samples (e.g., fruit juices and wines) and purified carbohydrates (e.g., concentrated syrups and high-purity carbohydrate powders such as inulin, maltodextrins, and commercially available oligosaccharides (fructooligosaccharides, xylooligosaccharides, galactooligosaccharides, etc.) can only be diluted and filtered before injection into the HPAEC-PAD system.
8. The samples and calibration points must be prepared in ultrapure water instead of the mobile phase starting the chromatographic run. Carbohydrates can undergo several reactions at high pH (e.g., epimerization, keto-enol tautomerization, de-acetylation of N-acetylated sugars, and β-elimination or peeling of 3-O-substituents on reducing sugars) that can potentially interfere with chromatography [15]. Although, in most cases, these reactions occur slowly at room temperature, long waiting times in the autosampler may compromise the quality of analytical curves (loss of linearity and peak area variation for the same standard concentration) and sample analysis (under-quantification of target carbohydrates and appearance of breakdown products) when prepared in alkaline solvents.
9. For colored samples, if the color remains intense even after the appropriate dilution, it must be purified before injection into the HPAEC-PAD system using specific filter cartridges. The choice of the filter cartridge will depend on the type of colored compound in the sample. For example, anthocyanin-rich samples can be purified by using polyvinylpyrrolidone (PVP) filter cartridges. Samples rich in other interferents, such as proteins,

phenolic compounds, and metals, must also be purified. Samples containing high levels of proteins or peptides can be purified by passing the sample solution through a hydrophobic filter cartridge (e.g., Thermo Scientific™ Dionex™ OnGuard™ II RP Cartridges). Samples containing high levels of humic acids or phenolics can be purified by passing the sample solution through a polyvinylpyrrolidone (PVP) filter cartridge (e.g., Thermo Scientific™ Dionex™ OnGuard™ II P Cartridges). Samples containing high levels of metals can be purified by passing the sample solution through an iminodiacetate filter cartridge (e.g., Thermo Scientific™ Dionex™ OnGuard™ II M Cartridges).

10. Filters composed of hydrophilic materials (e.g., polyvinylidene fluoride (PVDF) and polytetrafluoroethylene (PTFE)) can be used to filter aqueous samples. Cellulose acetate and other filters should be avoided because they may leach carbohydrates.

11. We recommend constructing the analytical curves using 0.25–12.50 μg/mL concentration range. However, the analyst may work in other concentration ranges depending on the concentrations of target analytes in the sample and/or detector sensitivity as long as the relationship between peak areas and analyte concentrations remains linear. However, we recommend using calibration concentrations that are as low as possible. This will allow the analyst to work with very diluted samples, enhancing compound separation and extending column and detector lifetime.

12. Stock solutions can be aliquoted and stored in a freezer (temperatures less than −18 °C) for over a year. We advise that the aliquots be made in sufficient volume for a single analysis, as once the aliquot is thawed, it must be used, and the rest discarded.

13. Working standard solutions and calibration points must always be freshly prepared before each analysis. Solutions at low concentrations, such as working standard solutions and calibration points, can be degraded during frozen storage, reducing analysis accuracy.

14. The sodium hydroxide eluent can be prepared by either weight or volume. We have used a plastic volumetric pipette to add the 50% (w/w) sodium hydroxide weight because it is more precise and convenient. Volume is less precise but more effective in preventing contamination than weighing. If the analyst decides to prepare the eluent by volume, they should rinse the pipette by drawing up the solution and dispensing it back into the bottle/cylinder (repeat it two- to three-fold). Minimize the time the solution is exposed to air to avoid carbon dioxide contamination.

15. The sodium acetate/sodium hydroxide eluent must be blanketed under helium or nitrogen during the storage (1–2 weeks). Indeed, it is a very important blanket for all eluents employed in the HPAEC-PAD.
16. Sodium hydroxide and sodium acetate/sodium hydroxide eluents in eluent bottles should be blanketed with helium or nitrogen immediately after preparation and during storage to avoid carbon dioxide contamination from the air (sodium hydroxide eluents may readily absorb carbon dioxide and produce carbonate at pH $\geq$ 12) [36]. Although carbonate cannot be eliminated from manually prepared eluents, it can be minimized by adopting this procedure.
17. Ultrapure water must be used in the injection bottles. They are responsible for washing the injection system and preparing samples to inject into the column. They should be degassed as previously presented (*see* Subheading 3.3).
18. Carbohydrate chromatographic separations can be carried out at room temperature. However, to ensure better analysis reproducibility, especially the retention time, it is recommended to set the oven temperature.
19. It is always recommended that the HPAEC-PAD system be cleaned with ultrapure water before starting the analyses to remove precipitates that can have formed (especially in a long-time stopped system). These precipitates can increase system pressure, cause the appearance of ghost peaks, interfere with the separation of target analytes, make compound retention time unstable, affect detector sensitivity, and clog up the column inlet. This procedure must be carried out before connecting the column to the system.
20. We recommend using 0.2 M sodium hydroxide for cleaning the CarboPac PA1 column and 0.5 M sodium acetate containing 0.1 M sodium hydroxide for cleaning the CarboPac PA100 column. However, if the column requires more severe cleaning, more concentrated solutions can be used (up to 1 M).
21. The choice of elution program will depend on several aspects related to the target carbohydrates (e.g., carbohydrate class, number of carbohydrates to be separated, degree of polymerization, and presence of charged groups). Figures 1 and 2 show different elution programs for distinct purposes, namely, mono- and oligosaccharides separation [27] and fructooligosaccharides separation [24]. Elution programs for purposes other than those presented here can be found in Pereira et al. [19], Rodrigues et al. [21], Barbosa et al. [30], and Silva et al. [33].
22. CarboPac PA1 column must be stored in 0.2 M sodium hydroxide, while 0.5 M sodium acetate containing 0.1 M

sodium hydroxide must be used as a storage solution for CarboPac PA100 column.

23. The mixture of eluents from the chromatographic runs is the main waste generated by the carbohydrate analysis by HPAEC-PAD. This waste must be disposed of after neutralizing it with some acid type.

Acknowledgments

Henrique Silvano Arruda thanks Coordenação de Aperfeiçoamento de Pessoal de Nível Superior (CAPES, grant number 88887.469390/2019-00) for his postdoctoral assistantship. Gustavo Araujo Pereira would like to thank the Federal University of Pará (UFPA).

References

1. Rahkila J, Saloranta T, Leino R (2015) Oligosaccharides for pharmaceutical applications. In: Murzin D, Simakova O (eds) Biomass sugars for non-fuel applications. The Royal Society of Chemistry, Cambridge, pp 205–227
2. Lemus-Mondaca R, Vega-Gálvez A, Zura-Bravo L, Kong AH (2012) *Stevia rebaudiana* Bertoni, source of a high-potency natural sweetener: a comprehensive review on the biochemical, nutritional and functional aspects. Food Chem 132:1121–1132. https://doi.org/10.1016/j.foodchem.2011.11.140
3. Arruda HS, Pereira GA, Almeida MEF, Neri-Numa IA, Sancho RAS, Molina G, Pastore GM (2017) Current knowledge and future perspectives of oligosaccharides research. In: Atta-ur-Rahman FRS (ed) Frontiers in natural product chemistry, vol 3. Bentham Science Publishers, Sharjah, pp 91–175
4. Sievenpiper JL (2020) Low-carbohydrate diets and cardiometabolic health: the importance of carbohydrate quality over quantity. Nutr Rev 78:69–77. https://doi.org/10.1093/nutrit/nuz082
5. Hardy MR, Rohrer JS (2007) High-pH Anion-Exchange Chromatography (HPAEC) and Pulsed Amperometric Detection (PAD) for carbohydrate analysis. In: Kamerling H, Boons G-J, Lee YC, Suzuki A, Taniguchi N, Voragen AGJ (eds) Comprehensive glycoscience: from chemistry to systems biology. Elsevier, Amsterdam, pp 303–327
6. Guimarães JT, Esmerino EA, Freitas MQ, Arruda HS, Pastore GM, Fernandes LM, Silva MC, Cruz AG, Pimentel TC, Balthazar CF (2021) Prebiotics. In: Toldrá F, Nollet LML (eds) Handbook of dairy foods analysis, 2nd edn. CRC Press, Boca Raton, pp 505–520
7. AOAC (1995) AOAC official method 985.29. Total dietary in foods - enzymatic-gravimetric method. In: AOAC AOAC official methods of analysis, 16th ed. AOAC International, Gaithersburg
8. Sun Z, Zou W, Huang J, Su Z, Bai Y (2019) The triple-wavelength overlapping resonance Rayleigh scattering method and the fluorescence quenching method for the determination of chitooligosaccharides using trisodium-8-hydroxypyrene-1,3,6-trisulfonate as a probe. Spectrochim Acta - Part A Mol Biomol Spectrosc 220:117100. https://doi.org/10.1016/j.saa.2019.05.005
9. Correia DM, Dias LG, Veloso ACA, Dias T, Rocha I, Rodrigues LR, Peres AM (2014) Dietary sugars analysis: quantification of fructooligossacharides during fermentation by HPLC-RI method. Front Nutr 1:11. https://doi.org/10.3389/fnut.2014.00011
10. Corradini C, Cavazza A, Bignardi C (2012) High-performance anion-exchange chromatography coupled with pulsed electrochemical detection as a powerful tool to evaluate carbohydrates of food interest: principles and applications. Int J Carbohydr Chem 2012:487564. https://doi.org/10.1155/2012/487564
11. Rohrer JS, Basumallick L, Hurum D (2013) High-performance anion-exchange chromatography with pulsed amperometric detection for carbohydrate analysis of glycoproteins.

Biochemist 78:697–709. https://doi.org/10.1134/S000629791307002X

12. Odjo S, Béra F, Jacquet N, Richel A, Malumba P (2016) Characterization of saccharides released during an *in vitro* pepsin-pancreatin digestion of corn flour using HPAEC-PAD. Starch - Stärke 68:691–699. https://doi.org/10.1002/star.201500281
13. Cataldi TRI, Campa C, De Benedetto GE (2000) Carbohydrate analysis by high-performance anion-exchange chromatography with pulsed amperometric detection: the potential is still growing. Fresenius J Anal Chem 368:739–758. https://doi.org/10.1007/s002160000588
14. Corradini C, Lantano C, Cavazza A (2013) Innovative analytical tools to characterize prebiotic carbohydrates of functional food interest. Anal Bioanal Chem 405:4591–4605. https://doi.org/10.1007/s00216-013-6731-6
15. Rohrer J (2013) Analysis of carbohydrates by High Performance Anion Exchange Chromatography with Pulsed Amperometric Detection (HPAE-PAD). In: Thermo Fisher Scientific Dionex technical note 20. Thermo Fisher Scientific, Sunnyvale, pp 1–12
16. Lee YC (1990) High-performance anion-exchange chromatography for carbohydrate analysis. Anal Biochem 189:151–162. https://doi.org/10.1016/0003-2697(90)90099-U
17. Bruggink C (2012) Oligosaccharide analysis by High-Performance Anion-Exchange Chromatography hyphenated to integrated pulsed amperometric detection and on-line ion-trap mass spectrometry. In: Bhattacharyya L, Rohrer JS (eds) Applications of ion chromatography for pharmaceutical and biological products. Wiley, Hoboken, pp 379–391
18. Rohrer J (2013) Optimal settings for Pulsed Amperometric Detection of carbohydrates using the Dionex ED40 electrochemical detector. In: Thermo Fisher Scientific Dionex technical note 21. Thermo Fisher Scientific, Sunnyvale, pp 1–4
19. Pereira GA, Arruda HS, de Morais DR, Eberlin MN, Pastore GM (2018) Carbohydrates, volatile and phenolic compounds composition, and antioxidant activity of calabura (*Muntingia calabura* L.) fruit. Food Res Int 108:264–273. https://doi.org/10.1016/j.foodres.2018.03.046
20. Pereira GA, Arruda HS, de Morais DR, Peixoto Araujo NM, Pastore GM (2020) Mutamba (*Guazuma ulmifolia* Lam.) fruit as a novel source of dietary fibre and phenolic compounds. Food Chem 310:125857. https://doi.org/10.1016/j.foodchem.2019.125857
21. Rodrigues AP, Pereira GA, Tomé PHF, Arruda HS, Eberlin MN, Pastore GM (2019) Chemical composition and antioxidant activity of monguba (*Pachira aquatica*) seeds. Food Res Int 121:880–887. https://doi.org/10.1016/j.foodres.2019.01.014
22. Pereira GA, Arruda HS, Molina G, Pastore GM (2018) Extraction optimization and profile analysis of oligosaccharides in banana pulp and peel. J Food Process Preserv 42:e13408. https://doi.org/10.1111/jfpp.13408
23. Arruda HS, Pereira GA, Pastore GM (2017) Oligosaccharide profile in Brazilian Cerrado fruit araticum (*Annona crassiflora* Mart.). LWT 76:278–283. https://doi.org/10.1016/j.lwt.2016.05.017
24. Guimarães JT, Silva EK, Arruda HS, Freitas MQ, Pastore GM, Meireles MAA, Cruz AG (2020) How does the degree of inulin polymerization affect the bioaccessibility of bioactive compounds from soursop whey beverage during *in vitro* gastrointestinal digestion? Food Hydrocoll 101:105511. https://doi.org/10.1016/j.foodhyd.2019.105511
25. Emerick LB, Silva EK, Arruda HS, Pastore GM, de Meireles MAA (2019) Non-thermal processing of inulin-enriched araticum juice using supercritical technology. Rev dos Trab Iniciação Científica da UNICAMP. https://doi.org/10.20396/revpibic262018872
26. de Bezerra YM, Silva EK, Strieder MM, Pastore GM, Meireles MAA, Arruda HS (2019) Processing of inulin-enriched apple juice using high-intensity ultrasound technology. Rev dos Trab Iniciação Científica da UNICAMP. https://doi.org/10.20396/revpibic2620181390
27. Pereira GA, Silva EK, Peixoto Araujo NM, Arruda HS, Meireles MAA, Pastore GM (2019) Obtaining a novel mucilage from mutamba seeds exploring different high-intensity ultrasound process conditions. Ultrason Sonochem 55:332–340. https://doi.org/10.1016/j.ultsonch.2019.01.010
28. Arruda HS, Silva EK, Pereira GA, Meireles MAA, Pastore GM (2020) Inulin thermal stability in prebiotic carbohydrate-enriched araticum whey beverage. LWT 128:109418. https://doi.org/10.1016/j.lwt.2020.109418
29. Geraldi MV, Betim Cazarin CB, Dias-Audibert FL, Pereira GA, Carvalho GG, Kabuki DY, Catharino RR, Pastore GM, Behrens JH, Cristianini M, Maróstica Júnior MR (2021) Influence of high isostatic pressure and thermal pasteurization on chemical composition, color,

antioxidant properties and sensory evaluation of jabuticaba juice. LWT 139:110548. https://doi.org/10.1016/j.lwt.2020.110548

30. Barbosa FC, Kendrick E, Brenelli LB, Arruda HS, Pastore GM, Rabelo SC, Damasio A, Franco TT, Leak D, Goldbeck R (2020) Optimization of cello-oligosaccharides production by enzymatic hydrolysis of hydrothermally pretreated sugarcane straw using cellulolytic and oxidative enzymes. Biomass Bioenergy 141: 105697. https://doi.org/10.1016/j.biombioe.2020.105697
31. Neri-Numa IA, Pessôa MG, Arruda HS, Pereira GA, Paulino BN, Angolini CFF, Ruiz ALTG, Pastore GM (2020) Genipap (*Genipa americana* L.) fruit extract as a source of antioxidant and antiproliferative iridoids. Food Res Int 134:109252. https://doi.org/10.1016/j.foodres.2020.109252
32. Silva EK, Bargas MA, Arruda HS, Vardanega R, Pastore GM, Meireles MAA (2020) Supercritical CO_2 processing of a functional beverage containing apple juice and aqueous extract of *Pfaffia glomerata* roots: Fructooligosaccharides chemical stability after non-thermal and thermal treatments. Molecules 25:3911. https://doi.org/10.3390/molecules25173911
33. Silva EK, Arruda HS, Pastore GM, Meireles MAA, Saldaña MDA (2020) Xylooligosaccharides chemical stability after high-intensity ultrasound processing of prebiotic orange juice. Ultrason Sonochem 63:104942. https://doi.org/10.1016/j.ultsonch.2019.104942
34. Silva EK, Arruda HS, Eberlin MN, Pastore GM, Meireles MAA (2019) Effects of supercritical carbon dioxide and thermal treatment on the inulin chemical stability and functional properties of prebiotic-enriched apple juice. Food Res Int 125:108561. https://doi.org/10.1016/J.FOODRES.2019.108561
35. Magnusson B, Örnemark U (2014) Eurachem guide: the fitness for purpose of analytical methods - a laboratory guide to method validation and related topics, 2nd edn. EURACHEM Working Group, Teddington
36. Rohrer J (2017) Eluent preparation for High-Performance Anion-Exchange Chromatography with Pulsed Amperometric Detection. In: Thermo Fisher Scientific Dionex technical note 71. Thermo Fisher Scientific, Sunnyvale, pp 1–6

Part V

Liposoluble Vitamins

Chapter 16

Analysis of Vitamin D and Its Respective Vitamers by Liquid Chromatography with Triple Quadrupole Mass Spectrometry

Cuicui Wang and Frank Vriesekoop

Abstract

Vitamin D is essential for maintaining a healthy mineral and bone metabolism. Currently, six forms of vitamin D are known. They range from vitamin D_2 to vitamin D_7 and are distinguished by their differing side chains. Only vitamin D_2 and vitamin D_3 have been found to be biologically active based on the elucidation of activation pathways. Deficiencies in Vitamin D cause rickets and many other health problems, including reduced muscle function, autoimmune diseases, cardiovascular disease, and some cancers. The following chapter describes a simple and reproducible method for the analysis of Vitamin D via liquid chromatography with triple quadrupole mass spectrometry.

Key words Vitamin, High-performance liquid chromatography, Mass spectrometry

1 Introduction

Compared with the other known vitamins essential to health, vitamin D is unique in its synthesis, because vitamin D is converted from the corresponding pre-vitamin by non-biological means, namely, ultraviolet radiation. Ergocalciferol (vitamin D_2) is generated from ergosterol by exposure to UV irradiation. Ergosterol is a steroid found in some plants, yeast, and other fungi [1]. Pre-vitamin D_3 is produced from 7-dehydrocholesterol via UV irradiation at UVB wavelengths of 290–320 nm and is then further thermally isomerized to form cholecalciferol (vitamin D_3) [2].

As a lipid-soluble vitamin, Vitamin D is a steroid derivative. There are several known forms of vitamin D, namely, D_2, D_3, D_4, D_5, D_6, and D_7, which vary from each other according to the structure of its side chain (Fig. 1). Only D_2 (ergocalciferol) and D_3 (cholecalciferol) are the active forms that play a role in animal nutrition. The hydroxylated metabolite of vitamin D_3, 25(OH)D_3, likely has a greater biological activity than vitamin D_3 itself [3, 4]. There is some evidence that vitamin D_3 taken orally is

Lilian Mariutti and Frank Vriesekoop (eds.), *Food Chemistry: Basic Protocols*, Methods and Protocols in Food Science,
https://doi.org/10.1007/978-1-0716-4678-6_16,

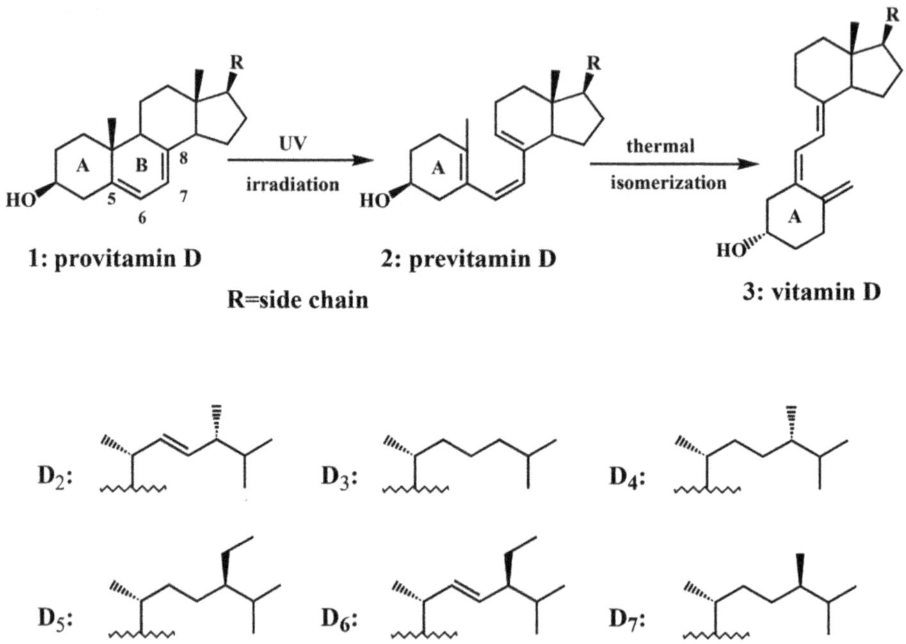

Fig. 1 Biosynthesis of vitamin D_3 from provitamin D_1 via pre-vitamin D_2 and side chain structure of vitamin D_2 through to vitamin D_7

more effective than vitamin D_2 at increasing levels of 25(OH)D [5, 6], as 25(OH)D may be between one and five times more bioactive than vitamin D_2 [4, 7, 8].

Vitamin D is a colorless crystal, soluble in fats, fatty solvents and organic solvents, chemically stable, heat resistant in neutral and alkaline solutions, not easily oxidized, but it gradually decomposes in acidic solutions. Vitamin D solutions are unstable because of dissolved oxygen, and its biological effect is reduced after double bond reduction. Therefore, vitamin D should generally be stored in low temperatures without light, acid, or oxygen.

Vitamin D promotes bone calcification by increasing calcium and phosphorus absorption in the gastrointestinal tract. Lack of vitamin D can lead to calcium absorption and bone mineralization disorders, leading to the occurrence of rickets [9, 10]. Vitamin D deficiency has also been linked to other health conditions besides bone health, including reduced muscle function, cardiovascular disease, autoimmune diseases, and some cancers [11–14].

Methodologies of Analysis

Vitamin D's sensitivity to light makes detection a challenge. The most commonly used method for vitamin D analysis is high-performance liquid chromatography [15, 16], but this method is not sensitive and specific and requires higher quality of pretreatment. The application of two-dimensional chromatography simplifies the preprocessing steps, but two-dimensional chromatography needs more stringent requirements for instruments, therefore, the application scope of two-dimensional chromatography is relatively limited [17]. High-performance liquid chromatography-ultraviolet (HPLC-UV) with mass spectrometry or liquid chromatography-tandem mass spectrometry (LC-MS/MS) has been used to measure vitamin D metabolites directly in the food due to its high sensitivity and specificity [18–26]. Recently, the Australian National Institute of Measurement (NMI) has developed a new sensitive and specific LC-QQQ method that allows accurate quantification of vitamin D_2, vitamin D_3, 25(OH)D_2, and 25(OH)D_3 at low concentrations in a single analytical run, with a throughput speed capable of generating data on a scale suitable for food composition analysis [27, 28].

2 Materials

All equipment and consumables mentioned in this chapter can be substituted by alternative products from other manufacturers. In this chapter, we describe a tried set of methodologies based on the equipment and consumables mentioned.

2.1 Basic Reagents

All reagents/standards mentioned in this chapter assume sourcing through Sigma-Aldrich (St. Louis, MO) and include vitamin D_2 (CAS No. 50-14-6, purity: ≤99%), Vitamin D_3 (CAS No. 67-97-0, purity: ≤99%), d6-vitamin D_2 (CAS No. 1311259–89-8, enrichment: ≤99%, purity: ≤99%), and d6-vitamin D_3 (CAS No. 118584-54-6, enrichment: ≤99%, purity: ≤99%) (*see* **Note 4**). LC-MS grade reagents include formic acid, ethanol, methanol, isooctane, acetone, and acetonitrile. 4-Phenyl-1,2,4-triazoline-3,5-dione (PTAD), potassium hydroxide, and pyrogallol are also used in this research (*see* **Note 3**). All solutions/standards are prepared with 18 MΩ water (Sartorius Arium 611VF water purification system—Sartorius AG, Göttingen, Germany). It is recommended that all glassware used in this analysis is cleaned via sonication and rinsed in 18 MΩ water.

2.2 Instrumentation

The LC-QQQ system described in this chapter consisted of an Agilent 1290 ultra high-pressure liquid chromatograph (Agilent, USA) and AB API 4000 triple quadrupole mass spectrometer (AB Sciex, USA);

Agilent XDB-C18 column (1.8 μm, 2.1 mm × 50 mm); UV spectrophotometer; digital display constant temperature water baths;1 mL disposable syringes; syringe filters (PTFE, 0.2 μm, 13 mm); filter membranes (0.45 μm nylon); centrifuge; 2 mL cryogenic vials; ultrasonic cleaning machine (SB-5200DT,

China); organomation nitrogen evaporators (N-EV AP, USA) (*see* **Note 2**).

The mobile phase gradient to be employed is as follows: Solvent A: formic acid (0.1%, v/v) in deionized water; solvent B: methanol (100%, v/v);

0–2 minutes 20:80 A: B;

2–10 minutes 0:100 A: B;

10–15 minutes 20:80 A: B; at a total run time of 15 minutes.

3 Methods

3.1 Sample Preparation

The weights of sample aliquots (powder sample: 1–2.5 g, Slurry sample: 19.0–21.0 g, liquid samples: 8.0–10.0 ml) should be determined by fat content, ensuring that the amount of saponified fat is ≤1 g. This is to ensure that the sample fat does not exceed the capacity of the saponification liquor, which is limited by the volume of the saponification vessel, in this instance 50 mL [27, 28].

All samples must be prepared under UV-shielded lighting (*see* **Note 1**), stored in amber glassware, and stored at refrigerated conditions. It is also recommended that samples and standards be prepared just prior to analysis but can be stored for up to 72 hours under refrigerated conditions to prevent bacterial growth [29].

3.2 Vitamin D Extraction from Prepared Samples

Vitamin D is sensitive to light; perform all steps under UV-shielded lighting.

- *Saponification*: The saponification mixture will consist of the sample and the chemically labelled internal standard solution, including the stable isotope-labeled (SIL) d6-vitamin D_2 and d6-vitamin D_3 internal standards, dissolved 1 g sodium ascorbate in 10 mL of deionized water, 30 mL ethanol, and 2 g potassium hydroxide. Then, make up to 50 mL with deionized water. The headspace is flushed with nitrogen gas and capped. Vitamin D analytes are then saponified in the ethanolic potassium hydroxide solution at 80 °C for 30 min. After which, the saponified mixture is cooled down to room temperature.
- *Extraction*: The saponified samples undergo centrifugation, and 10 mL of the liquid fraction is extracted onto diatomaceous earth solid phase extraction tubes (Agilent Technologies Chem Elut™ 10 mL unbuffered SPE cartridges [part # 12198007]). The organic soluble compounds will be washed off with two 30 mL aliquots of an organic solvent like petroleum spirits or hexane. The organic fractions are collected into 80 mL glass EPA vials and then evaporated to dryness under high-purity nitrogen gas. The residue is then resolvated into 400 μL heptane

Table 1
MS instrument settings. These settings are suitable for the AB API 4000 triple quadrupole MS. Optimal settings on alternative instruments may differ

Instrument parameter	Value
Ionization mode	ESI+
Curtain gas, psi	30
Nebulizer gas GS1, psi	55
Heater gas GS2, psi	45
Collision gas	N_2
Source temperature, °C	500
Ion spray voltage, V	5500

and transferred to a liquid chromatography vial and evaporated to dryness under nitrogen.

- *Derivatization*: The dried residue is resolvated into a solution of 200uL anhydrous acetonitrile containing 1 mg/mL 4-phenyl-1,2,4-triazoline- 3,5-dione (PTAD) to form high-molecular weight, easily ionizable vitamin D-PTAD derivatives. After 10 min at room temperature, the derivatization reaction is stopped by the addition of 100 μL water. Using a syringe filter, transfer extracts to amber micro-vials for analysis by LC-QQQ [19–30].

3.3 Separation and Analysis of Vitamin D

Form gradients by high-pressure mixing of the two mobile phases, A and B, using the gradient details described in Subheading 2.2. Set up the MS system with the instrument settings shown in Table 1.

The calculation of vitamin content refers to the previous method [29, 30]; the method detail is as follows:

- Concentration of d6-vitamin D_2/d6-vitamin D_3/vitamin D_2/vitamin D_3 in the stock standard (μg/mL) needs to be corrected by UV absorbance.

Take 100 μL d6-vitamin D_2, d6-vitamin D_3, vitamin D_2, and vitamin D_3 stock standard solutions into 10 ml brown volumetric flask, respectively. Add absolute ethanol, constant volume to the scale line and shake, use 1 cm quartz colorimetric cups, take absolute ethanol as blank control, and measure the absorbance. The concentration of d6-vitamin D_2 /d6-vitamin D_3/vitamin D_2/vitamin D_3 in the stock standard (μg/mL) is calculated according to the following formula.

- $$SS = \frac{UVA}{E_{1cm}^{1\%}} \times 10000$$

SS: Respective concentration of d6-vitamin D_2 /d6-vitamin D_3/ vitamin D_2/vitamin D_3 in the stock standard (μg/mL).

UV A: UV absorbance of the stock standard at 265 nm (1/cm);

$E_{1cm}^{1\%}$: extinction coefficient for vitamin D_2 (461 dL/g.cm)/vitamin D_3 (485 dL/g.cm) in ethanol.

10,000: concentration conversion factor (g/dL to μg/mL).

Dilute the stock standard with methanol as required and prepare it into a mixed standard solution with the appropriate concentration.

Making a standard curve:

Determine the linear regression curve y = mx + c, the ratio of peak areas (non-labeled vitamin D/SIL d6-vitamin D) is the vertical coordinate, and the concentration of the ratio of concentrations (non-labeled vitamin D/SIL d6-vitamin D) is the horizontal coordinate.

- The concentration of vitamin D_2 or vitamin D_3 in samples is calculated as follows:
- $$\textbf{Result D} = \frac{\rho \times V \times f \times 100}{m}$$

- Result D: vitamin D_2 or vitamin D_3 concentration in the sample (μg/100 g).
- P: The concentration of vitamin D2(or vitamin D3) in the sample was calculated according to the standard curve (μg/ml).
- V: The volume of the extract.
- *f*: Dilution ratio of the sample solution to be tested.
- 100: The conversion coefficient for the amount per 100 g in the sample.
- m: Weight of sample (g).

- Determine the limit of quantitation (LOQ) and the limit of detection (LOD).

For each vitamer analyzed, each standard and sample should be tested in triplicate, and those triplicate values to be averaged to obtain one mean value for each sample. An additional sample, spiked with the same vitamer, can then be analyzed for each sample tested to provide quality control data. The mean percentage

Table 2
Quantifier and qualifier ions for liquid chromatography with triple quadrupole mass -spectrometry analysis of vitamin D in foods

	Product ions (Da)	
Compound	**Quantifier**	**Qualifiers**
Vitamin D_2–PTAD adduct	572.2–298.0	572.2–280.0
D6-vitamin D_2–PTAD adduct	578.2–298.0	578.2–280.0
Vitamin D_3–PTAD adduct	560.2–298.0	560.2–280.0
D6-vitamin D_3–PTAD adduct	566.2–298.0	566.2–280.0

Table 3
Accuracy, precision, LOQ, LOD, range of linearity, and LOC of the described *vitamin D* methods for sample analysis

Analyte	Accuracy (av. % spike recovery)	Precision (% RSD)	LOQ (μg/ 100 g)	LOD (μg/ 100 g)	Range of linearity (ng/mL)	LOC (R^2 value)
Vitamin D_2	97.6	3.2	0.10	0.01	5–500	0.9999
Vitamin D_3	98.4	3.8	0.10	0.05	5–500	0.9999

recovery and mean relative percentage difference can then be calculated for each vitamer.

Table 2 depicts the compound parameters. Table 3 shows the accuracy, precision LOD, and RSD of this analysis. The described protocol provides an accurate, precise, rapid, and robust method for the analysis of vitamin D that is suitable both for routine compliance testing and as a reference method to demonstrate product compliance against global regulations.

4 Notes

1. Vitamin D is sensitive to light; perform all steps under UV-shielded lighting. It is recommended that all work is carried out away from direct sunlight, utilizing amber or covered glassware for both standards and samples.
2. Precolumn filters are recommended for all samples to prevent excessive column fouling.
3. Use appropriate safety equipment when handling acids, bases and solvents and refer to the material safety data sheet for detailed safety instructions for each chemical.

4. Stable isotope-labeled (SIL) d6-vitamin D_2 and d6-vitamin D_3 internal standards used for quantitation have the advantage of correcting for extraction inefficiencies and potential MS matrix enhancement or suppression effects.

References

1. Calman K (ed) (1998) Nutrition and bone health with particular reference to calcium and vitamin D: report of the subgroup on bone health (working group on the nutritional status of the population) of the committee on medical aspects of food and nutrition policy. The Stationary Office, London
2. Ross CA, Taylor CL, Yaktime AL, Del Valle HB (eds) (2010) Dietary reference intakes for calcium and vitamin. Committee to review dietary reference intakes for vitamin D and calcium; Institute of Medicine. National Academies Press, Washington, DC
3. Cashman K, Seamans K, Lucey A, Stöcklin E, Weber P, Kiely M, Hill T (2012) Relative effectiveness of oral 25-hydroxyvitamin D3 and vitamin D3 in raising wintertime serum 25-hydroxyvitamin D in older adults. Am J Clin Nutr 95:1350–1356
4. Ovesen L, Brot C, Jakobsen J (2003) Food contents and biological activity of 25-hydroxyvitamin D: a vitamin D metabolite to be reckoned with? Ann Nutr Metab 47:107–113
5. Lehmann U, Hirche F, Stangl GI, Hinz K, Westphal S, Dierkes J (2013) Bioavailability of vitamin D(2) and D(3) in healthy volunteers, a randomized placebo-controlled trial. J Clin Endocrinol Metab 98:4339–4345
6. Itkonen ST, Skaffari E, Saaristo P, Saarnio EM, Erkkola M, Jakobsen J, Cashman KD, Lamberg-Allardt C (2016) Effects of vitamin D2-fortified bread v. supplementation with vitamin D2 or D3 on serum 25-hydroxyvitamin D metabolites: an 8-week randomised-controlled trial in young adult Finnish women. Br J Nutr 115:1232–1239
7. Cashman KD, Seamans KM, Lucey AJ, Stocklin E, Weber P, Kiely M, Hill TR (2012) Relative effectiveness of oral 25-hydroxyvitamin D3 and vitamin D3 in raising wintertime serum 25-hydroxyvitamin D in older adults. Am J Clin Nutr 95(6):1350–1356
8. Jakobsen J, Melse-Boonstra A, Rychlik M (2019) Challenges to quantify total vitamin activity: how to combine the contribution of diverse vitamers? Curr Dev Nutr 3(10):nzz086
9. Prentice A (2013) Nutritional rickets around the world. J Steroid Biochem Mol Biol 136: 201–206
10. Uday S, Hogler W (2017) Nutritional rickets and osteomalacia in the twenty-first century: revised concepts, public health, and prevention strategies. Curr Osteoporos Rep 15:293–302
11. Wintermeyer E, Ihle C, Ehnert S, Stockle U, Ochs G, de Zwart P, Flesch I, Bahrs C, Nussler AK (2016) Crucial role of vitamin D in the musculoskeletal system. Nutrients 8:319
12. Lucas RM, Norval M, Neale RE, Young AR, de Gruijl FR, Takizawa Y, van der Leun JC (2015) The consequences for human health of stratospheric ozone depletion in association with other environmental factors. Photochem Photobiol Sci 14:53–87
13. Bais AF, Lucas RM, Bornman JF, Williamson CE, Sulzberger B, Austin AT, Wilson SR, Andrady AL, Bernhard G, McKenzie RL (2018) Environmental effects of ozone depletion, UV radiation and interactions with climate change: UNEP environmental effects assessment panel, update 2017. Photochem Photobiol Sci 17:127–179
14. Binks MJ, Smith-Vaughan HC, Marsh R, Chang AB, Andrews RM (2016) Cord blood vitamin D and the risk of acute lower respiratory infection in indigenous infants in the Northern Territory. Med J Aust 204(6):238
15. Qian H, Sheng M (1998) Simultaneous determination of fat-soluble vitamins A, D and E and pro-vitamin D2 in animal feeds by one-step extraction and high-performance liquid chromatography analysis. J Chromatogr A 825(2):127–133
16. Higashi T, Shibayama Y, Fuji M, Shimada K (2008) Liquid chromatography-tandem mass spectrometric method for the determination of salivary 25-hydroxyvitamin D3: a noninvasive tool for the assessment of vitamin D status. Anal Bioanal Chem 391(1):229–238
17. Ye J, Wang J, Ye L (2020) Rapid determination of vitamin D in milk powder and rice flour by online solid phase extraction and two-dimensional chromatography. Food Fermentat Industr 46(7):239–243

18. Bilodeau L, Dufresne G, Deeks J (2011) Determination of vitamin D3 and 25-hydroxyvitamin D3 in foodstuffs by HPLC UV-DAD and LC-MS/MS. J Food Composit Analys 24(3):441–448
19. Adamec J, Jannasch A, Huang J, Hohman E, Fleet JC, Peacock M, Weaver CM (2011) Development and optimization of an LC-MS/MS-based method for simultaneous quantification of vitamin D2, vitamin D3, 25-hydroxyvitamin D2 and 25-hydroxyvitamin D3. J Sep Sci 34(1):11–20
20. Strobel N, Buddhadasa S, Adorno P, Stockham K, Greenfield H (2013) Vitamin D and 25-hydroxyvitamin D determination in meats by LC–IT-MS. Food Chem 138(2–3): 1042–1047
21. Bilodeau L, Dufresne G, Deeks J, Cl'ement G, Bertrand J, Turcotte S, Fouquet A (2011) Determination of vitamin D3 and 25-hydroxyvitamin D3 in foodstuffs by HPLC UV-DAD and LC-MS/MS. J Food Compos Anal 24(3):441–448
22. Bussche JV, Decloedt A, Van Meulebroek L, De Clercq N, Lock S, Stahl-Zeng J, Vanhaecke L (2014) A novel approach to the quantitative detection of anabolic steroids in bovine muscle tissue by means of a hybrid quadrupole time-of-flight-mass spectrometry instrument. J Chromatogr A 1360:229–239
23. Svarc PL, Barnkob LL, Jakobsen J (2021) Quantification of vitamin D3 and 25-hydroxyvitamin D3 in food – the impact of eluent additives and labelled internal standards on matrix effects in LC-MS/MS analysis. Food Chem 357:129588
24. Jakobsen J, Clausen I, Leth T, Ovesen L (2004) A new method for the determination of vitamin D3 and 25-hydroxyvitamin D3 in meat. J Food Compos Anal 17:777–787
25. Trenerry VC, Plozza T, Caridi D, Murphy S (2011) The determination of vitamin D3 in bovine milk by liquid chromatography mass spectrometry. Food Chem 125(4):1314–1319
26. Phillips KM, Craig Byrdwell W, Exler J, Harnly JM, Holden JM, Holick MF, Wolf WR (2008) Development and validation of control materials for the measurement of vitamin D3 in selected US foods. J Food Compos Anal 21(7):527–534
27. Dunlop E, James AP, Cunningham J, Strobel N, Lucas RM, Kiely M, Black LJ (2021) Vitamin D composition of Australian foods. Food Chem 358:e129836
28. Dunlop E, Shepherd C, Cunningham J, Strobel N, Lucas RM, Black LJ (2022) Vitamin D composition of Australian game products. Food Chem 387:132965
29. Gill BD, Indyk HE (2018) Analysis of vitamin D_2 and vitamin D_3 in infant and adult nutritional formulas by liquid chromatography-tandem mass spectrometry: a multilaboratory testing study. J AOAC Int 101(1):256–263
30. Hughes LJ, Black LJ, Sherriff JL, Dunlop E, Strobel N, Lucas RM, Bornman JF (2018) Vitamin D content of Australian native food plants and Australian-grown edible seaweed. Nutrients 10(7):876

Chapter 17

Tocopherol Analysis in Animal Tissues

Ligia Manoel Martins, Lilian Regina Barros Mariutti, and Neura Bragagnolo

Abstract

This chapter describes the analysis of tocopherols in edible animal tissues undergoing saponification. First, a survey of the main techniques to extract tocopherols is presented, and their advantages and disadvantages are discussed. Next, we explain the steps involving saponification followed by solvent extraction. Finally, the method is described with all the steps for quantification, calibration curve construction, extraction, and analysis using RP-HPLC with a fluorescence detector. Tocopherol analysis is delicate and requires attention and care at various steps, starting with sample preparation. Thus, a good understanding of the steps and critical points of the analysis guarantees a more representative and reliable result.

Key words Vitamin E, Tocotrienols, Extraction, Saponification, RP-HPLC-FLD

1 Introduction

Vitamin E was first discovered in 1922 by Evans and Bishop and was considered an essential factor for reproduction in rats. Firstly called "factor X," it was named vitamin E in 1923 by Sure, as a sequence of vitamins A, B, C, and D previously described [1, 2]. Vitamin E is a fat-soluble vitamin comprising eight compounds derived from tocol (2-methyl-2-(4,8,12-trimethyltridecyl) chroman-6-ol), including tocopherols (from the Greek *Tokos*, which means "childbirth," and *Phero*, "to bear") [3] and tocotrienols. The tocopherol structure consists of a polar chromanol ring with 16 carbons saturated phytyl side chain, while the tocotrienols have an unsaturated side chain with three *trans* double bonds at positions 3′, 7′, and 11′. Tocopherols and tocotrienols comprise all mono-, di-, and trimethyl tocols. Four isomers, alpha (α), beta (β), gamma (γ), and delta (δ), differ in the substitution of the methyl radical on the chromanol ring (Fig. 1) [4].

Vitamin E has antioxidant activity, and α-tocopherol is the most biologically active form. The biological activity of the other isomers

Lilian Mariutti and Frank Vriesekoop (eds.), *Food Chemistry: Basic Protocols*, Methods and Protocols in Food Science, https://doi.org/10.1007/978-1-0716-4678-6_17,

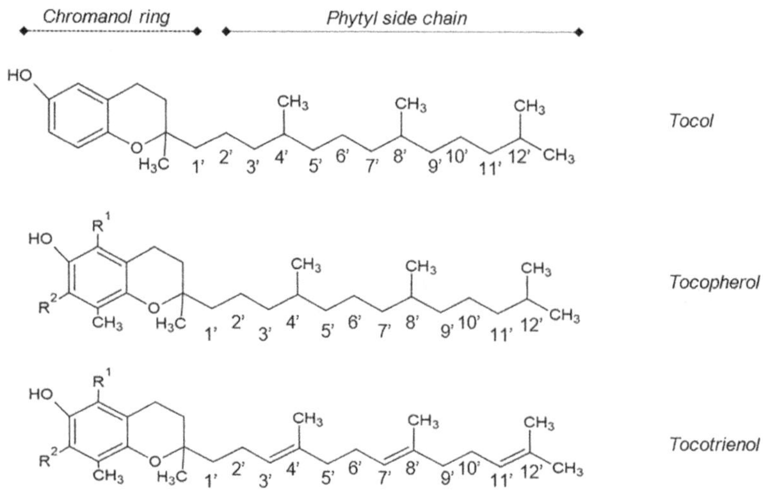

	R1	R2
α	CH_3	CH_3
β	CH_3	H
γ	H	CH_3
δ	H	H

Fig. 1 Tocol, tocopherol, and tocotrienol structures

is usually expressed as α-tocopherol equivalents (%), of which β-, γ-, and δ-tocopherol account for 50%, 10%, and 3%, respectively, while α-, β-, and γ-tocotrienol account for 30%, 5%, and 8%, respectively, of α-tocopherol equivalents [5].

Photosynthetic organisms produce tocopherols, and the main isomers found in plants are α- and γ-tocopherol. Despite not producing tocopherols, animals can store tocopherols obtained through the diet in the muscles [6]. Adding tocopherols in feed, usually as acetate ester (tocopheryl acetate), increased the lipid stability of meat, reducing its oxidation [7, 8]. Although the tocopheryl acetate does not present antioxidant activity because the hydroxyl group in C6 is esterified, it is hydrolyzed during digestion. Thus, the intestinal mucosa can absorb the released tocopherol [9]. Furthermore, its absorption is influenced by digestion and dietary fat absorption; medium-chain triglycerides increase the absorption, while polyunsaturated fatty acids inhibit it [10]. Several studies have already evidenced and quantified tocopherols in edible portions of animal tissue, such as chicken [11–14], pork [15, 16], and fish meat [17, 18]. The tocopherol levels in meat samples vary widely, depending on animal feed and even from study to study.

Vitamin E analysis in food begins with sampling, which must be representative and follow standardized methods. Sample handling must prevent exposure to light, prolonged contact with oxygen, and high temperatures. Thus, the samples must be kept protected from light, stored at low temperatures (freezer), and under vacuum or modified atmosphere (e.g., nitrogen) to avoid vitamin degradation [19]. Another precaution is to carry out the sample processing on the analysis day and reduce the sample particle size for better extraction. The influence of sample particle size on vitamin E extraction was evidenced in traditional and alternative extraction methods with higher yield in powdered samples than in whole samples [20, 21].

Among the extraction methods described for tocopherols and tocotrienols, direct solvent extraction, saponification, and Soxhlet extraction are the most widely used in different food samples, such as grains, oilseeds, and tissues [22]. Nevertheless, alternative approaches have also been developed, for instance, ultrasonic-assisted extraction (UAE) [23, 24], extraction with matrix solid-phase dispersion (MSPD) [25–27], supercritical fluid extraction (SCFE) [28–30], pressurized liquid extraction (PLE) [31–33], and supercritical carbon dioxide (SC-CO_2) [20, 21, 34].

Vitamin E is a fat-soluble compound extracted with organic solvents. Thus, for foods with a high vitamin E concentration, the direct extraction with apolar solvents (such as hexane, petroleum ether, and chloroform) or solvent mixtures in different proportions is a reliable and straightforward procedure. However, when the sample has high lipid and low vitamin E concentration, direct solvent extraction may not be effective, so saponification is indicated in this case [35, 36]. Saponification is an alkaline hydrolysis, usually using KOH to improve extraction from complex food matrices, releasing tocopherol from interactions with proteins, carbohydrates, and triacylglycerols [36]. The vitamin E released in saponification is then extracted with an organic solvent, such as hexane.

Alternative methods (UAE, MSPD, SCFE, PLE, and SC–CO_2) have been optimized and shown effectiveness, reducing cost, sample size, and solvent volume and mainly use environmentally friendly solvents (e.g., SCFE). Furthermore, these methods allow simultaneous sample analysis and greater extraction efficiency. However, some limitations can be cited, such as the sample nature dependence for adequate extraction by UAE, the high cost of the equipment used in SCFE, and the time consumption for MSPD analysis [22]. Thus, the extraction method should be chosen considering the specific food matrices studied and the available instrumental laboratory conditions.

Chromatographic methods are the most widely used method for vitamin E identification and quantification. Gas chromatography (GC) has been used since early 1960, with flame ionization detection (FID). However, the compounds require derivatization with trimethylsilyl before analysis, and since vitamin E is temperature-sensitive, GC can cause losses. Therefore, high-performance liquid chromatography (HPLC) is the most used method nowadays, and either normal (NP) or reversed phase (RP) can be applied. The NP-HPLC consists of a polar stationary phase and a non-polar mobile phase. The affinity between the hydroxyl group of the chromanol ring with the silica in the stationary phase allows the separation of the four isomers (α, β, γ, and δ), responsible for this good method selectivity. By using the non-polar mobile phase, oil samples can be directly analyzed.

On the other hand, RP-HPLC uses a non-polar stationary phase and a polar mobile phase. This method results in faster analysis, less toxic solvents, and greater reproducibility. However, β- and γ- isomers are generally not separated when a C18 column (stationary phase most used in reversed phase) is used because both have identical substituents on the chromanol ring, varying only the position. In the meantime, applying columns with a longer carbon chain (such as C30) showed good selectivity, allowing the separation of these isomers [37, 38]. Therefore, depending on the analysis goal, identifying all the isomers or not, the HPLC method (reversed or normal phase) is chosen. The main detectors used are ultraviolet (UV) and fluorescence (FLD) since the tocopherol and tocotrienol absorbance in UV light ranges from 290 to 300 nm [4]. The FLD detector features better selectivity and sensitivity.

2 Material

2.1 Apparatus

1. Analytical balance—must be capable of measuring four decimal places to the right of the decimal point.
2. UV spectrophotometer—capable of performing precision measurements in the absorbance range required for tocopherols (292–298 nm).
3. Ultra-turrax.
4. Vortex-mixer.
5. Shaker.
6. Water bath—must reach the temperature required for analysis (70 °C) and keep it constant during the experiment.
7. Centrifuge—equipment that allows rotation parameter configuration required in the method (1,500 g).
8. HPLC system—equipped with a high-pressure pump, automatic injector (or 20 μL manual injector loop), column oven, analytical column, and fluorescence detector (FLD).
9. HPLC analytical column—250 × 4 mm, packed with octadecylsilane reverse phase (C18).

2.2 Reagents

1. Tocopherol (α-, β-, γ-, δ-) standards.
2. Potassium chloride—KCl.
3. Potassium hydroxide—KOH.
4. n-Hexane.
5. Butylated hydroxytoluene—BHT.
6. Pyrogallol.

7. Ethanol HPLC grade.
8. Methanol HPLC grade.

2.3 Preparation of Solvents

1. *1.15% KCl solution*: Prepare the solution by weighing 11.5 g of KCl per liter. Use ultrapure water and keep the solution at room temperature.
2. *KOH saturated solution*: Prepare the solution by weighing 1 g of KOH per milliliter. Use ultrapure water, store it at room temperature, and protect it from light. The potassium hydroxide must be stored in plastic bottles (polyethylene or polypropylene) as it can react with glass containers in prolonged contact (over a few days). KOH saturated is highly corrosive. Use safety goggles and gloves.
3. *n-Hexane with 0.001% BHT*: Prepare the solution by weighing 0.001 g of BHT and adding it to 100 mL of n-hexane. Store the solution at room temperature and protect it from light. Wear gloves when working with hexane, and always work in a fume hood.
4. *0.5% Pyrogallol in ethanol*: Prepare the solution by weighing 0.5 g of pyrogallol and adding it to 100 mL ethanol. This solution cannot be stocked. Calculate the volume needed for the analysis day. It must be prepared in a fume hood, and use gloves.
5. *96% Ethanol with 0.001% BHT*: Prepare the solution by weighing 0.001 g of BHT into a 100 mL volumetric flask. Add 4 mL of ultrapure water, and make up the volume with HPLC-grade ethanol. Keep the solution at room temperature and protected from light.

All waste from the sample glasses, which can contain hexane, must be discarded in the waste section. Only the KCl portion can be discarded in the sink.

3 Methods

This chapter describes a method for analyzing tocopherol in edible animal tissues with saponification and extraction steps. The tocopherol separation, identification, and quantification are described using RP-HPLC.

Saponification before solvent extraction has already been applied to several food matrices, such as meat, tissues, oilseeds, and grains. Saponification involves complex lipid (triacylglycerides, glycerophospholipids, and sphingolipids) alkaline hydrolysis, leading to soap formation and raising polarity. Tocopherol is a non-saponifiable compound, so it remains in the non-polar fraction. In addition to releasing interfering lipids, saponification helps separate protein complexes and release tocopherols complexed to esters.

The method described in this chapter was initially developed for tocopherol analysis in tissues by Buttriss and Diplock [39], with modifications by Juncher et al. [40]. First, the sample is solubilized in a KCl solution, which helps tocopherol extraction from the food matrix, especially proteins. Then, the pyrogallol is added to a basic solution (KOH). Under these conditions, the pyrogallol auto-oxidation occurs in contact with oxygen, preventing the tocopherol in the sample from oxidation. Saponification occurs through the reaction with KOH in a water bath at 70 °C. After cooling the tubes, ultra-pure water is added, followed by hexane with butylated hydroxytoluene (BHT), so that phase separation occurs. BHT, like pyrogallol, has an antioxidant role and should be used cautiously and in low concentrations as it may have UV absorbance and fluorescence similar to tocopherol, which may interfere with the analysis. Finally, tocopherol is transferred to the hexane portion due to its polarity. This fraction is separated and prepared for analysis.

After tocopherol extraction from samples, the isomers will be separated by RP-HPLC. As previously described, the reversed-phase column (C18) does not allow the β- and γ-tocopherol isomer separation due to the two isomers' molecular characterization, which only differs the methyl and hydrogen group substitution position on the chromanol ring radicals. Despite this limitation, the separation of the isomers through a reversed-phase column occurs in a shorter time, using less aggressive solvents, and with better reproducibility in the analysis.

3.1 Procedures

3.1.1 Preparation of Tocopherol Standard Solution

1. Prepare a stock solution for α-, β-, γ-, and δ-tocopherol standard, using 96% ethanol with 0.001% BHT. As previously discussed, the conditions used in the RP-HPLC do not allow β and γ separation. So, consider only the isomer in higher concentration in the analysed sample (regarding literature).
2. The prepared solution will be quantified using a spectrophotometer. For spectrophotometer calibration, 96% ethanol with 0.001% BHT will be used as a blank solution.
3. Each tocopherol must be analyzed (in triplicate) at its respective wavelength according to Table 1.
4. The tocopherol concentration will be calculated according to Lambert-Beer law as follows:

$$A = \varepsilon \times c \times b$$

where A = absorbance

ε = molar absorption coefficient (M^{-1} cm^{-1})
c = molar concentration (M)
b = optical path length (cm)

Table 1
Wavelength, molecular weight, and $E_{1cm}^{1\%}$ of α-, β-, γ-, and δ-tocopherol

	λ max (nm)	$E_{1cm}^{1\%}$ (ε)[a]	Molecular weight (MW)
α-tocopherol	292	75.8 (3265)	430.71
β-tocopherol	296	89.4 (3725)	416.69
γ-tocopherol	298	91.4 (3809)	416.69
δ-tocopherol	298	87.3 (3515)	402.66

[a]Values in ethanol solution [41]

To obtain the tocopherol concentration in μg/mL, consider the molecular weight and a factor to convert milligrams to micrograms (multiply per 1,000), as follows:

$$\text{Tocopherol concentration } (\mu\text{g/mL}) = \frac{A \times \text{MW} \times 1000}{\varepsilon \times b}$$

where A = absorbance

ε = molar absorption coefficient ($\text{M}^{-1}\ \text{cm}^{-1}$)
MW = molecular weight
b = optical path length (cm)

The molar absorption coefficient and molecular weight for each tocopherol isomer are given in Table 1.

5. The stock solution of each tocopherol standard must be kept in the freezer (−20 °C). As described above, measuring the solution concentration each time the solution is used is necessary.

3.2 Tocopherol Standard Curve

1. A calibration curve is prepared using six to eight different concentrations from the standard solutions appropriately quantified at the spectrophotometer. The calibration curve can be prepared individually for each standard or through a mix containing all the standards.
2. Calculate the tocopherol stock solution volume for each concentration, and pipette it into a volumetric flask. Complete the volume with 96% ethanol with 0.001% BHT. Table 2 illustrates the calibration curve points considering a 200 μg/mL stock solution of α-tocopherol and a 10 mL final solution.
3. Each point on the curve must be carried out in triplicate and injected into the HPLC under the following conditions:
   (a) Vol. of injection: 20 μL
   (b) Column: reversed phase (125 mm × 4.0 mm, 5 μ, C18)
   (c) Mobile phase: methanol/water (ultrapure) (98:2 v/v)
   (d) Flow: 1 mL/min
   (e) Detector: fluorescence (FLD)

Table 2
Example of a calibration curve of α-tocopherol from a 200 μg/mL[a] stock solution

Concentration (μg/mL)	Stock solution vol. (mL)	96% Ethanol with 0.001% BHT vol (mL)
0.5	0.025	9.975
1.0	0.050	9.950
2.5	0.125	9.875
5.0	0.250	9.750
7.5	0.375	9.625
10	0.50	9.50
15	0.75	9.25
25	1.25	8.75

[a]Concentration verified and adjusted after spectrophotometer quantification

α-Tocopherol conc. (μg/mL)	α-Tocopherol (μg)	Area (LU)
0.5	0.01	22.1
1	0.02	44.8
2.5	0.05	114.2
5	0.1	228.1
7.5	0.15	346.2
10	0.2	442.2
15	0.3	730.5
25	0.5	1462.3

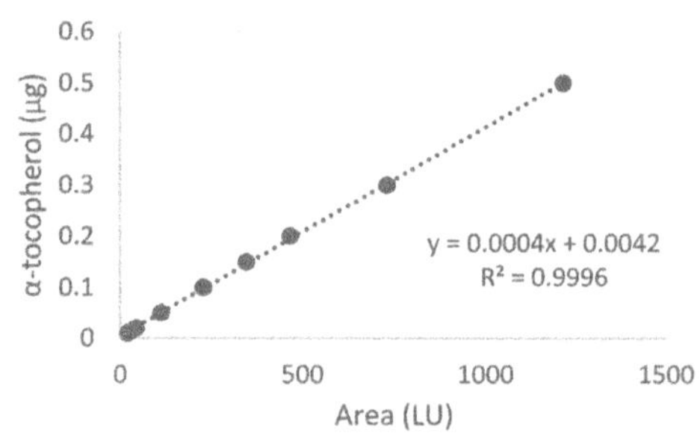

Fig. 2 Calibration curve of α-tocopherol

(f) λ excitation: 295 nm

(g) λ emission: 325 nm

4. Linear regression is done by plotting the area by tocopherol (μg). To calculate the tocopherol mass (μg) in the 20 μL injected on the HPLC, multiply the concentration by 0.02. Figure 2 shows an example of the α-tocopherol calibration curve and the data obtained after analysis.
5. The regression equation obtained will be used to calculate the tocopherol (μg) in food samples.

3.3 Sampling

Samples must be clearly described, including all possible information such as origin, production site, collection period, production size, batch size, and expiry date. If the samples undergo heat treatment, the entire process must be described in detail, specifying temperature, pressure, and cooking time. Likewise, if samples will be stored, it is necessary to report all the storage conditions, such as temperature, light, and humidity.

The sample amount must be representative and commensurate with the research. For meat, purchase at least 1 kg of fresh meat or fish or three to five packs of a single processed food batch in an appropriate marketplace. At the laboratory, remove connective, cartilage, connective, and adipose tissues, and chop and mince the meat. The sample portion selected for analysis must not contain inedible parts. The meat or fish should be ground in a food processor until an homogenous mass is obtained. For chicken, the sample should be hand-boned and have the skin and apparent external fat removed, and the lean part must be then finely minced in a food processor until obtaining an homogenous mass [12–14]. Depending on the research objective, external fat may be included and adequately described in the sample section.

Sample processing and fractionation must be performed on the same day of sample acquisition to begin the analysis. If the analysis will be carry out later, samples must be kept under a vacuum or modified atmosphere in a freezer to avoid oxidation.

3.4 Tocopherol Extraction from Tissues and Meat

1. Pipette 2 mL of the 0.5% pyrogallol solution into a tube, and add 200 uL of the KOH solution. Close the tube and reserve it.
2. Weigh 4 g of the sample (meat samples) in a separate tube, and add 20 mL of the 1.15% KCL solution.
3. Homogenize the sample solution (**item 2**) for 30 s in an ultra-turrax, and then weigh 2 g of this solution into the tube containing pyrogallol and KOH solution (**item 1**). Caution: This step is essential to ensure a homogeneous extraction.
4. Homogenize this mixture for 10 s by vortexing, and place it in a water bath at 70 °C for 30 min for saponification.
5. After saponification, remove the tubes from the water bath, and place them in an ice bath for 5 min. Remember to protect the samples from light with an aluminum foil.
6. Add 1 mL of ultrapure water to the tube and 4 mL of n-hexane with BHT.
7. Homogenize the tube on the vortex for 15 s, and centrifuge the sample at 1,500 g for 5 min.
8. After centrifugation, collect the upper phase (corresponding to n-hexane), and place it in another tube with a cover. This part is essential because only the n-hexane phase and nothing else must be transferred. Keep the samples on ice and cover them with aluminum foil.
9. Repeat the procedure, adding 3 mL of n-hexane with BHT to the extraction tube, and centrifuge again under the same conditions.
10. Then, the upper phase (which contains the n-hexane) is collected and added to the n-hexane fraction collected in the first

centrifugation kept in the covered tube. If the samples are not immediately analyzed, the sample tube should be wrapped in aluminum foil and stored in a freezer. The samples must be analyzed within 2 days.

11. Evaporate the hexane portion under nitrogen up to 40 °C.
12. Resuspend the extract in 96% ethanol with 0.001% BHT. The volume can range from 250 to 1000 μL of solvent, depending on the sample concentration. The volume will be considered in the calculation.
13. Filter the extract with a 0.22 μm syringe filter (preferably, but a 0.45 μm filter can be used). Attention to the composition of the filter is necessary due to compatibility with the solvent used. In this extraction, ethanol filters, regenerated cellulose (RC), polytetrafluoroethylene (PTFE), and nylon (NY) can be used.

3.5 Chromatographic Separation and Calculations

1. The HPLC must be turned on and stabilized before use.
2. Set the HPLC to the same conditions as described previously:
   (a) Vol. of injection: 20 μL
   (b) Column: reversed phase (125 mm × 4.0 mm, 5 μ, C18)
   (c) Mobile phase: methanol/water (ultrapure) (98:2 v/v)
   (d) Flow: 1 mL/min
   (e) Detector: fluorescence (FLD)
   (f) λ excitation: 295 nm
   (g) λ emission: 325 nm
3. Start the analysis with a blank injection of solvent or mobile phase to verify the occurrence of system contamination.
4. Inject the filtered extract into the HPLC. A standard should be injected for every ten samples injected into the HPLC to ensure the method runs correctly.
5. The mean area obtained from each sample will be considered for tocopherol quantification using the linear regression equation resulting from the standard curve, expressed in μg (*see* **item 4.2, step 5**). Figure 3 illustrates the elution order of the tocopherol isomers.
6. After determining the tocopherol content (μg) in the sample, it is necessary to calculate the sample mass in the 20 μL injected in the analysis. Use the formula below:

$$\text{Sample mass injected (g)} = \frac{\text{SW1} \times \text{SW2} \times \text{VI}}{\text{KCl vol}}$$

where

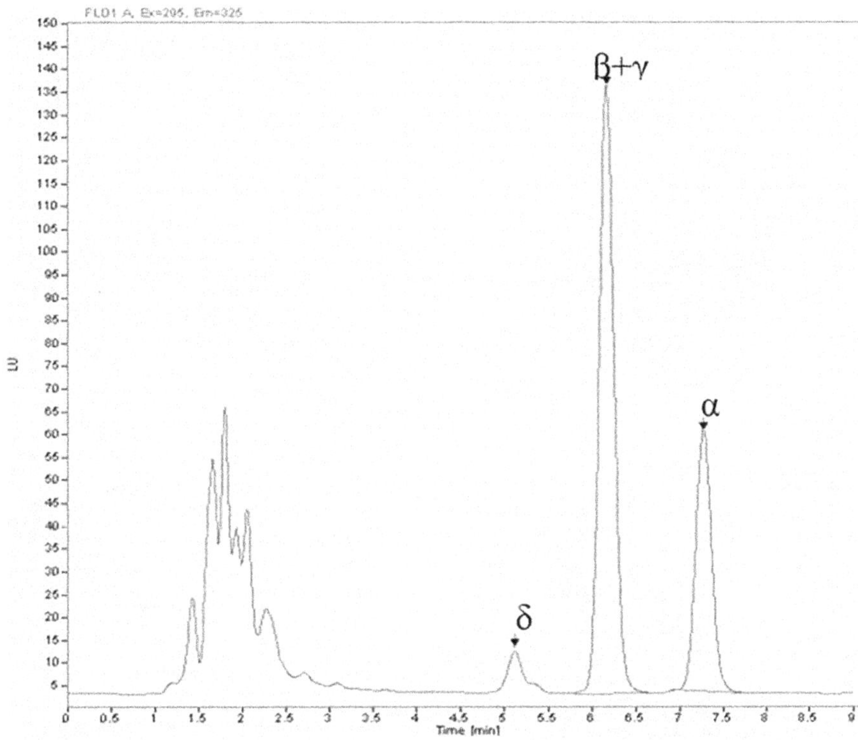

Fig. 3 Chromatogram of tocopherol isomer elution order (retention time of δ-, β + γ-, and δ-tocopherol was approximately 7–7.5, 6–6.5, and 5 min, respectively)

SW1 = sample weight—**item 4.4**, **step 2** (use the precise weight from the 4 g samples)

SW2 = solution weight—**item 4.4**, **step 3** (use the precise weight from the 2 g solution)

VI = volume of injection (mL) (for 20 μL use 0.02 mL)

KCL vol = volume of KCl solution = 20 mL

7. The tocopherol concentration (μg/g) is then calculated by dividing the tocopherol content found in the linear regression equation (μg) by the sample mass injected (g).

4 Notes

This chapter covered the method for determining and quantifying tocopherols in edible animal tissues. Despite not being produced by animals, tocopherols can be found in meat samples because they are acquired from the animal's diet. Tocopherol quantification is a susceptible analysis that requires care and attention to detail from the beginning. Therefore, before starting the investigation, it is essential to establish the analysis objectives to choose the best extraction method and option for determining and quantifying the tocopherol. Also, it is indispensable to validate the method. The validation must consider specificity, linearity, detection limit, quantification limit, precision, and robustness.

Acknowledgments

The authors are grateful to the Brazilian agencies CAPES (Finance code 001), CNPq (grant 304131/2019-9 and 305778/2022-6), and FAPESP (grant 2013/07914-8).

References

1. Evans HM, Bishop KS (1922) On the existence of a hitherto unrecognized dietary factor essential for reproduction. Science 56:650–651
2. Sure B (1924) Dietary requirements for reproduction. J Biol Chem 58(3):693–709
3. Evans HM, Emerson OH, Emerson GA (1936) The isolation from wheat germ oil of an alcohol, α-tocopherol, having the properties of vitamin E. J Biol Chem 113(1):319–332
4. Eitenmiller RR, Lee J (2004) Vitamin E: food chemistry. In: Composition and analysis. Marcel Dekker, New York
5. Szewczyk K, Chojnacka A, Górnicka M (2021) Tocopherols and tocotrienols—bioactive dietary compounds; what is certain, what is doubt? Int J Mol Sci 22(12). https://doi.org/10.3390/ijms22126222
6. Lushchak VI, Semchuk NM (2012) Tocopherol biosynthesis: chemistry, regulation and effects of environmental factors. Acta Physiol Plant 34(5):1607–1628. https://doi.org/10.1007/s11738-012-0988-9
7. Morrissey PA, Buckley DJ, Sheehy PJA, Monahan FJ (1994) Vitamin E and meat quality. Proc Nutr Soc 53(2):289–295. https://doi.org/10.1079/pns19940034
8. Sales J, Koukolová V (2011) Dietary vitamin E and lipid and color stability of beef and pork: modeling of relationships. J Anim Sci 89(9):2836–2848. https://doi.org/10.2527/jas.2010-3335
9. Reboul E (2017) Vitamin e bioavailability: mechanisms of intestinal absorption in the spotlight. Antioxidants 6(4). https://doi.org/10.3390/antiox6040095
10. Sayago A, Marín MI, Aparicio R, Morales MT (2007) Vitamina E y aceites vegetales. Grasas Aceites 58(1):74–86
11. Bou R, Guardiola F, Grau A, Grimpa S, Manich A, Barroeta A, Codony R (2001) Processing and products: influence of dietary fat source, α-tocopherol, and ascorbic acid supplementation on sensory quality of dark chicken meat. Poult Sci 80(6):800–807. https://doi.org/10.1093/ps/80.6.800
12. Bragagnolo N, Danielsen B, Skibsted LH (2005) Effect of rosemary on lipid oxidation in pressure-processed, minced chicken breast during refrigerated storage and subsequent heat treatment. Eur Food Res Technol 221:610–615
13. Mariutti LRB, Nogueira GC, Bragagnolo N (2011) Lipid and cholesterol oxidation in chicken meat are inhibited by sage but not by garlic. J Food Sci 76:C909–C915
14. Castro WF, Mariutti LRB, Bragagnolo N (2011) The effects of colorifico on lipid oxidation, color and vitamin E in raw and grilled chicken patties during frozen storage. Food Chem 124:126–131
15. Pfalzgraf A, Frigg M, Steinhart H (1995) A-Tocopherol contents and lipid oxidation in pork muscle and adipose tissue during storage. J Agric Food Chem 43(5):1339–1342. https://doi.org/10.1021/jf00053a039
16. Morrissey PA, Buckley DJ, Sisk H, Lynch PB, Sheehy PJA (1996) Uptake of α-tocopherol in porcine plasma and tissues. Meat Sci 44(4):275–283. https://doi.org/10.1016/S0309-1740(96)00033-2
17. Pazos M, Sánchez L, Medina I (2005) α-Tocopherol oxidation in fish muscle during chilling and frozen storage. J Agric Food Chem 53(10):4000–4005. https://doi.org/10.1021/jf047843k
18. Secci G, Parisi G (2016) From farm to fork: lipid oxidation in fish products. A review. Ital J

Anim Sci 15(1):124–136. https://doi.org/10.1080/1828051X.2015.1128687

19. Bartosińska E, Buszewska-Forajta M, Siluk D (2016) GC–MS and LC–MS approaches for determination of tocopherols and tocotrienols in biological and food matrices. J Pharm Biomed Anal 127:156–169
20. Leo L, Rescio L, Ciurlia L, Zacheo G (2005) Supercritical carbon dioxide extraction of oil and α-tocopherol from almond seeds. J Sci Food Agric 85(13):2167–2174
21. Peng WL, Mohd-Nasir H, Setapar SHM, Ahmad A, Lokhat D (2020) Optimization of process variables using response surface methodology for tocopherol extraction from Roselle seed oil by supercritical carbon dioxide. Ind Crop Prod 143(April 2019):111886
22. Saini RK, Keum YS (2016) Tocopherols and tocotrienols in plants and their products: a review on methods of extraction, chromatographic separation, and detection. Food Res Int 82:59–70. https://doi.org/10.1016/j.foodres.2016.01.025
23. Xiao S, Zhang Y, Xie J, Wen Z (2018) Ultrasonic-assisted extraction of squalene and vitamin E based oil from Ziziphi Spinosae semen and evaluation of its antioxidant activity. J Food Meas Charact 12(4):2844–2854
24. Ofori-Boateng C, Lee KT (2014) Ultrasonic-assisted extraction of α-tocopherol antioxidants from the fronds of Elaeis guineensis Jacq.: optimization, kinetics, and thermodynamic studies. Food Anal Methods 7(2):257–267
25. Lu D, Yang Y, Sun C, Wang S (2019) Determination of tocopherols and tocotrienols in cereals and nuts by dispersive solid-phase microextraction-gas chromatography-mass spectrometry. Anal Methods 11(42):5439–5446
26. Köseoğlu K, Ulusoy Hİ, Yilmaz E, Soylak M (2020) Simple and sensitive determination of vitamin A and E in the milk and egg yolk samples by using dispersive solid phase extraction with newly synthesized polymeric material. J Food Compos Anal 90(March):103482
27. Tsochatzis ED, Tzimou-Tsitouridou R (2015) Validated RP-HPLC method for simultaneous determination of tocopherols and tocotrienols in whole grain barley using matrix solid-phase dispersion. Food Anal Methods 8(2):392–400
28. Ibáñez E, Palacios J, Señoráns FJ, Santa-María G, Tabera J, Reglero G (2000) Isolation and separation of tocopherols from olive by-products with supercritical fluids. J Am Oil Chem Soc 77(2):187–190
29. Przygoda K, Wejnerowska G (2015) Extraction of tocopherol-enriched oils from Quinoa seeds by supercritical fluid extraction. Ind Crop Prod 63:41–47
30. de Lucas A, Martinez de la Ossab E, Rincóna J, Blancob MA, Graciaa I (2002) Supercritical fluid extraction of tocopherol concentrates from olive tree leaves. J Supercrit Fluids 22(3):221–228
31. Delgado-Zamarreño MM, Bustamante-Rangel M, Sierra-Manzano S, Verdugo-Jara M, Carabias-Martínez R (2009) Simultaneous extraction of tocotrienols and tocopherols from cereals using pressurized liquid extraction prior to LC determination. J Sep Sci 32(9):1430–1436. https://doi.org/10.1002/jssc.200800707
32. Viñas P, Bravo-Bravo M, López-García I, Pastor-Belda M, Hernández-Córdoba M (2014) Pressurized liquid extraction and dispersive liquid-liquid microextraction for determination of tocopherols and tocotrienols in plant foods by liquid chromatography with fluorescence and atmospheric pressure chemical ionization-mass spectrometry detection. Talanta 119:98–104
33. Bustamante-Rangel M, Delgado-Zamarreño MM, Sánchez-Pérez A, Carabias-Martínez R (2007) Determination of tocopherols and tocotrienols in cereals by pressurized liquid extraction-liquid chromatography-mass spectrometry. Anal Chim Acta 587(2):216–221
34. Kraujalis P, Venskutonis PR (2013) Supercritical carbon dioxide extraction of squalene and tocopherols from amaranth and assessment of extracts antioxidant activity. J Supercrit Fluids 80:78–85
35. Craft NE (2016) Tocopherols: properties and determination. In: Encyclopedia of food and health, 1st edn. Academic Press
36. Parrish DP, Waltking AE (1980) Determination of vitamin e in foods—a review. Crit Rev Food Sci Nutr 13(2):161–187
37. Rimmer CA, Sander LC, Wise SA (2005) Selectivity of long-chain stationary phases in reversed-phase liquid chromatography. Anal Bioanal Chem 382(3):698–707
38. Britz SJ, Prasad PVV, Moreau RA, Allen LH, Kremer DF, Boote KJ (2007) Influence of growth temperature on the amounts of tocopherols, tocotrienols, and γ-oryzanol in brown rice. J Agric Food Chem 55(18):7559–7565
39. Buttriss J, Diplock A (1984) High-performance liquid chromatography methods for vitamin E in tissues. Methods Enzymol 105(1981):131–138

40. Juncher D, Vestergaard CS, Søltoft-Jensen J, Weber CJ, Bertelsen G, Skibsted LH (2000) Effects of chemical hurdles on microbiological and oxidative stability of a cooked cured emulsion type meat product. Meat Sci 55(4): 483–491

41. Kofler M, Sommer PF, Bolliger HR, Schmidli B, Vecchi M (1962) Physicochemical properties and assay of the tocopherols. Vitam Horm 20(C):407–440. https://doi.org/10.1016/S0083-6729(08)60727-X

Part VI

Hydrosoluble Vitamins

Chapter 18

Vitamin C

Bin Jiang and Peng Dong

Abstract

Vitamin C (aka ascorbate acid) is an essential micronutrient abundant in diverse types of food, which is required in humans for various biological processes. There are many analytical techniques to determine vitamin C. Here we describe five common methods, titrimetry, spectrophotometry, fluorometry, HPLC, and electrochemical methodologies. Their principles, materials, procedures, and applications are explained in detail.

Key words Vitamin C, Titrimetry, Spectrophotometry, Fluorometry, HPLC, Electrochemical method

1 Introduction

Vitamin C, also known as L-ascorbic acid (AA), is a polyhydroxy compound with a molecular formula as $C_6H_8O_6$. Its chemical structure is shown in Fig. 1a. AA is a water-soluble vitamin which is vital in the biosynthesis of blood vessels, muscle, cartilage, collagen in bones, and certain neurotransmitters [1]. The antioxidant activity is one of the important properties of AA. It neutralizes free radicals by sacrificing one of its own electrons, ending the oxidative reaction, which can protect the body against the damage of free radicals [2]. AA also helps to prevent certain diseases such as vascular diseases, cancer, and infections [3].

Because the human body does not synthesize AA, it can only be supplied through diet. Fruits and vegetables are the best sources of AA. The oxidized form of AA is dehydroascorbic acid (Fig. 1b). A total ascorbate refers to the sum of the two substances. The analytical methods of ascorbic acid in food products have been receiving overwhelming attention.

Lilian Mariutti and Frank Vriesekoop (eds.), *Food Chemistry: Basic Protocols*, Methods and Protocols in Food Science, https://doi.org/10.1007/978-1-0716-4678-6_18,

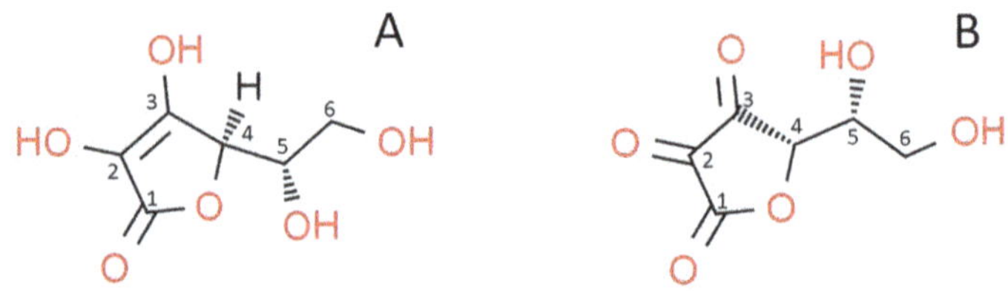

Fig. 1 Chemical structures of (**a**) ascorbic acid and of (**b**) dehydroascorbic acid

Many analytical techniques such as titrimetry, spectrophotometry, fluorometry, high-performance liquid chromatography, and electrochemical method are used for the determination of AA in different food matrices.

1.1 Titrimetric Methods

Titrimetric methods are a relatively simple, low-tech, low-cost, and widely used approach to detect and analyze AA in many types of food, both from solid and liquid matrices. Two main types of titration methods/reagents are described in this chapter.

The first is 2,6-dichloroindophenol titration (Subheadings 2.1 and 3.1), which is a well-established method for determination of AA based on its reducibility. The official method is AOAC 967.21 [4]. A blue basic dye (2,6-dichloroindophenol (DCIP)) can be reduced to a colorless compound by ascorbic acid in an acidic medium. During that process ascorbic acid is oxidized to dehydroascorbic acid (DHAA). At the end point of the titration, excess DCIP turns a rose pink color. AA content of the sample is calculated from the consumption of DCIP. The second titrimetric method is iodine titration (Subheadings 2.2 and 3.2). Iodine serves as a widely used redox reaction reagent in determination of AA. AA reduces potassium iodate (KIO_3) in the presence of potassium iodide (KI). When all AA is oxidized, the excess iodine is free to react with starch indicator, forming a bluish-black starch–iodine complex, which indicates the end point of the titration.

1.2 Spectrophotometric Methods

Low cost, versatility, and easy implementation are in general inherent to all spectrophotometric methods [5] (Subheadings 2.3 and 3.3). Generally, a complex formed by ascorbic acid and specific reagents has a specific/unique maximum absorption wavelength (λ_{max}) in the ultraviolet-visible range (200–760 nm) which can be determined employing a spectrophotometer. The absorbance related to the concentration of AA in sample solutions follows the Beer–Lambert law.

A wide spectrum of reagents, analytes, applications, systems, and their combinations involved in spectrophotometric methods have been investigated in numerous studies for the determination of ascorbic acid. A ferric ion–ferrous ion redox system is a conventional application for the qualitative determination of ascorbic acid, dehydroascorbic acid, and total ascorbate. AA is reduced by ferric ion to the ferrous ion, forming a complex with λ_{max} of 525 nm in the presence of α-α′-bipyridyl [6]. DHAA is reduced to AA by dithiothreitol (DTT). Excess DTT is removed by reacting with *N*-ethylmaleimide (NEM). Total ascorbate is subsequently assayed using the α-α′-bipyridyl method, and DHAA is calculated by subtracting the reduced AA from the total pool of ascorbate [6–9].

1.3 Fluorescent Methods

The molecules of some substances can absorb energy and emit fluorescence when exposed to specific wavelengths. According to the spectrum and intensity of fluorescence, substances could be qualitatively or quantitatively analyzed. AA is oxidized to dehydroascorbic acid (DHAA) in the presence of Nortrit, and then DHAA reacts with *o*-phenylenediamine to form a fluorescent quinoxaline compound with the activation wavelength at 350 nm and emission length at 430 nm. The fluorescence intensity at 430 nm is determined by fluorometer and is proportional to the concentration of AA (Subheadings 2.4 and 3.4).

The formation of fluoresce compound is prevented by boric acid-dehydrovitamin acid chelate and *o*-phenylenediamine. Therefore, any other fluorescence from materials besides AA serves as blank. The procedure described in this chapter is adapted from AOAC 967.22 [10].

1.4 High-Performance Liquid Chromatography (HPLC) Methods

High-performance liquid chromatography (HPLC) is a chromatographic technique of great versatility and analytic power used in food component research. HPLC can separate a large number of compounds both rapidly and at high sensitivity, reduce separation time, and reduce the volume of sample needed [11]. HPLC is widely used for the simultaneous qualitative and quantitative analysis of AA in food (Subheadings 2.5 and 3.5).

1.5 Electrochemical Methods

Electrochemical analysis (Subheadings 2.6 and 3.6) has been widely used due to its high sensitivity, simple procedure, and low cost. Polarographic analysis, voltammetric analysis, potentiometric titration, and coulometry titration are developed for ascorbic acid determination. Voltammetry are increasingly popular in real-time samples, and coulometry titration is a conventional method [12, 13].

Quantitative or qualitative analysis of an electrolyte is determined by the current–voltage curve under specific condition. Voltammetric methods include cyclic voltammetry, stripping voltammetry, differential pulse method, etc. Cyclic voltammetry has been previously used for antioxidant content assessment and in particular low-molecular-weight antioxidants [12]. During the redox reaction of ascorbic acid and its oxidized form, the concentration of AA and DHAA varies, which change the corresponding current. Consequently, a cyclic voltammogram curve can be obtained.

1.6 Coulometric Methods

Coulometric titration is produced by electrolyzing reaction in the cell (Subheadings 2.7 and 3.7). The quantity of electricity consumed demonstrates the content of the substance titrated. Comparing to the classical chemical titration, the end point is indicated by magnitude of current in coulometric titration, which is applicable for colored and turbid solution. Besides, the preparation and calibration of standard solution are not required, electrolyte is employed for coulometric titration, and the analysis data is gained by measuring the magnitude of current, which is simplified and accurate.

2 Materials

2.1 2,6-Dichloroindophenol Titration

1. Solutions
   (a) Metaphosphoric acid–acetic acid solution: Dissolve 15 g HPO_3 in 40 mL CH_3COOH (36%) and 200 mL of water; dilute to 500 mL. Filter through Whatman #1 paper into a glass-stoppered bottle, which can then be kept in a refrigerator for up to 7–10 days.
   (b) Metaphosphoric acid–acetic acid–sulfuric acid solution: Processed as in Subheading 2.1, **item 1(a)**, except substitute H_2O for 0.15 M H_2SO_4.
2. AA standard solution (1 mg/mL): Weigh 0.050 g ascorbic acid reference standard, and dissolve in HPO_3–CH_3COOH solution (Subheading 2.1, **item 1(a)**). Transfer to a 50 mL volumetric flask, and dilute to volume with HPO_3-CH_3COOH solution (Subheading 2.1, **item 1(a)**). Prepare immediately before use.
3. Indophenol standard solution: Dissolve 50 mg DCIP in 50 mL distilled water containing 42 mg $NaHCO_3$. Shake well. After dissolving completely, dilute to 200 mL with distilled water. Filter through Whatman #1 paper into a brown glass-stoppered bottle and store at 4–8 °C. The titer (T, mg/mL, represents mg of ascorbic acid equivalent to 1 mL DCIP solution) needs to be calibrated with a standard ascorbic acid solution before each use.
4. Thymol blue pH indicator: 0.04%. Dissolve 0.1 g thymol blue by triturating in agate mortar with 10.75 mL 0.02 M NaOH, and dilute to 250 mL with distilled water. Transition range: pH 1.2 (red)–2.8 (yellow).
5. *Calibration*: Transfer 2.0 mL ascorbic acid standard solution into three 50 mL Erlenmeyer flasks. Add 5.0 mL HPO_3–CH_3COOH solution (Subheading 2.1, **item 1(a)**) to each Erlenmeyer flask, and shake vigorously. Quickly titrate with the DCIP solution until a rose pink color develops that should last for 15 s (each titration should require 15 mL indophenol

solution, and titrations should be within 0.1 mL). Similarly, titrate three 7.0 mL of HPO_3–CH_3COOH solution (Subheading 2.1, **item 1(a)**) as blank test. Calculate by the following formula:

$$T = \frac{c \times v}{v_1 - v_0.}$$

where T is titer of DCIP solution, the concentration of indophenol solution as mg ascorbic acid equivalent to 1 mL reagent, mg/mL; c is the concentration of ascorbic acid standard solution, mg/mL; V is the volume of ascorbic acid standard solution, mL; V_1 is the volume of DCIP solution consumed for titrating ascorbic acid standard solution, mL; and V_o is the volume of DCIP solution consumed for titrating bland solution, mL. The value of the titer of DCIP solution is the average of three Ts.

2.2 Iodine Titration

1. Iodide solution (0.025 mol/L): Dissolve 18 g solid iodide in about 100 mL distilled water, and then add 6.4 g iodine. Make up to 1 L with distilled water in a volumetric flask.
2. Primary standard arsenic trioxide solution: Dissolve 1.2364 g arsenic trioxide to 100 mL distilled water. Dropwise add 5.0 M NaOH until arsenic trioxide fully dissolves. Adjust pH to 6.0 with 5 M HCl. Transfer the solution to a 250 mL volumetric flask, and make up to the volume with distilled water.
3. Starch indicator solution (1%): Add 1 g soluble starch to 100 mL boiling water in a conical flask. Stir to dissolve and cool before use.

2.3 AA Analysis by Spectrophotometry

1. Trichloroacetic acid (TCA, 6% w/v, 10% w/v): Keep at room temperature.
2. Potassium phosphate, monobasic (KH_2PO_4).
3. Potassium phosphate, dibasic (K_2HPO_4).
4. Phosphate buffer (75 mM, pH 7.0): Prepare from potassium phosphate, monobasic and potassium phosphate, dibasic.
5. Dithiothreitol solution (DTT, 10 mM): Keep in an ice bath.
6. *N*-ethylmaleimide solution (NEM, 0.5% w/v): Keep in an ice bath.
7. Orthophosphoric acid solution (H_3PO_4, 43% v/v): 1:1 dilution of 85% reagent-grade H_3PO_4. Keep at room temperature.
8. Ferric chloride solution ($FeCl_3$, 3% w/v): Keep at room temperature.
9. α-α′-bipyridyl solution (4%, w/v): Keep in an ice bath.
10. Ascorbic acid standard solution (0.15–19 mM ascorbate standards in 6% w/v TCA). Keep in an ice bath.

2.4 Fluorometric Analysis of AA

1. Metaphosphoric acid–acetic acid solution: *See* Subheading 2.1, **item 1(a)**.
2. Metaphosphoric acid–acetic acid–sulfuric acid solution: *See* Subheading 2.1, **item 1(b)**.
3. AA standard solution (100 μg/mL): Make 1 mg/mL AA standard solution 2.1.1.1–2. Dilute 10.0 mL of the solution, and fix the volume to 100 mL with HPO_3–CH_3COOH solution (a) immediately before use.
4. Acid-washed activated carbon (Norit): Weigh about 200 g Norit (chemically pure, 80 mesh to 200 mesh), add 1 L of 10% (v/v) HCl, heat to bp, and filter with vacuum. Remove the agglomerate into a large beaker, and wash with water until there is no ferric iron in the filtrate. Dry at 110 °C–120 °C for about 10 h.

   Method for detecting ferric iron–Prussian blue reaction. Mix 20 g/L potassium ferrocyanide and 1% (v/v) HCl in equal volume. Drip the filtrate in the mixed solution, and a blue precipitate demonstrates the presence of ferric iron.
5. Sodium acetate solution: Dissolve 500 g $CH_3COONa{\cdot}3H_2O$ in water, and dilute to 1 L.
6. Boric acid–sodium acetate solution*:* Dissolve 3.0 g H_3BO_3 in CH_3COONa solution (4), and dilute to 100 mL. Make fresh before use.
7. o-Phenylenediamine solution (*400 mg/L*): Dissolve 40 mg *o*-phenylenediamine·2HCl in water, and dilute to 100 mL. Make fresh before use.
8. Thymol blue pH indicator: *See* Subheading 2.1, **item 4**.

2.5 High-Performance Liquid Chromatography

1. Metaphosphoric acid (20 g/L) solution.

   Dissolve 200 g metaphosphoric acid in 1 L of water; the solution can be stored at 4 °C for 1 month. Take 50 mL of the solution and dilute it with water to 500 mL.
2. Ascorbic acid standard solution.

   AA standard solution (1 mg/mL): Accurately weigh 0.050 g ascorbic acid reference standard, and dissolve in metaphosphoric acid solution (2.5.1). Transfer to a 50 mL volumetric flask, and dilute to volume with metaphosphoric acid solution (2.1). Prepare immediately before use.

2.6 Voltammetry

1. Stock solution of AA (100 mmol/L): Prepare before use.
2. Electrolyte: 0.34 mol/L KCl solution.
3. Standard solution of AA: Dilute stock solution in 0.34 mol/L KCl solution to a concentration range between 0.1–10 mmol/L.

4. Standard solution of odium benzoate solution (1 mol/L): Dissolve the respective amount of odium benzoate in 0.34 mol/L KCl solution.
5. Standard solution of glucose solution (1 mol/L): Dissolve the respective amount of glucose in 0.34 mol/L KCl solution.
6. Standard solution of tartaric acid solution (1 mol/L): Dissolve the respective amount of tartaric acid in 0.34 mol/L KCl solution.
7. Standard solution of citric acid solution (1 mol/L): Dissolve the respective amount of citric acid in 0.34 mol/L KCl solution.
8. Phosphate buffer (300 mL, 0.066 mol/L, pH = 6.98): Mix the respective volumes of potassium dihydrogen phosphate and sodium monohydrogen phosphate solutions in 2:3.
9. Dichlorophenol indophenol (DCPIP) solution (5×10^4 mol/L): Dissolve 145 mg DCPIP and sodium salt in 100 mL hot distilled water. Add 300 mL phosphate buffer and homogenize. Keep from light and filter.
10. Oxalic acid solution (2.0%).

2.7 Coulometric Titration

1. Iodine solution (0.05 mol/L): Weigh 6.6 g I_2 pellet and 10 g KI in a mortar, grinding with water. Transfer to a 500 mL brown volumetric flask after fully dissolved.
2. Potassium iodide solution (1 mol/L).
3. Phosphate solution (1 mol/L).
4. Standard solution of ascorbic acid (10^{-3} mol/L).

3 Methods

3.1 2,6-Dichloroindophenol Titration

1. Preliminary test for appreciable amount of basic substances.

   Add 25 mL HPO_3–CH_3COOH solution to a finely ground solid sample, or dilute liquid samples with twofold HPO_3–CH_3COOH solution. Test pH by thymol blue pH indicator. (pH >1.2 represents an appreciable amount of basic substances). The appropriate treatment process is chosen according to the content of basic substances.
2. Sample treatment
   (a) Fruit and vegetable juices. Mix thoroughly by shaking to ensure uniform distribution, and filter through absorbent cotton or rapid paper. Prepare fresh juices by pressing well-pulped fruit and filtering. Press juice of citrus fruits by any commercial device and filter. Add aliquots of ≥100 mL prepared juice to equal volumes of HPO_3–CH_3COOH solution. Designate total volume as *V* mL. Mix and filter through fluted Whatman #1 paper.

(b) Dry products containing no appreciable quantities of basic substances. Pulverize test portion by gentle grinding, add HPO_3–CH_3COOH solution, and triturate until test portion is in suspension. Dilute with HPO_3–CH_3COOH solution to measured volume. Designate this volume as V mL. (Use 10 mL extracting solution/g dry test sample. Final test solution should contain 10–100 mg ascorbic acid/100 mL).

(c) Dry products containing appreciable of basic substances. Pulverize test portion by gentle grinding, add HPO_3–CH_3COOH–H_2SO_4 solution to adjust pH to 1.2, and triturate until test portion is in suspension. Dilute with HPO_3–CH_3COOH solution to measured volume. Designate this volume as V mL. (Use 10 mL extracting solution/g dry test sample. Final test solution should contain 10–100 mg ascorbic acid/100 mL).

(d) Liquid products. For liquid products such as non-alcoholic and alcoholic beverages, take amount of test portion containing an estimated 100 mg ascorbic acid. If appreciable amounts of basic substances are present, adjust pH to 1.2 with HPO_3–CH_3COOH–H_2SO_4 solution. Dilute with HPO_3–CH_3COOH solution to measured volume containing 10–100 mg ascorbic acid/100 mL. Designate this volume as V mL.

3. Titration

Transfer 10 mL filtered sample to a 50 mL Erlenmeyer. Titrate with DCIP solution until rose pink color develops lasting at least 5 s. Titrate a blank solution similarly. Repeat all titrations three times.

4. Calculation

mg ascorbic acid/g, mL, etc. $= (V_1 - V_0) \times \frac{T}{E} \times \frac{V}{\Upsilon}$,

where V_1 is average mL for test solution titration; V_0 is average mL for test blank titration; T is mg ascorbic acid equivalent to 1.0 mL indophenol standard solution; E is number of g, mL, etc. assayed; V is volume initial test solution; and Υ is volume test solution titrated.

3.2 Iodine Titration

1. Calibration of iodine solution

The procedure is modified from the study of Ciancaglinia et al. [14] as follows: transfer exactly 20 mL (designated as V) of 0.025 mol/L (designated as c) standard solution of As_2O_3 in an Erlenmeyer flask, and add 0.1 g $NaHCO_3$ and 1 mL starch solution indicator. Fill the 50 mL burette with the iodine solution, and slowly titrate the As_2O_3 solution until the solution turns blue-black. Mix 20 mL distilled water, 0.1 g $NaHCO_3$, and 1 mL starch solution in an Erlenmeyer, serving as blank solution. Titrate the blank solution by the iodine solution similarly. Repeat this procedure three times, and calculate the correct concentration of the iodine solution.

The concentration of iodine solution is calculated according to the following chemical equation and formula:

$$As_2O_3 + 2H_2O + 2I_2 \leftrightarrow As_2O_5 + 4HI,$$

the concentration of iodine solution, mol/L $= \frac{2(c \times V)}{V_{iodine} - V_0}$,

where V_{iodine} is the volume of iodine solution consumed in titrating As_2O_3 solution, mL; V_0 is the volume of iodine solution consumed in titrating blank solution, mL; V is the volume of As_2O_3 solution used in titration, mL; and c is the concentration of As_2O_3 solution, mol/L.

2. Sample treatment
   (a) Liquid food. Strain through cheesecloth if it contains pulp or seeds [15]. It should be noted that depending on the liquid viscosity, dilution may be necessary.
   (b) Solid food. Cut a 100 g sample into small pieces, and grind in a mortar and pestle. Add 10 mL portions of distilled water several times while grinding the sample, each time decanting off the liquid extract into a 100 mL volumetric flask. Finally, strain the ground pulp through cheesecloth, rinsing the pulp with a few 10 mL portions of water and collecting all filtrate and washings in the volumetric flask. Make the extracted solution up to 100 mL with distilled water [15].
3. Titration

   Transfer 40 mL sample solution to a 250 mL Erlenmeyer, adding 60 mL distilled water, 1 mL HCl (15%, v/v), and 1 mL starch indicator solution. Titrate the sample solution mixture with standardized iodine solution until it turns blue-black and persists >5 s. The mixture of 100 mL distilled water, 1 mL HCl (15%, v/v), and 1 mL starch solution serves as blank and is titrated similarly. Repeat three times.
4. Calculation

   The amount of ascorbic acid is calculated as the following formula:

   g ascorbic acid/L, g $= \frac{c_{iodine} \times (V_{iodine} - V_0)}{V} \times M_{vc}$

   where c_{iodine} is the concentration of standardized iodine solution, mol/L; V_{iodine} is the volume of standardized iodine solution consumed in titrating sample solution, mL; V_0 is the volume of standardized iodine solution consumed in titrating blank solution, mL; V is the volume of sample of sample solution, mL; and M_{vc} is the molecular weight of ascorbic acid, g/mol.

3.3 AA Analysis by Spectrophotometry

The procedure is modified from the study of Gillespie and Ainsworth [7].

1. Extraction
   (a) Juice, milk, and liquid products: A relevant portion of sample will be mixed with 6% TCA at a volume ratio of 1:1. Mix well. Centrifuge at 4 °C. The volume of the supernatant is made up to 100 mL with distilled water, serving as the sample solution.
   (b) Fruit, vegetable, and other solid products: Weigh a representative amount of sample (~ 1 g). Homogenize with a mortar and pestle. Add 1 mL 6% TCA and transfer to a centrifugal tube. Wash the mortar and pestle with another 1 mL 6% TCA, and collect in the same tube. Centrifuge at 4 °C, and the supernatant is transferred into a 100 mL volumetric flask, and the volume is made up to the mark with distilled water, serving as the sample solution.
2. Detection
   (a) Prepare blank, standard, and sample solutions in duplicate for both reduced AA and total AA (four assays per sample). Add 1 mM phosphate buffer and 200 mL of either 6% TCA (blank), AA standards (range, 0.15–10 mM), or sample to a 2 mL tube.
   (b) Add 100 μL of 10 mM DTT to the total AA tubes, and incubate at room temperature for 10 min. This will reduce the pool of oxidized AA.
   (c) Add 100 μL 0.5% NEM to the total AA tubes to remove the excess DTT, and incubate at room temperature for at least 30 s.
   (d) Add 200 μL distilled water to the reduced AA assay tubes to account for the volume of DTT and NEM added to the total AA assay tubes.
   (e) Add 500 μL 10% TCA, 400 μL 43% H_3PO_4, 400 mL 4% α-α′-bipyridyl solution, and 200 mL 3% $FeCl_3$ to all assay tubes. After adding $FeCl_3$, shake the mixture vigorously to avoid formation of a white precipitate that could interfere with the absorbance reading.
   (f) Incubate the assay tubes at 37 °C for 1 h.
   (g) Transfer 200 μL sample, standard, or blank reaction solution from the assay tube to a cuvette, and read the absorbance of each at 525 nm.
3. Calculation
   (a) Standard curve. Draw a curve with the concentrations of AA standard solutions as *X* and the corresponding absorbance as *Y*.

(b) The concentrations of AA and total ascorbate in samples can be calculated by the regression equation of the standard curve.

(c) DHAA is calculated as the difference between the total pool and the reduced pool.

3.4 Fluorometric Analysis of AA

1. Preliminary test for appreciable amount of basic substances
   See Subheading 3.1, **item 1**.
2. Sample treatment
   (a) Fruit and vegetable juices. *See* Subheading 3.1, **item 2(a)**.
   (b) Dry products containing no appreciable of basic substances. Pulverize test portion by gentle grinding, add HPO_3–CH_3COOH solution, and triturate until test portion is in suspension. Dilute with HPO_3–CH_3COOH solution to 100 μg ascorbic acid/mL. Designate this volume as *V* mL. Filter solutions containing large amounts of suspended solids through filter paper.
   (c) Dry products containing appreciable of basic substances. Pulverize test portion by gentle grinding, add HPO_3–CH_3COOH–H_2SO_4 solution to adjust pH to 1.2, and triturate until test portion is in suspension. Dilute with HPO_3–CH_3COOH solution to measured volume. Designate this volume as *V* mL. Dilute with HPO_3–CH_3COOH solution to 100 μg ascorbic acid/mL. Designate this volume as *V* mL. Filter solutions containing large amounts of suspended solids through filter paper. Designate as test solution.
   (d) Liquid products. Take amount of test portion containing 100 mg ascorbic acid. If appreciable amounts of basic substances are present, adjust pH to 1.2 with HPO_3–CH_3COOH–H_2SO_4 solution. Dilute with HPO_3–CH_3COOH solution to measured volume containing 10–100 mg ascorbic acid/100 mL. Designate this volume as *V* mL. Filter solutions containing large amounts of suspended solids through filter paper. Dilute with HPO_3–CH_3COOH solution to 100 μg ascorbic acid/mL. Designate this volume as *V* mL. Filter solutions containing large amounts of suspended solids through filter paper. Designate as test solution.
3. Preparation of test solution
   (a) Transfer 100 mL sample solution and AA standard solution to a 250 mL Erlenmeyer. Add 2 g acid-washed Norit, shaking vigorously. Filter and discard the first few mL (about 5 mL). Transfer 5.0 mL of each filtrate to separate 100 mL volumetric flasks containing 5.0 mL H_3BO_3–

CH_3COONa solution, and shake well. After standing for 30 min, fix the volume with distilled water. Designate as the blank solution for the corresponding sample and standard solution.

(b) During the 30 min, transfer 5.0 mL of each filtrate to separate 100 mL volumetric flasks containing 5.0 mL CH_3COONa and 75 mL H_2O. Fix the volume with distilled water. Designate as the sample and standard solutions.

(c) Transfer 2.0 mL sample solution and blank solution for sample to separate 10 mL test tubes. Add 5.0 mL *o*-phenylenediamine solution to all tubes. Shake well, and protect from light for 60 min at room temperature before testing.

4. Fluorometric detection

Transfer appropriate volume of sample solution, corresponding blank solution and standard AA solution to separate fluorescence reading tubes. Fluorescence is determined at $\lambda_{Ex} = 350$ nm, $\lambda_{Ex} = 430$ nm.

5. Calculation

mg Ascorbic acid/g, mL, etc. $= \frac{\text{average } X - \text{average } D}{\text{average } C - \text{average } B} \times (20 \times S \times \frac{V}{E})$,

where V is initial test solution volume; E is number of g, mL, etc.; and S is concentration of standard in mg/mL added to reading tube.

3.5 High-Performance Liquid Chromatography

1. Sample treatment

(a) Liquid or powder foods

Mix thoroughly before and during weighing aliquots for analysis to ensure representative subsampling.

(b) Fruits, vegetables, and other solid food

Cut into small pieces (1.5 cm), grind in a mortar, and/or homogenize. Prepare immediately before use. If that is not possible, freeze the samples in liquid nitrogen, and store at <-18 °C.

2. Extraction and preparation of sample solution

(a) Weigh an appropriate amount of sample (designated as m) to make sure the concentration of AA in test sample solution lies in a reasonable range.

(b) Add metaphosphoric acid solution (20 g/L) to the sample, and then vortex or homogenize. Centrifuge the extraction mixture at 4–8 °C and collect the supernatant. Transfer the supernatant to a volumetric flask, and fix to the volume (designated as V) with the metaphosphoric acid (20 g/L) solution.

(c) Filter the sample solution through 0.45 μm polyvinylidene fluoride (PVDF) membrane. The sample solution is kept away from light at 4–8 °C before use.

3. Detection

(a) Separation of AA in sample solution is conducted by a C_{18} reversed phase column (250 mm × 4.6 mm, 5 μm particle size) or one of equivalent performance.

(b) The mobile phase is as follows: A—dissolve 6.8 g potassium dihydrogen phosphate and 0.91 g cetyltrimethylammonium bromide in ultrapure water, and adjust pH to 2.5–2.8 with phosphoric acid. Then transfer to a 1 L volumetric flask and fill to the volume. B—100% methanol. Mix A and B in ratio 98:2 (v:v). Filter through 0.45 μm PDVF filter and degas. The flow rate is 0.7 mL/min. The mobile phase and flow rate are not fixed. It can be modified to suit the food matrix, extractant, and the form of AA. Acid solutions like 0.05% (w/v) aqueous formic acid, 0.02% (w/v) aqueous orthophosphoric acid, and 0.025% trifluoroacetic acid have been used as an alternative mobile phase.

(c) The temperature of the column is set at 25 °C. A 20 μL injection volume is used.

(d) Detection of AA is performed employing a UV detector at 245 nm.

4. Calculation

(a) Detect standard solutions first, and draw a standard curve or calculate a regression equation by taking concentration of the standard solution (μg/mL) as *X* while the peak height or the peak area as *Y*.

(b) Detect sample solution, and the quantification of AA of interest is conducted by the workstation of the equipment.

(c) The amount of AA (L(+)-ascorbic acid/D(−)-ascorbic acid) is calculated by the following equation:

$$\text{mg AA}/100\text{ g} = \frac{(c_1 - c_0) \times V}{m \times 1000} \times F \times 100,$$

where c_1 is the concentration of AA of sample solution, μg/mL; c_1 is the concentration of AA of blank solution, μg/mL; V is the volume of sample solution, mL; m is g of the sample, g; and F is dilution ratio during the extraction and sample solution preparation.

3.6 Voltammetry

1. Sample treatment

(a) Fruit and vegetable juice: Shake well.

(b) Fruit, vegetable, and other solid products: Weigh 100 g, and cut into small pieces. Add 100 mL 2.0% oxalic acid solution and homogenize. Centrifugate 5 min at 8500 r/min. Collect the supernatant as sample solution. Prepare before use.

2. Detection

The progress is followed with Pisoschi et al. [12]. All measurements were performed at 295.5 K using 0.34 mol/L KCl solution as supporting the electrolyte. Before each determination, the Pt working electrode was cleaned mechanically, by polishing it on alumina (63–200 μm granularity), and electrochemically, by applying a − 1.5 V potential pulse for 3 s. For each measurement, the potential was scanned within the range 100–1000 mV, with a 50 mV/s scan rate, and the value of the background current obtained for the KCl 0.34 mol/L solution was subtracted from the current corresponding to the analyzed solution/sample. For investigating the potential scan rate influence, this parameter varied from 50 mV/s to 250 mV/s.

Detect the standard solutions of AA in different concentrations. Draw a calibration graph with *X* representing the current *I* (μA) and *Y* concentration of AA solution (mmol/L).

Detect sample solutions similarly. The concentration of AA in the sample solution can be calculated by substituting into the standard curve equation, and then multiply the dilution ratio (if there is a dilution step in sample solution preparation, otherwise ×1) to get the amount of AA in the food product.

3.7 Coulometric Titration

1. Sample treatment
   (a) Fruit and vegetable juice: Shake well.
   (b) Fruit, vegetable, and other solid products: Homogenize a certain amount of sample. Weigh 10–30 g pulp in a beaker, and add 4.2 mL 1 mol/L H_3PO_4 solution and some water with the total volume no more than 40 mL. Sonicate 5 min, and vacuum filter through a rapid paper plus four-layer gauze twice. Transfer the filtrate in a 50 mL volumetric flask, and fill up with water to the volume, serving as the sample solution. Prepare before use.
2. Detection

Mix 5 mL 1 mol/L KI solution, 5 mL 1 mol/L H_3PO_4 solution, and 45 mL water as the electrolyte in the cell. Electrodes are connected to their electric circuits. The magnetic stirrer is set at a relatively high speed and the pH-meter set to READ. The generator current is adjusted to 10 mA, and the measuring range set to 100 mV.

The KI generation is initiated simultaneously as the recorder pen crosses a vertical line on the chart after a stable baseline has been achieved. Record the magnitude of current. Add 2.00 mL standard solution of AA in the cell and goes through the process as the blank. A 1000 μL sample solution is titrated the same way. Iodine is generated by the generation electrode (anode) of KI and reacts with AA in the sample solution quantitatively. When the AA is fully oxidized, the indicator is on.

Electrode reaction: $2I^- - 2e = I_2$

Chemical reaction: $C_6H_8O_6$ (AA) + I_2 = $C_6H_6O_6$ (DHAA) + $2I^-$ + $2H^+$

3. Calculation

The content of AA in sample is calculated by Faraday's law, mg AA of the sample $= \frac{QM}{nF}$,

where Q is magnitude of current, mC; M is molecular weight of ascorbic acid, g/mol; n is number of electron; and F is Faraday constant, 96,487 C/mol.

4 Notes

4.1 2,6-Dichloroindophenol Titration

(a) The testing process shall be carried out in the dark because AA is reversibly oxidized into DHAA upon exposure to light.

(b) As dehydroascorbic acid (DHAA) has no reducibility, this method only quantifies the oxidized form (ascorbic acid) in samples. If a proportion of AA in its oxidized form is already present in the food, the results will underestimate the total amount of AA. The method can be conducted as titrimetric or colorimetric where L-dehydroascorbic acid is further converted into oxazone with 2,4-dinitrophenylhydrazine, recolored with sulfuric acid and measured at 520 nm [16].

(c) Colored samples can cause difficulties in determining the titration end points, especially reddish-purplish colors. Add 0.4 g white pottery clay/g sample to sample solution, and then filter to eliminate any innate color. Another approach is to employ repeated ether/dimethylbenzene/ethyl acetate extractions to remove lipid soluble pigments such as lycopene in food to eliminate the color, thus facilitating the visual detection of the titration end point [17, 18].

(d) Since this method is based on the redox reaction, a number of reducing substances in foods could interfere with the determination, i.e., phenols, sulfhydryls, triose reductones, and ions such as ferrous, cuprous, and sulfite. Their presence can lead to high false-positive results [17]. Interferences can be overcome by adjusting the pH and other reaction conditions so that most other materials react only very much more slowly than does ascorbate acid [17, 19].

(e) Products containing ample ferrous (Fe^{2+}), stannous (Sn^{2+}), and/or cuprous (Cu^+) could give values in excess of their actual ascorbic acid content by this method due to their reducibility.

Following are simple tests to determine whether these reducing ions are present in such amounts as to invalidate test: Add 2 drops 0.05% aqueous solution of methylene blue to 10 mL freshly prepared mixture (1 + 1) of test solution and HPO_3–CH_3COOH reagent and mix. Disappearance of methylene blue color in 5–10 s indicates presence of interfering substances.

Stannous (Sn^{2+}) does not show up in this test and may be tested for as follows: To another 10 mL test solution to which 10 mL HCl (1 + 3) has been added, add 5 drops 0.05% aqueous solution of indigo carmine and mix. Disappearance of color in 5–10 s indicates presence of stannous (Sn^{2+}) or other interfering substance [4].

(f) In detection of AA in ready-to-feed milk-based infant food, interfering Fe and Cu can be removed by means of the chelating agent EDTA [20].

4.2 Iodine Titration

(a) DHAA is not detected by this method.

(b) As AA is susceptible to oxidation by air, freshly prepared samples should be titrated immediately and rapidly. However, if the samples have to be prepared several hours earlier, oxidation can be minimized by the addition of a small amount of oxalic acid (e.g., 1 g oxalic acid per 100 mL of sample solution) [15].

(c) Wear a mask to weigh I_2 in a fume hood for safety.

(d) Food sample solution with dark color could cause an interference to determine the end point. Solvent and precipitant of the dark pigments could be attempted to improve the accuracy of determination.

4.3 AA Analysis by Spectrophotometry

(a) The absorbance range between 0.2 and 0.8 is favorable in order to reduce any measurement errors. Therefore, the concentrations of AA standard and sample solutions should be adjusted to fall within this range.

(b) The colored complex of Cu(I) with 2,2-biquinoline (λ_{max} = 540 nm) [21], rhodamine (λ_{max} = 473 nm) [22], and 2,9-dimethyl-1,10-phenanthroline [23] (λ_{max} = 450 nm) was tested AA in different samples [24]. However, in Cu(II)-Cu(I) redox system, Fe(II), cysteine, and sodium thiosulfate interfered seriously which caused a reduction of Cu(II).

(c) Potassium permanganate can also act as a reducing agent in the spectrophotometric determination of AA in fruit juice, which induced higher results than values labeled on the products. Potassium permanganate was concluded to be applicable to less complex matrices like AA solution and pills [25].

(d) A spectrophotometric method for the determination of AA in citrus (lemon, orange, and grapefruit) juices is based on the reaction with a 0.02% methanolic peri-naphthindan-2,3,4-trione hydrate solution at pH 3, which yields dihydroxy-peri-naphthindenone which was measured at 460 nm [26].

(e) Bromine water oxidized ascorbic acid to dehydroascorbic acid in presence of acetic acid. After coupling with 2,4-dinitrophenyl hydrazine at 37 °C, the solution was treated with 85% H_2SO_4 to produce a red color complex, and the absorbance is spectrophotometrically measured at 521 nm [23].

(f) Dimethoxydiquinone gives a violet-colored product with AA in a phosphate buffer (pH 6.6). The absorbance of the colored solution in dioxane is measured at 510 nm which can be applied to the analysis of citrus fruits [24, 27].

(g) Ascorbic acid is determined at 336 nm by using tetrachloro-benzoquinone reagent [28].

(h) Methylene blue and ninhydrin turn colorless after reducing AA, which is used for the assay of AA in food at 653 nm [29].

(i) Molybdenum blue is applied for the determination of AA in fruit and vegetables, pastries, and beverages, forming on reduction of phosphomolybdate complex [30], ammonium molybdate, or molybdic acid [31, 32].

(j) Aniline diazonium salts like diazotized-4-methoxy-2-nitroaniline, 4-nitrobenzene diazonium, and diazotized 4-nitroaniline-2:5-dimethoxy-aniline are found to couple with ascorbic acid to form a colored compound which can be detected at different λ_{max} [29].

(k) Folin phenol reagent reacts with AA to show a blue complex with λ_{max} of 730 nm [32, 33]. But this method is time-consuming because it takes 40–50 min for a full color intensity to develop.

(l) Ascorbate oxidase can also be used to catalyze the oxidation of AA to DHAA. Then DHAA was reacted with *o*-phenylenediamine to produce a 2,2-anhydro quinoxaline which was observed at 358 nm in fruit and vegetable [34, 35].

4.4 Fluorometric Analysis of AA

(a) This method detects the total AA in food, including both AA and DHAA.

(b) For certain food such as pet foods and some vegetables, fruits, and cereals, 2.0 g Norit is excessive, which could cause a longer filter time (>15 min) and cloudy filtrates which might not be eliminated by repeatedly filtering or centrifuging [36].

(c) Acid-washed Norit is not a unique candidate for oxidant of AA. *N*-bromosuccinimide is a substitution in an investigation by Dunmire et al. [36]. Roy et al. [37] have reported products containing cocoa resulted in higher AA values when oxidized with *N*-bromosuccinimide rather than Norit, which probably resulted from the oxidation of cocoa.

(d) As metaphosphoric acid is insoluble, it is time-consuming to make the solution. Besides, its storage time could not be too long (stored in brown volumetric flasks for 7–10 days). Metaphosphoric acid was replaced by oxalic acid in AA determination in yogurt [38], fruit juice [39], and tea [40], serving as the extractant of AA and the precipitant of protein.

4.5 High-Performance Liquid Chromatography

(a) Extractant should stabilize AA in its original forming in food, preventing interference such as protein, enzymes, carbohydrate, fat, Fe(II), Cu(II), etc., and provides an acid solution with pH <4 [41–45].

(b) EDTA chelates metals efficiently [46, 47]; 10 mL 0.1 M citric acid and an equal volume of methylene chloride were added to extract AA from fresh fruit and vegetable, juice, and canned products [48]; 8 mL extraction buffer mixed with 5% metaphosphoric acid, 1 mM ethylenediaminetetraacetic acid (EDTA), and 5 mM Tris (2-carboxyethyl) phosphine (TCEP) were added to a 50 mL centrifuge tube containing 2 g sample [49].

(c) Besides chemical extraction, microextraction by packed sorbent (MEPS) is also applied to separate AA from beverages [50]. In a typical experiment, 300 μL of sample (ice-tea or fruit juice) is passed through the silica sorbent and the entrapped AA quantitatively recovered by eluting with 60 μL methanol-water (10%, v/v) [41]. Then the elution is filtered through 0.45 μm PVDF membrane and is ready to be detected by HPLC. The procedure is simple with higher cost than chemical extraction.

(d) This method is applicable for individual determination of L(−)-ascorbic acid, D(+)-ascorbic acid, and DHAA by their standard counterparts. The total ascorbate (AA+DHAA) is therefore gained.

(e) UV detector is much used for L-ascorbic acid detection [51].

(f) Fluorescence has also been applied for identification purposes by the replacement of the fluorometric detector instead of the UV one. AA is oxidized to DHAA, which forms highly fluorescent quinoxaline derivatives with substrate like benzene-1,2-diamine and *o*-phenylene diamine. The derivatives would be detected by the fluorometric detector [52]. This procedure demands the use of additional reagents and is more time-consuming, but the sensitivity to DHAA increased than that of the UV detector according to some research [51].

(g) HPLC-UV analysis can also be conducted at other conditions. AA and DHAA from food are determined at 254 nm and 210 nm, respectively, by Rose and Nahrwold [53]; ion exchange column is also feasible [54].

(h) Different extractant is also widely applied. Metaphosphoric acid is a common alternative [54].

4.6 Voltammetry

In this method, a potentiostat galvanostat is necessary where Pt disc electrode serves as working electrode, saturated calomel electrode (SCE) serves as reference electrode, and Pt strip serves as counter electrode.

4.7 Coulometric Titration

Ultrapure water is used throughout the procedure.

References

1. Doseděl M, Jirkovský E, Macáková K, Krčmová LK, Javorská L, Pourová J, Mercolini L, Remião F, Nováková L, Mladěnka PJN (2021) Vitamin C—sources, physiological role, kinetics, deficiency, use, toxicity, and determination. Food Secur 13(2):615
2. Schlueter AK, Johnston CS (2011) Vitamin C: overview and update. J Evidence-Based Integr Med 16(1):49–57
3. Jacob RA, Sotoudeh G (2002) Vitamin C function and status in chronic disease. Nutr Clin Care 52:66–74
4. AOAC Official Method 967.21 (1967) Ascorbic acid in vitamin preparations and juices. https://vsip.info/aoac-method-ascorbic-ac-967-21-pdf-free.html
5. Zezzi-Arruda MA, Poppi RJ (2005) Spectrophotometry | inorganic compounds. In: Worsfold P et al (eds) Encyclopedia of analytical science, 2nd edn. Elsevier
6. Masato O (1980) An improved method for determination of L-ascorbic acid and L-dehydroascorbic acid in blood plasma. Clin Chim Acta 103(3):259–268
7. Gillespie KM, Ainsworth EA (2007) Measurement of reduced, oxidized and total ascorbate content in plants. Nat Protoc 2(4):871–874
8. Kampfenkel K, Vanmontagu M, Inze D (1995) Extraction and determination of ascorbate and dehydroascorbate from plant tissue. Anal Biochem 225(1):165–167
9. Stevens R, Buret M, Garchery C et al (2006) Technique for rapid, small-scale analysis of vitamin C levels in fruit and application to a tomato mutant collection. J Agri Food Chem 54(17): 6159–6165
10. AOAC Official Method 967.22 (1968) Vitamin C (Total) in vitamin preparations. https://www.wdfxw.net/doc96206270.htm
11. Aqel A, Yusuf K, Al-Rifai et al (2014) Vitamin analysis in food by UPLC-MS. In: Naushad M (ed) Ultra performance liquid chromatography mass spectrometry, 1st edn. CRC Press, Boca Raton
12. Pisoschi AM, Danet AF, Kalinowski S (2008) Ascorbic acid determination in commercial fruit juice samples by cyclic voltammetry. J Autom Methods Manag Chem 2008:1–8
13. Wawrzyniak J, Ryniecki A, Zembrzuski W (2005) Application of voltammetry to determine vitamin C in apple juices. Acta Sci Pol Technol Aliment 42:5–16
14. Ciancaglini P, Santos HI, Daghastanli KRP et al (2001) Using a classical method of vitamin C quantification as a tool for discussion of its role in the body. Biochem Mol Biol Educ 29: 110–114
15. College of Science, University of Canterbury (2011) Determination of vitamin C concentration by titration. http://www.outreach.canterbury.ac.nz/chemistry/vitamin_C_iodine.shtml

16. Šaponjac VT, Čanadanović-Brunet J, Ćetković G et al (2016) Detection of bioactive compounds in plants and food products. In: Nedović V et al (eds) Emerging and traditional technologies for safe, healthy and quality food, food engineering series. Springer International Publishing, Switzerland
17. Tee ES, Young SI, Ho SK et al (1988) Determination of vitamin C in fresh fruits and vegetables using the dye-titration and microfluorometric methods. Pertanika 11:39–44
18. Zhao X, Jiang Y, Liu K et al (2006) Assay research on VC in fruit and vegetable. Shipin Kexue 27:197–199
19. Bender DA (1982) Vitamin C. In: Barker BM et al (eds) Vitamins in medicine, 4th edn. William Heinemann Medical Books Ltd, London
20. AOAC Official Method 985.33 (1988) Vitamin C (reduced ascorbic acid) in ready-to-feed milk-based infant formula. http://www.aoacofficialmethod.org/index.php?main_page=product_info&products_id=227
21. Shieh H-H, Sweet TR (1979) Spectrophotometric determination of ascorbic acid. Anal Biochem 96(1):1–5
22. Tütem E, Ülküseven B, Apak R (1992) Spectrophotometric determination of ascorbic acid in pharmaceutical products with the copper (II)-rhodanine reagent. Anal Lett 25:471–483
23. Baker WL, Lowe T (1985) Sensitive ascorbic acid assay for the analysis of pharmaceutical products and fruit juices. Analyst 110:1189–1192
24. Arya SP, Mahajan M, Jain P (1998) Photometric methods for the determination of vitamin C. Anal Sci 14:889–895
25. Zanini DJ, Silva MH, Aguiar-Oliveira E et al (2018) Spectrophotometric analysis of vitamin C in different matrices utilizing potassium permanganate. Eur Int J Sci Technol 7:70–84
26. Hassan SSM, El Fattah MMA, MTM Z (1975) Spectrophotometric determination of vitamin C in citrus fruits using *peri*-naphthindan-2,3,4-trione. Z Anal Chem 277:369–371
27. Martens M, Martens F, Maenhout P et al (1975) Systematic identification of unknown drugs in powder form by means of ultraviolet spectrophotometry in forensic toxicology. Anal Chem 47:458–461
28. Pandey NK (1982) Spectrophotometric and titrimetric determinations of ascorbic acid. Anal Chem 54:793–796
29. Hosain MM, Khan MMR, Rahman MM (1970) Analysis of vitamin C (ascorbic acid) contents in various fruits and vegetables by UV-spectrophotometry. Bangladesh J Sci Ind Res 42(4):417–424
30. Wang Z, Chen Y (1991) Spectrophotometric determination of ascorbic acid in medicines and foods with phosphorus-molybdenum heteropoly acid. Fenxi Huaxue 19:357–361
31. Elnenaey ES, Soliman R (1979) A sensitive colorimetric method for estimation of ascorbic acid. Talanta 26:1164–1166
32. Bajaj KL, Kaur G (1981) Spectrophotometric determination of L-ascorbic acid in vegetables and fruits. Analyst 106:117–120
33. Jagota SK, Dani IIM (1982) A new colorimetric technique for the estimation of vitamin C using folin phenol reagent. Anal Bioanal Chem 127:178–182
34. Esteban MR, Ho C-N (1997) Enzymatic spectrophotometric determination of ascorbic acid in commercial vitamin C tablets. Microchem J 56:122–129
35. Casella L, Gullotti M, Marchesini A et al (1989) Rapid enzymatic method for vitamin C assay in fruits and vegetables using peroxidase. J Food Sci 54:374–375
36. Dumier DL, Reeze JD, Bryan R et al (1979) Automated fluorometric determination of vitamin C in foods. J Assoc Off Anal Chem 62: 648–652
37. Roy RBCA, Salpeter J (1976) Automated fluorometric method for the determination of total vitamin C in food products. J Assoc Off Anal Chem 59:1244–1250
38. Wang M, Liu Z (1999) Measurement of vitamin C in yogurt. J Shanghai Fisher U 8:279–281
39. Wang J, Li S, Wang X (2018) Fluorimetric determination of vitamin C in pomegranate. Shandong Chem Ind 47:44–46
40. Zheng Y, Xie D, Zhang W (2005) Spectrofluorometry determination for vitamin C in tea. Fujian Anal Test 14:2307–2317
41. Alessandra G, Fulvia C (2017) Analysis of vitamins by liquid chromatography. In: Fanali S et al (eds) Liquid chromatography. Elsevier
42. Bognár A, Daood HG (2000) Simple in-line postcolumn oxidation and derivatization for the simultaneous analysis of ascorbic and dehydroascorbic acids in foods. J Chromatogr Sci 38(4):162–168
43. Castro RN, Azeredo LC, Azeredo MAAA et al (2001) HPLC assay for the determination of ascorbic acid in honey samples. J Liq Chromatogr Relat Technol 24:1015–1020
44. Rizzolo A, Brambilla A, Valsecchi S et al (2002) Evaluation of sampling and extraction

procedures for the analysis of ascorbic acid from pear fruit tissue. Food Chem 77:257–262
45. Chebrolu KK, Jayaprakasha GK, Yoo KS et al (2012) An improved sample preparation method for quantification of ascorbic acid and dehydroascorbic acid by HPLC. LWT Food Sci Technol 47:443–449
46. Pérez AG, Olías R, Ezpada J (1997) Rapid determination of sugars, nonvolatile acids, and ascorbic acid in strawberry and other fruits. J Agric Food Chem 45:3545–3549
47. Barros AIRNA, Silva AP, Gonçalves B (2010) A fast, simple, and reliable hydrophilic interaction liquid chromatography method for the determination of ascorbic and isoascorbic acids. Anal Bioanal Chem 396:1863–1875
48. Vanderslice JT, Higgs DT (1984) HPLC analysis with fluorometric detector of vitamin C in food samples. J. Chromatogr Sci 22:485–489
49. Tarrago-Trani MT, Phillips KM, Cotty M et al (2012) Matrix-specific method validation for quantitative analysis of vitamin C in diverse foods. J Food Compos Anal 26:12–25
50. Adam M, Pavlíková P, Čížková A et al (2012) Microextraction by packed sorbent (MEPS) as a suitable selective method for l-ascorbic acid determination in beverages. Food Chem 135: 1613–1618
51. Spinola V, Llorent-Martinez EJ, Castilho PC (2014) Determination of vitamin C in foods: current state of method validation. J Chromatogr A 1369:2–17
52. Burini G (2007) Development of a quantitative method for the analysis of total l-ascorbic acid in foods by high-performance liquid chromatography. J Chromatogr A 1154(1):97–102
53. Rose RC, Nahrwold DL (1981) Quantitative analysis of ascorbic acid and dehydroascorbic acid by high-performance liquid chromatography. Anal Biochem 114(1):140–145
54. Graham WD, Annette D (1992) Determination of ascorbic and dehydroascorbic acid in potatoes (*Solanum tuberosum*) and strawberries using ion-exclusion chromatography. J Chromatogr A 594:187–194

Chapter 19

Analysis of Thiamin and Its Respective Vitamers by High-Performance Liquid Chromatography

Barry Hucker and Frank Vriesekoop

Abstract

Thiamin and its respective vitamers, thiamin mono- and di-phosphate, can be easily analyzed using high-performance liquid chromatography-fluorescence methods. By utilizing a 1% w/w solution of potassium ferricyanide, thiamin vitamers can be easily converted to their respective thiochrome esters and analyzed to very low levels in samples. The method described in this chapter can be easily utilized as a base for a variety of samples with high reproducibility and repeatability.

Key words Thiamin, Vitamers, High-performance liquid chromatography, Thiochrome, Fluorescence

1 Introduction

Thiamin (aka aneurin or vitamin B_1) is an essential micro-nutrient to support metabolism. It occurs naturally in multiple phosphorylated forms (vitamers) plus as free thiamin. The most common biological form is thiamin-pyrophosphate, the diphosphorylated form. Thiamin pyrophosphate is most typically found as an enzymatic cofactor in decarboxylase enzymes [1, 2], notably pyruvate dehydrogenase in most cellular systems and pyruvate decarboxylase in ethanol fermentation pathways [3].

The thiamin molecule is readily soluble in water and relatively stable under acidic conditions, even when heated. In neutral or alkaline environments, thiamin is unstable and rapidly degraded when heated. Thiamin is sensitive to oxygen and ultraviolet light and is susceptible to cleavage by sulfite [4].

In the absence of sufficient quantities of thiamin, the condition known as Beriberi can occur, which, when left untreated, can be fatal. Worldwide, thiamin deficiency is primarily due to inadequate dietary intake, often linked to diets high in polished rice and processed grains. In Western countries, thiamin deficiency is most

Lilian Mariutti and Frank Vriesekoop (eds.), *Food Chemistry: Basic Protocols*, Methods and Protocols in Food Science, https://doi.org/10.1007/978-1-0716-4678-6_19,

commonly presenting in patients suffering from alcoholism or chronic illness or in people that have undergone a gastric bypass [5]. In people suffering from alcoholism, the main cause of thiamin deficiency is due to impaired thiamin absorption from the gut [6], while the ability to take up thiamin following a gastric bypass is significantly reduced due to the altered intestinal physiology. This is argued to be either caused by prolonged nausea and emesis or induced by an altered gut ecology [7]. Regardless of the cause of thiamin deficiency, there are two main types of Beriberi: dry Beriberi and wet Beriberi. Dry Beriberi typically affects the central nervous system resulting in confusion and numbness of the hands and feet often resulting in difficulties moving one's legs and arms. Wet Beriberi typically affects the cardiovascular system resulting in an increased heart rate, with shortness of breath and swelling of the legs [8].

Methodologies of Analysis

Over the past 100 years testing methods for thiamin have developed significantly from their basic microbiological methods [9, 10] and evolved to spectrophotometric methods [11, 12]. With the invention of chromatography and in particular high-performance liquid chromatography (HPLC), methods have been created for a variety of matrices [13–18]. Unlike their predecessors (microbial and fluorometric methods), these chromatographic methods have the ability to differentiate between the various forms of thiamin found in nature, as well as providing more accurate results as matrix interferences can be removed [14]. The advances in separation techniques have led to the more accurate separation and quantitation of the various vitamers of thiamin (thiamin; thiamin mono-phosphate (TMP); and thiamin di-phosphate (TDP)). Most modern chromatographic systems utilize a variety of detectors including ultraviolet/visible (UV/Vis), photodiode array (PDA), refractive index (RI), and mass spectrometer (MS) to increase the potential detection limits. Due to its high sensitivity, relatively low cost, and the use of specific excitation and emission wavelengths, the fluorescence detector has been typically employed for the analysis of low-level (ppb range) thiamin vitamers [19–22]. Unfortunately, thiamin vitamers do not naturally fluoresce, and this can be overcome by specific derivatization of the base compound in order to induce artificial fluorescence.

Thiamin and its vitamers can be derivatized into thiochrome esters (Fig. 1) prior to detection by methods that have been well described in the past, by a variety of authors for fluorimetric assays [23–27]; capillary electrophoresis [28]; and HPLC analysis [13, 18, 29–35].

The derivatization (via an oxidation reaction) of thiamin to thiochrome generally occurs under alkaline conditions due to

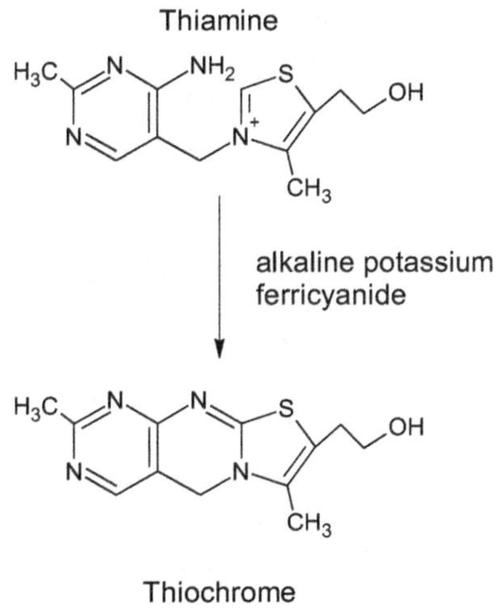

Fig. 1 Reaction of thiamin to the fluorescent thiochrome in the presence of alkaline potassium ferricyanide

strongly enhanced fluorescence of the derivatized product at pH >8 [20, 36–38]. However, due to the alkaline nature of the derivatized sample, it is not possible to directly introduce a derivatized sample into a HPLC system without it first being neutralized [39]. This is mainly due to the fact that the vitamers are best separated by reverse-phase HPLC or non-polar C18 columns, which degrade extremely rapidly under alkaline conditions [40–42]. Strengthened alkali-silica columns have been utilized in the past for alkaline separations [40, 42, 43] that could be employed for this work; however there is no information on whether the strong oxidative potential of potassium ferricyanide would destroy a column. Alternatively, the derivatization can be facilitated after the compounds have been separated on the column (post-column) [18]. This method employs a secondary pump and a mixing coil system that is connected to the effluent of the column, which allows the effective conversion of thiamin to thiochrome without risk of dissolving the silica gel. This method too has its drawbacks as the HPLC effluent has to be neutralized before release. Pre-column derivatization [33, 44] may be time-consuming; however it is more advantageous over post-column methods as the analyst can control the reaction more thoroughly; there is no risk of column degradation in case the flow of mobile phase is halted during analysis for whatever reason, and it has a lower capital cost.

2 Materials

All equipment and consumables mentioned in this chapter can be substituted by alternative products from other manufacturers. In this chapter we describe a tried set of methodologies based on the equipment and consumables mentioned. Any variations from this method may cause elution timing changes and different elution profiles. Also matrix interferences may need to be accounted for resulting in changes to the elution profiles.

2.1 Basic Reagents

All reagents/standards used describe the methodologies were sourced through Sigma-Aldrich (St. Louis, MO) and were of 95% purity or greater with the exception of methanol which was supplied through Scharlau (Barcelona, Spain). In order to achieve an optimal detection limit, it is recommended that only 18 MΩ water is used. It is also recommended that any reused equipment should be cleaned via sonication and washed in 18 MΩ water. It is possible to get artifacts in the chromatograms if glassware is not well cleaned. For the sake of the methodology described here, we used a Sartorius Arium 611VF water purification system (Sartorius AG, Göttingen, Germany).

The derivatization reagent used consisted of 1% w/w potassium ferricyanide in a 15% w/w sodium hydroxide solution. This must be prepared weekly and stored in amber bottle (out of direct sunlight) at 4 °C until used.

The mobile phase used was a 10 mM KH_2PO_4/K_2HPO_4; pH 6.5 solution and 100% methanol were used in a gradient combination as described in Subheading 2.2.

The neutralizing solution was 1.33 M phosphoric acid solution.

2.2 Instrumentation

The HPLC system consisted of a Varian Prostar 230 solvent delivery system (flow rate 1 mL/min), a Prostar 410 autosampler (10 μL injection), a Prostar 500 column valve module (30 °C), and a Prostar 363 fluorescence detector 360/425 nm excitation/emission for thiamin and its vitamers.

The column utilized was a Varian Pursuit C18 5 μm (250 mm × 4.6 mm) fitted with a Securityguard Cartridge (C18 4 × 3.0 mm) (Phenomenex, Torrance, CA).

The mobile phase gradient employed was as follows: solvent A, 10 mM potassium phosphate buffer; solvent B, 100% methanol; 0–0.5 min 95:5 A:B; 0.5–10 min 65:35 A:B; 10–15 min 65:35 A:B; 15–16 min 95:5 A:B; at a total run time of 25 min.

3 Methodology

3.1 Sample Preparation

A variety of sample preparation techniques can be used to extract thiamin vitamers from a sample matrix. Due to the unstable nature of thiamin, it is always recommended to keep the extraction methods simple. Due to the water solubility of the vitamers, simply vortexing or mixing a sample in the phosphate buffer can be effective. Hucker et al. [45] discussed simply degassing and filtering samples of alcoholic beverages, while Liddicoat et al. [46] utilized a 100 g/L polyvinylpolypyrrolidone (PVPP) solution to remove polyphenolic compounds via precipitation in wine prior to derivatization. Solid phase extraction can also be employed; however this will increase the cost of analysis and may influence the recovery efficiency of the vitamers.

3.1.1 Thiamin Extraction from Beverages

Liquid samples such as beers, ciders, and wines require to be degassed, which can be facilitated by passing the sample through a 0.45 μm RC filter. Other methods of degassing could be used, but in order to preserve the HPLC column, it is best for all samples to be passed through a 0.45 μm RC filter anyway. Directly following filtration, the samples can be processed for thiochrome derivatization.

3.1.2 Thiamin Extraction from Beverages with High Antioxidant Content (Red Wine)

Using the preparation step presented in Subheading 3.1.2, a significant interference in the analysis of red wine matrix can occur in a number of red wines [45] with no detectable thiamin vitamers being reported. It is possible that no thiamin vitamers exist in red wine; however spike recoveries of only ~20% suggest a sequestering potential of the ferricyanide oxidation of thiamin to thiochrome. This is likely due to the antioxidant activity of polyphenolic compounds in red wines. It has previously been reported that red wines contain substantial concentrations of total polyphenolic compounds (1700–1800 mg/L) compared to white wines (200–300 mg/L) [47]. It has also been reported by several authors [48–51] that potassium ferricyanide solutions can be used to measure the reducing power of anthocyanins in a variety of fruits. It can therefore be concluded that polyphenols such as anthocyanins and flavonoids which exhibit antioxidant activity could easily interfere in the essential oxidation of thiamin to thiochrome as described by Hucker et al. [45]. Without the effective removal of antioxidant inhibitors such as polyphenols in red wine, thiamin vitamer measurements will always be underestimated. Polyvinylpolypyrrolidone (PVPP) has been shown to remove unwanted polyphenols from beverages [52–55]. Bulk polyphenol removal by PVPP is due to preferential hydrogen bonding with haze active polyphenols over other components in beers and wines. The most effective reagent to remove polyphenols from red wine while preserving all thiamin vitamers is PVPP [46].

Using a PVPP stock solution (100 g/L), dilute with the wine sample so that the final concentration of PVPP is 50 g/L, in a 15 mL centrifuge tube. Mix thoroughly and continuously agitate the samples for 15 min [46]. A dense and fluffy precipitate will form. Centrifuge at 2000 rcf and transfer the supernatant (~5 mL) into a clean test tube. The treated sample can then be filtered through a 0.45 μm regenerated cellulose (RC) filter prior to analysis. Directly following filtration, the samples can be processed for thiochrome derivatization.

3.1.3 Thiamin Extraction from Solid Matrices and High Protein Containing Samples

When analyzing samples that present a solid matrix such as grains, these samples have to be processed to a fine powder. Many solid matrices will contain relatively high concentrations of proteins, which might have a range of vitamers incorporated into them. In order to dislodge the vitamers from the proteins, and to be able to remove proteins from the sample, a trichloroacetic acid (TCA) can be used.

To a 0.5 g of powdered sample, add 5 mL of 0.5 M (TCA). Vortex to mix and continuously agitate for 15 min. This extraction should not exceed 30 min as it has been reported that losses in thiamine vitamer recovery have been observed [56]. After 15 min, centrifuge at 2000 g, and remove 2 mL of supernatant (2 mL) into a clean tube. Add 1 mL of 10 mM phosphate buffer (K_2HPO_4/KH_2PO_4) solution to the supernatant, and further adjust the pH to 6.5–6.6 with 2 M KOH. Dilute the sample to 10 mL with 10 mM phosphate buffer, and filter through a 0.45 μm regenerated cellulose (RC) filter prior to analysis [56]. Directly following filtration, the samples can be processed for thiochrome derivatization.

3.1.4 Thiamin Extraction by Solid Phase Extraction

Solid phase extraction (SPE) is a useful technique for removing interfering artifacts from a variety of sample matrix. Due to the various possible matrices, each extraction will need to be modified specifically for the matrix. A detailed method for separating the thiamine vitamers in salmonid eggs has been described [57] and has possible applications with other matrices. This method utilized the elution with a phosphate buffer at pH 2 in combination with 10% w/w methanol.

3.2 Conversion of Thiamin to Thiochrome

To 1 mL of extracted and prefiltered (0.45 μm) sample/standard, add 300 μL of alkaline potassium ferricyanide solution, and vortex for 15 s.

Neutralize with 600 μL of 1.33 M phosphoric acid (Sigma-Aldrich, St Louis, MO, USA) solution, and make up to 2 mL with mobile phase buffer (10 mM KH_2PO_4/K_2HPO_4; pH 6.5). This should result in a sample/standard pH of 6.5 and allows for easy injection into the HPLC system. This entire conversion can take place within a HPLC vial to allow for ease of operation and analysis. It is recommended that converted esters are analyzed as soon as possible; however they are stable in solution for up to 3 days at 20 °C.

3.3 Separation and Analysis of Thiamin Vitamers

Figure 2 depicts the elution order of the described vitamers. Table 1 shows the accuracy, precision LOD, and RSD of this analysis. This method also achieved ppm detection limits of the target analytes.

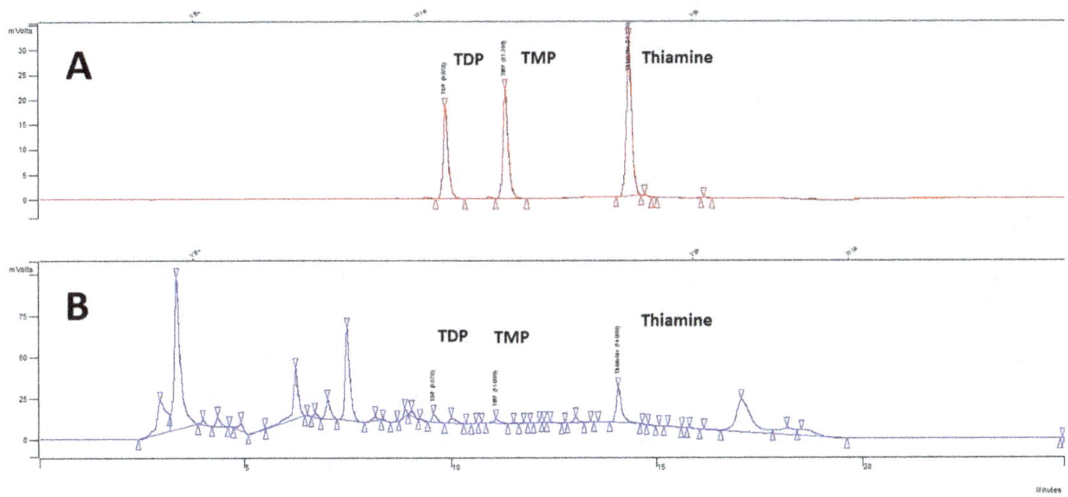

Fig. 2 Chromatographic analysis of thiamin vitamers by HPLC: (**a**) standard solution of thiamin vitamers; (**b**) sample (beer) containing thiamin vitamers

Table 1
Accuracy, precision, limit of quantitation (LOQ), linearity of calibration (LOD), range of linearity, and LOC of described thiamin vitamer methods for liquid sample analysis

Analyte	Accuracy (% spike recovery)	Precision (% RSD)	LOQ (ng/mL)	LOD (ng/mL)	Range of linearity (ng/mL)	LOC (R^2 value)
Thiamin diphosphate (TDP)	97.8	2.4	5.0	1.5	5–500	0.9998
Thiamin monophosphate (TMP)	98.2	2.5	5.0	1.5	5–530	0.9997
Thiamin	98.4	3.5	2.0	0.6	2–550	0.9998

4 Notes

1. Both thiamin vitamer standards and potassium ferricyanide solution are light sensitive. It is recommended that amber glassware is used throughout all of the analysis.
2. Potassium ferricyanide solution is a hazardous chemical and should be handled and disposed of with respect to local environmental requirements.
3. As noted in Subheading 3.1, a variety of sample preparation techniques may need to be investigated depending on matrix type, especially if extraction is required. All samples should be evaluated for matrix interferences.

4. Where possible all analysis should be performed as quickly as possible post derivatization, or the samples refrigerated until used. Storing derivatized samples greater than 72 h can result in bacterial growth in the neutralized sample that can foul the column.

References

1. Hawkins CF, Borges A, Perham RN (1989) A common structural motif in thiamin pyrophosphate-binding enzymes. FEBS Lett 255(1):77–82
2. Bettendorff L (2020) Thiamine. In: Marriott B, Birt D, Stalling V, Yates A (eds) Present knowledge in nutrition, 11th edn. Academic Press, Cambridge
3. Raj KC, Talaric LA, Ingram LO, Maupin-Furlow JA (2002) Cloning and characterization of the Zymobacter palmae pyruvate decarboxylase gene (pdc) and comparison to bacterial homologues. Appl Environ Microbiol 68(6):2869–2876
4. Leichter J, Joslyn MA (1969) Kinetics of thiamin cleavage by sulphite. Biochem J 113(4): 611–615
5. Wilson RB (2020) Pathophysiology, prevention, and treatment of beriberi after gastric surgery. Nutr Rev 78(12):1015–1029
6. Thomson AD (2002) Mechanisms of vitamin deficiency in chronic alcohol misusers and the development of the Wernicke–Korsakoff syndrome. Alcohol Alcohol 35:658–667
7. Lakhani SV, Shah HN, Alexander K, Finelli FC, Kirkpatrick JR, Koch TR (2008) Small intestinal bacterial overgrowth and thiamin deficiency after Roux-en-Y gastric bypass surgery in obese patients. Nutr Res 28(5):293–298
8. Smith TJ, Johnson CR, Koshy R, Hess SY, Qureshi UA, Mynak ML, Fischer PR (2021) Thiamine deficiency disorders: a clinical perspective. Ann NY Acad Sci 1498(1):9–28
9. Hoover AA, Jayasuriya GCN (1950) Microbiological assay of vitamins. 1. Thiamin Ceylon. J Med Sci 7:66–72
10. International AOAC (2005) Official methods of analysis. 45.2.01. AOAC official method 960.46 vitamin analysis – microbiological methods. American Organisation of Analytical Chemists International
11. Hennesey DJ, Cerecedo LR (1939) The determination of free and phosphorylated thiamin by a modified thiochrome assay. J Am Chem Soc 61(1):179–183
12. Loofbourow JR (1943) Physical methods for the identification and assay of vitamins and hormones. In: Harris RS, Thiman KV (eds) Vitamins & hormones, vol 1. Academic Press, pp 190–155
13. Batifoulier F, Verny MA, Besson C, Demigné C, Rémésy C (2005) Determination of thiamin and its phosphate esters in rat tissues analyzed as thiochromes on a RP-amide C16 column. J Chromatogr B 816(1–2):67–72
14. Lynch P, Young T (2000) Determination of thiamin by high-performance liquid chromatography. J Chromatogr A 881:267–284
15. Hong-Zhi H, Hua-Bin L, Feng C (2005) Determination of vitamin B1 in seawater and microalgal fermentation media by high-performance liquid chromatography with fluorescence detection. In: Analytical & bioanalytical chemistry, vol 383. Springer Science & Business Media B.V.
16. Esteve M-J, Farre R, Frigola A, Garcia-Cantabella J-M (2001) Simultaneous determination of thiamin and riboflavin in mushrooms by liquid chromatography. J Agric Food Chem 49(3):1450–1454
17. Vinas P, Balsalobre N, Lopez-Erroz C, Hernandez-Cordoba M (2004) Liquid chromatographic analysis of riboflavin vitamers in foods using fluorescence detection. J Agric Food Chem 52(7):1789–1794
18. Vinas P, Lopez-Erroz C, Balsalobre N, Hernandez-Cordoba M (2003) Determination of thiamin and its esters in beers and raw materials used for their manufacture by liquid chromatography with post column derivatization. J Agric Food Chem 51:3222–3227
19. Rizzolo A, Polesello S (1992) Chromatographic determination of vitamins in foods. J Chromatogr 624:103–152
20. Kawasaki T, Egi Y (2000) Thiamin. In: De Leenheer A, Lambert W, Van Bocxlaer J (eds) Modern chromatographic analysis of vitamins, 3rd edn. Marcel Dekker Inc, pp 365–391
21. Finglas PM, Faulks R (1987) Critical review of HPLC methods for the determination of thiamin, riboflavin and niacin in food. J Micronutr Anal 3:251–283
22. Finglas P (1993) Thiamin. Int J Vitam Nutr Res 63(4):270–274

23. Das Gupta V, Cadwallader DE (1968) Acid dye method for the analysis of thiamin. J Pharm Sci 57(1):112–116
24. Mickelsen O, Condiff H, Keys A (1945) The determination of thiamin in urine by means of the thiochrome technique. J Biol Chem 160: 361–370
25. Uber FM, Verbrugge F (1940) Photochemistry of the thiazole component of vitamin B1. J Biol Chem 136(1):81–86
26. Uber FM, Verbrugge F (1940) Photochemistry and absorption spectroscopy of the pyrimidine component of vitamin B1. J Biol Chem 134(1):273–282
27. Slater EC (1941) The thiochrome method of estimating thiamin (vitamin B1) in urine, milk and cereal products. Aust J Exp Biol Med Sci 19:29–32
28. Su S-C, Chou S-S, Hwang D-F, Chang P-C, Liu C-H (2001) Capillary zone electrophoresis and micellar electrokinetic capillary chromatography for determining water-soluble vitamins in commercial capsules and tablets. J Food Sci 66:10–14
29. Backermann S, Poel C, Ternes W (2008) Thiamin phosphates in egg yolk granules and plasma of regular and embryonated eggs of hens and in five- and seven-day-old embryos. Poult Sci 87(1):108–115
30. Pegg RB, Landen WO, Eitenmiller RR (2010) Vitamin analysis. In: Nielsen SS (ed) Food analysis, 4th edn. Springer, New York, pp 179–200
31. Bettendorff L (1991) Application of high-performance liquid chromatography to study thiamin metabolism and in particular thiamin triphosphatase. J Chromatogr 566:397–408
32. Bettendorff L, Peeters M, Jouan C, Wins P, Schoffeniels E (1991) Determination of thiamin and its phosphate esters in cultures neurons and astrocytes using an ion-pair reverse-phase high performance liquid chromatographic method. Anal Biochem 198:52–59
33. Bontemps J, Philippe P, Bettendorff L, Lombet J, Dandrifosse G, Schoffeniels E (1984) Determination of thiamin and thiamin phosphates in excitable tissues as thiochrome derivatives by reversed-phase high-performance liquid chromatography on octadecyl silica. J Chromatogr 307:283–294
34. European Committee for Standardisation (2003) BS EN 14152:2003 food stuffs – determination of vitamin B2 by HPLC. British International Standards: Technical Committee CEN/TC 275, UK
35. European Committee for Standardisation (2003) BS EN 14122:2003 food stuffs – determination of vitamin B1 by HPLC. British International Standards: Technical Committee CEN/TC 275, UK
36. Barger G, Bergel F, Todd AR (1935) A crystalline fluorescent dehydrogenation product from vitamin B1. Nature 136:259
37. Barger G, Bergel F, Todd AR (1935) Über das Thiochrom aus vitamin B1 (Antineurin). Ber Dtsch Chem Ges (A and B Series) 68(12): 2257–2262
38. Kuhn R, Vetter H (1935) Zur Kenntnis des Thiochroms. Ber Dtsch Chem Ges (A and B Series) 68(12):2375–2385
39. Fellman JK, Artz WE, Tassinari PD, Cole CL, Augustin J (1982) Simultaneous determination of thiamin and riboflavin in selected foods by high-performance liquid chromatography. J Food Sci 47:2048–2050
40. Lambert WE, Meyer E, Leenheer A, De P (1995) Systematic toxicological analysis of basic drugs by gradient elution of an alumina-based HPLC packing material under alkaline conditions. J Anal Toxicol 19(2):73–78
41. Kiel J, Morgan SL, Abramson RK (1985) Effects of amine modifiers on retention and peak shape in reversed-phase high-performance liquid chromatography. J Chromatogr A 320(2):313–323
42. Hill DW (1990) Evaluation of alkyl bonded silica and solvent phase modifiers for the efficient elution of basic drugs on HPLC. J Liq Chromatogr 13(16):3147–3175
43. Hansen S, Helboe P, Thomsen M (1988) Separation of basic drugs by high-performance liquid chromatography using dynamically modified silica. Curr Trends Anal Chem 7(10):389–393
44. Brunnekreeft JWI, Eidhof H, Gerrits J (1989) Optimized determination of thiochrome derivatives of thiamin and thiamin phosphates in whole blood by reversed-phase liquid chromatography with precolumn derivatization. J Chromatogr 491:89–96
45. Hucker B, Wakeling L, Vriesekoop F (2011) The quantitative analysis of thiamin and riboflavin and their respective vitamers in fermented alcoholic beverages. J Agric Food Chem 59(23):12278–12285
46. Liddicoat C, Hucker B, Liang H, Vriesekoop F (2015) Thiamin analysis in red wine by fluorescence reverse phase-HPLC. Food Chem 177: 325–329
47. Waterhouse A (2002) Wine Phenolics. Ann N Y Acad Sci 957(1):21–36
48. Duan X, Jiang Y, Su X, Zhang Z, Shi J (2007) Antioxidant properties of anthocyanins extracted from litchi (Litchi chinensis Sonn.)

fruit pericarp tissues in relation to their role in the pericarp browning. Food Chem 101:1365–1371

49. Gulcin I (2005) Antioxidant activity of caffeic acid (3, 4-dihydroxycinnamic acid). Toxicology 217:213–220
50. Oyaizu M (1986) Studies on products of browning reaction: antioxidant activity of products of browning reaction prepared from glucosamine. J Nutr 44:307–315
51. Veigas JM, Narayan MS, Laxman PM, Neelwarne B (2007) Chemical nature, stability and bioefficacies of anthocyanins from fruit peel of Syzygium cumini Skeels. Food Chem 105: 619–627
52. Briggs DE, Boulton CA, Brookes PA, Stevens R (2004) Brewing: science and practice. CRC Press
53. Pignatelli B, Scriban R, Descotes G, Bartsch H (1984) Modifying effects of polyphenols and other constituents of beer on the formation of N-nitroso compounds. J Am Soc Brew Chem 42(1):18–23
54. Rehmanji M, Gopal C, Mola A (2002) A novel stabilization of beer with Polyclar Brewbrite. MBAA Tech Quart 39(1):24–28
55. Whittle N, Eldridge H, Bartley J (1999) Identification of the polyphenols in barley and beer by HPLC/MS and HPLC/electrochemical detection. J Inst Brew 105(2):88–99
56. Hucker B, Wakeling L, Vriesekoop F (2012) Investigations into the thiamine and riboflavin content of malt and the effects of malting and roasting on their final content. J Cereal Sci 56(2):300–306
57. Zajicek JL, Tillitt DE, Brown SB, Brown LR, Honeyfield DC, Fitzsimons JD (2005) A rapid solid-phase extraction fluorometric method for thiamine and riboflavin in salmonid eggs. J Aquat Anim Health 17:95–105

Chapter 20

Analysis of Riboflavin and Its Respective Vitamers by High-Performance Liquid Chromatography

Barry Hucker and Frank Vriesekoop

Abstract

Riboflavin (vitamin B_2) is an essential vitamin which is utilized in a variety of biological functions. Intracellularly riboflavin is phosphorylated to riboflavin 5′phospate (FMN) or converted to flavin adenine dinucleotide (FAD) to play crucial coenzyme roles in the respiration pathways. Deficiencies in this vitamin (known as ariboflavinosis) generally lead to fatigue, skin disorders, and hyperemia.

The following chapter describes a simple and reproducible method for the analysis of riboflavin via high-performance liquid chromatography utilizing fluorescence detection.

Key words Riboflavin, Vitamers, High-performance liquid chromatography, Fluorescence

1 Introduction

Riboflavin is a yellow fluorescent compound that was first isolated from whey as lactoflavin or lactochrome [1–5]. It consists of a tricyclic dimethyl-isoalloxazine ring and ribitol moieties, and it is biologically active as riboflavin 5′-phosphate (flavin mononucleotide—FMN) and flavin adenine dinucleotide (FAD) (Fig. 1). It must be noted that technically FMN and FAD are not nucleotides, but these generic names are used to generally describe these vitamers.

Riboflavin is highly unstable and becomes biologically inactive when exposed to UV light, being easily degraded to lumiflavin in basic solutions (pH > 7) and lumichrome under acidic conditions (pH < 7) [6–12]. This photodegradation forms various free radicals that cause irreversible changes in food products as shown in the generation of 3-methyl-2-butene-1-thiol (MBT) (skunky odor) in beer and other light struck off-odors in milk and wines [13–18]. However under acidic conditions and in the absence of light riboflavin is heat stable (up to 100 °C) but is easily degraded in alkaline pH solutions [19, 20].

Lilian Mariutti and Frank Vriesekoop (eds.), *Food Chemistry: Basic Protocols*, Methods and Protocols in Food Science, https://doi.org/10.1007/978-1-0716-4678-6_20,

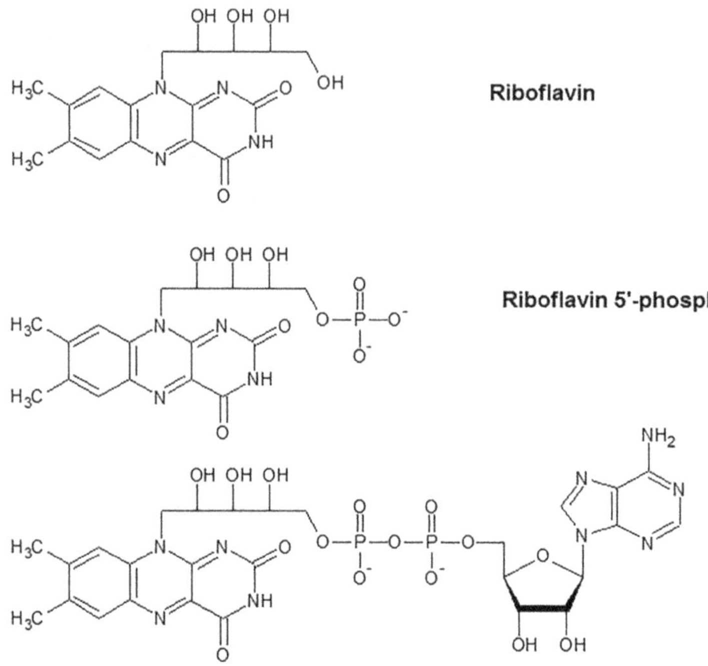

Fig. 1 Structure of the riboflavin vitamers

In biological systems, FMN and FAD are utilized as cofactors in a variety of flavoproteins and as electron acceptors in oxidation–reduction reactions involved in the tricarboxylic acid (TCA) cycle, electron transport chain, β-oxidation, and tryptophan metabolism [21–29]. Deficiencies in this vitamin result in a condition known as ariboflavinosis and can lead to skin disorders and hyperemia. However, this condition is quite rare due to its abundance in food and is often associated with other deficiencies and medical conditions.

1.1 Methodologies of Analysis

Riboflavin is traditionally easy to analyze for. Typically, HPLC methods coupled with UV/vis, fluorescence, or mass spectrometry are used to determine the riboflavin content of various matrices. However due to the natural fluorescence of riboflavin, fluorescence detection is the cheapest easiest way to analyze this vitamin. Fluorescence methods have been able to easily detect within the ppb range with high specificity and recovery [30–36]. Riboflavin's natural fluorescence also means that specific excitation and emission wavelengths can be used reducing artifact interferences. Being highly water soluble, extractions in water are possible with the use of trichloroacetic acid (TCA) being used for improved total riboflavin recovery [37].

When analyzing for riboflavin or its respective vitamers, it is crucial that all manipulations are performed in amber glassware (*see* **Note 1**). Failure to do so may lead to an underestimation in the results. Exposure to wavelengths in the range of 415–455 nm are particularly destructive to all forms of riboflavin. For the best fluorescence response, ideally the pH of the sample should remain around 6.5 to 7.0 [36–39] with excitation and emission wavelengths set at 270 and 516 nm, respectively [36, 37].

2 Materials

All equipment and consumables mentioned in this chapter can be substituted by alternative products from other manufacturers. In this chapter we describe a tried set of methodologies based on the equipment and consumables mentioned.

2.1 Basic Reagents

All reagents/standards used in this research were sourced through Sigma-Aldrich (St. Louis, MO) and were of 95% purity or greater. All solutions/standard are made with 18 MΩ water (Sartorius arium 611VF water purification system, Sartorius AG, Göttingen, Germany). It is recommended that all glassware used in this analysis is cleaned via sonication and rinsed in 18 MΩ water. As this method uses very specific wavelengths for low-level detection, interferences can be observed in the analysis. A phosphate buffer: methanol mobile phase was utilized for the analysis. A 10 mM KH_2PO_4/K_2HPO_4, pH 6.5 solution, and 100% methanol were used in a gradient combination up to 35:65 buffer:methanol concentration (~25-minute profile).

2.2 Instrumentation

The HPLC system consisted of a Varian Prostar 230 solvent delivery system (flow rate 1 mL/min), a Varian Prostar 410 autosampler (10 μL injection), a Varian Prostar 500 column valve module (30 ° C), and a Varian Prostar 363 fluorescence detector at 270/516 nm excitation/emission [36].

A Varian Pursuit C18 5 μm (250 mm × 4.6 mm) fitted with a Securityguard Cartridge (C18 4 × 3.0 mm) (Phenomenex, Torrance, CA) was used for the analysis (*see* **Note 2**).

The mobile phase gradient employed was as follows: solvent A, 10 mM potassium phosphate buffer; solvent B, 100% methanol; 0–0.5 min 95:5 A:B; 0.5–10 min 65:35 A:B; 10–15 min 65:35 A:B; and 15–16 min 95:5 A:B; at a total run time of 25 min.

3 Methods

3.1 Sample Preparation

If analyzing for total riboflavin, it is recommended that taka-diastase or Clara-diastase is used to extract riboflavin from samples. These enzymes can cleave the phosphate bonds in the vitamers into the free forms [40–42]. High recovery has been reported using these enzymes of the free forms, but for vitamer speciation, it is recommended that these enzymes are not used. This method describes the analysis of the speciated vitamers and therefore does not include the use of the diastase.

All prepared samples must be stored in amber glass wear and stored at refrigerated conditions. It is also recommended to prevent bacterial growth in samples that they are only prepared prior to analysis but can be stored for up to 72 h under refrigerated conditions [37].

3.1.1 Riboflavin Extraction from Liquid Samples

Due to the high solubility of the riboflavin vitamins, liquid sample analysis is reasonably straight forward. Unless sample cleanup is required due to matrix interferences, the only sample preparation required is particulate removal. In matrices such as beers, ciders, and wines, samples will be required to be degassed, which can be facilitated by passing the sample through a 0.45 μm RC filter. Other methods of degassing could be used, but to preserve the HPLC column, it is best for all samples to be passed through a 0.45 μm RC filter anyway.

It is recommended that all samples are at least diluted 1:1 with 10 mM phosphate buffer to allow for a smoother flow onto the column gradient in the HPLC. It is also possible that due to sensitive nature of fluorescence detection dilutions will need to be made.

3.1.2 Riboflavin Extraction from Solid Samples

Free riboflavin vitamers can easily be analyzed from solid samples by dissolving them in water. To ensure stability of the phosphorylated form, it is recommended that a 10 mM phosphate buffer is used. This will help to stabilize the pH during extraction and make it easier for later filtration and injection into the HPLC system. For total riboflavin vitamer testing, using trichloroacetic acid (TCA) can aid in the breakdown on binding proteins releasing the vitamers. In matrix such as grains, sample must be processed to fine powder, where 0.5 g is added to 5 mL of 0.5 M (TCA) and continuously for 15 minutes [38, 43]. Longer extraction times can lead to a reduction in recovery [38], and the mixing should be always shielded from light. Once mixed through, the sample is centrifuged at 2000 ×*g*, and the 2 mL supernatant is removed into a clean test tube. To this supernatant 1 mL of 10 mM phosphate buffer solution is added, and the pH is adjusted to 6.5–6.6 with 2 M KOH [38, 42]. Dilute the sample to 10 mL with phosphate buffer, and filter into a HPLC amber vial.

In sample matrix with interfering polyphenols present, the use of compounds such as polyvinylpolypyrrolidone (PVPP) is useful in removing the interferences. Liddicoat et al. [44] (2015) described the effective use of PVPP for polyphenol removal in wines from thiamin analysis. While this may not be required for riboflavin vitamer analysis, it is an effective method in removing polyphenols that may foul the column. This method [44] employed dosing the PVPP at a rate of 50 g/L in the sample, mixing for 15 min, and centrifuging at 2000 rcf. The samples are then filtered through a 0.45 μm RC filter and diluted as required in phosphate buffer.

3.2 Separation and Analysis of Riboflavin Vitamers

The elution of vitamers occurs on a buffer to methanol gradient with a final composition of 65:35 phosphate buffer:methanol as described in Subheading 2.2.

Figure 2 depicts the elution order of the described vitamers. Table 1 shows the accuracy, precision LOD, and RSD of this analysis. This method also achieved ppm detection limits of the target analytes.

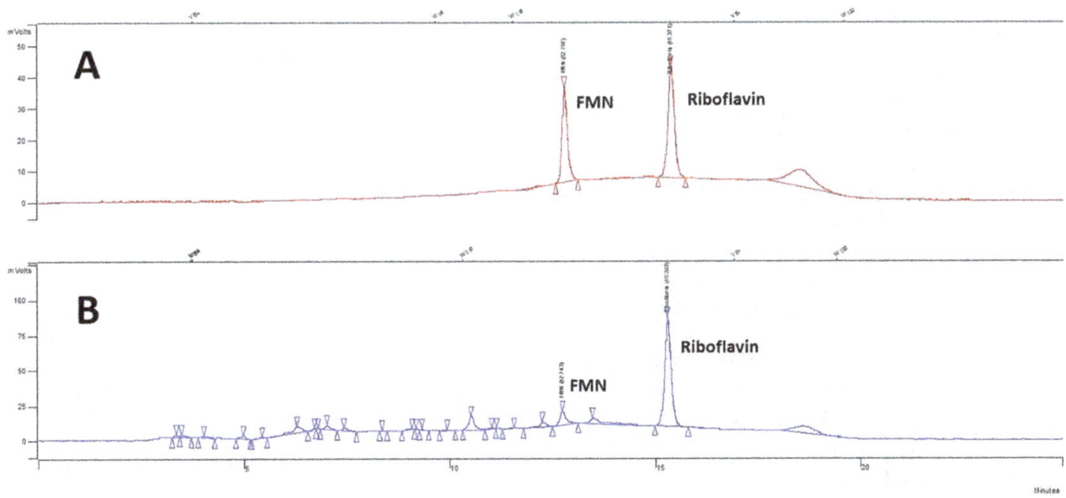

Fig. 2 Chromatographic analysis of riboflavin vitamers by HPLC: (**a**) standard solution of riboflavin vitamers; (**b**) sample (beer) containing riboflavin vitamers

Table 1
Accuracy, precision, LOQ, LOD, range of linearity, and LOC of described riboflavin vitamer methods for liquid sample analysis

Analyte	Accuracy (av. % spike recovery)	Precision (% RSD)	LOQ (ng/mL)	LOD (ng/mL)	Range of linearity (ng/mL)	LOC (R^2 value)
Flavin Mononucleotide (FMN)	97.6	2.7	5.0	1.5	5–520	0.9999
Riboflavin	98.4	2.8	2.0	0.6	2–550	0.9999

4 Notes

1. Riboflavin and FMN are not stable in the presence of light. It is recommended that all work is carried out away from direct sunlight utilizing amber or covered glassware for both standards and samples.
2. Precolumn filters are recommended for dirtier samples to prevent excessive column fouling.

Due to the high specificity of the fluorescence detection, it is critical that glassware use is thoroughly cleaned using sonication and washed in 18 MΩ water.

References

1. Karrer P, Schöpp K, Benz F (1935) Synthesen von Flavinen IV. Helv Chim Acta 18(1): 426–429
2. Booher LE (1938) Chemical aspects of riboflavin. JAMA 110(14):1105–1111
3. Warburg O, Christian W (1932) Ein zweites sauerstoffübertragendes Ferment und sein Absorptionspektrum. Naturwissenschaften 20:688–700
4. Warburg O, Christian W (1938) Co-ferment der d-aminosaure-deaminase. Biochem Z 295: 261
5. McCollum E, Kennedy C (1916) The dietary factors operating in the production of polyneuritis. J Biol Chem 24:491–502
6. Sattar A, deMan JM, Alexander CS (1977) Light-induced degradation of vitamins I. Kinetic studies on riboflavin decomposition in solution. Can Inst Food Sci Technol J 10(1): 61–64
7. Sattar A, deMan JM, Alexander JC (1977) Light-induced degradation of vitamins II. Kinetic studies on ascorbic acid decomposition in solution. Can Inst Food Sci Technol J 10(1):65–68
8. Parks OW, Allen C (1977) Photodegradation of riboflavin to lumichrome in milk exposed to sunlight. J Dairy Sci 60(7):1038–1041
9. Stull JW (1953) The effect of light on activated flavor development and on the constituents of milk and its products: a review. J Dairy Sci 36(11):1153–1164
10. Toba T, Adachi S, Arai I (1980) Sunlight and sodium hypochlorite induced color changes in milk. J Dairy Sci 63(11):1796–1801
11. Zandomeneghi M, Carbonaro L, Zandomeneghi G (2007) Biochemical fluorometric method for the determination of riboflavin in milk. J Agric Food Chem 55:5990–5994
12. Furuya E, Warthesen J, Labuza T (1984) Effects of water activity, light intensity, and physical structure of food on the kinetics of riboflavin photodegradation. J Food Sci 49(2):526–528
13. Foster R II, Samp E, Patino H (2001) Multivariate modelling of sensory and chemical data to understand staling in light beer. J Am Soc Brew Chem 59(4):201–210
14. Goldsmith M, Rogers R, Cabral N, Ghiggino K, Roddick F (2005) Riboflavin triplet quenchers inhibit lightstruck flavor formation in beer. J Am Soc Brew Chem 63(4): 177–184
15. Gray P, Stone I, Rothchild H (1940) The action of sunlight on beer. Wallerstein Lab Commun 4:29–40
16. Heyerick A, Zhao Y, Sandra P, Huvaere K, Roelens F, De Keukeleire D (2003) Photolysis of hop-derived trans-iso-α-acids and trans-tetrahydriso-α-acids: product identification in relation to the lightstruck flavour of beer. Photochem Photobiol Sci 2:306–314
17. Huvaere K, Andersen M, Storme M, Van Bocxlaer J, Skibsted L, De Keukeleire D (2006) Flavin-induced photodecomposition of sulfur-containing amino acids is decisive in the formation of beer lightstruck flavor. Photochem Photobiol Sci 5:961–969
18. Vendramin V, Primerano S, Leserri G, Paniccia G, Vincenzi S (2021) Riboflavin removal by commercial bentonites and charcoals in white and red wines. Italian J Food Sci 33(2):13–23
19. Surrey AR, Nachod FC (1951) Alkaline hydrolysis of riboflavin. J Am Chem Soc 73(5):

2336–2338. https://doi.org/10.1021/ja01149a128

20. Bender DA (2003) Nutritional biochemistry of the vitamins, 2nd edn. Cambridge University Press, Cambridge
21. Bacher A, Eberhardt S, Fischer M, Kis K, Richter G (2000) Biosynthesis of vitamin B2 (riboflavin). Annu Rev Nutr 20(1):153
22. Choe E, Huang R, Min DB (2005) Chemical reactions and stability of riboflavin in foods. J Food Sci 70(1):R28–R36
23. Fischer M, Römisch W, Saller S, Illarionov B, Richter G, Rohdich F, Eisenreich W, Bacher A (2004) Evolution of vitamin B2 biosynthesis. J Biol Chem 279(35):36299–36308
24. Kearney EB (1952) The interaction of yeast flavokinase with riboflavin analogues. J Biol Chem 194(2):747–754
25. Lehmann M, Degen S, Hohmann H-P, Wyss M, Bacher A, Schramek N (2009) Biosynthesis of riboflavin. FEBS J 276(15): 4119–4129
26. Roje S (2007) Vitamin B biosynthesis in plants. Phytochemistry 68(14):1904–1921
27. Shavlovsky GM, Logvinenko EM, Benndorf R, Koltun LV, Kashchenko VE, Zakalsky AE, Schlee D, Reinbothe H (1980) First reaction of riboflavin biosynthesis – catalysis by a guanosine triphosphate cyclohydrolase from yeast. Arch Microbiol 124(2):255–259
28. Ball GFM (2005) Vitamins in foods: analysis, bioavailability, and stability. CRC Press, Boca Raton
29. Ball GFM (2004) Vitamins: their role in the human body. Blackwell Publishing Ltd, India
30. Vinas P, Balsalobre N, Lopez-Erroz C, Hernandez-Cordoba M (2004) Liquid chromatographic analysis of riboflavin vitamers in foods using fluorescence detection. J Agric Food Chem 52(7):1789–1794
31. Vinas P, Lopez-Erroz C, Balsalobre N, Hernandez-Cordoba M (2003) Determination of thiamine and its esters in beers and raw materials used for their manufacture by liquid chromatography with post column derivatization. J Agric Food Chem 51:3222–3227
32. Becker ME, Christensen J, Frederiksen CS, Haugaard VK (2003) Front-face fluorescence spectroscopy and chemometrics in analysis of yogurt: rapid analysis of riboflavin. J Dairy Sci 86(8):2508–2515
33. Hong-Zhi H, Hua-Bin L, Feng C (2005) Determination of vitamin B1 in seawater and microalgal fermentation media by high-performance liquid chromatography with fluorescence detection. In: Analytical & bioanalytical chemistry, vol 383. Springer
34. Sikorska E (2007) Analysis of vitamin B2 using front-face intrinsic beer fluorescence. Eur Food Res Technol 225(1):43–48
35. Wang L-H, Hung H (2006) Simultaneous determination of water-soluble vitamins in human urine by fluorescence in a flow-injection analysis. J Liq Chromatogr Relat Technol 29: 329–338
36. Hucker B, Wakeling L, Vriesekoop F (2011) The quantitative analysis of thiamine and riboflavin and their respective vitamers in fermented alcoholic beverages. J Agric Food Chem 59:12278–12285
37. Hucker B, Wakeling L, Vriesekoop F (2012) Investigations into the thiamine and riboflavin content of malt and the effects of malting and roasting on their final content. J Cereal Sci 56(2):300–306
38. Ishii K, Sari K, Sanemori H, Kawasaki T (1979) Analysis of thiamine and its phosphate esters by high-performance liquid chromatography. Anal Biochem 97:191–195
39. Kawasaki T, Egi Y (2000) Thiamine. In: De Leenheer A, Lambert W, Van Bocxlaer J (eds) Modern chromatographic analysis of vitamins, 3rd edn. Marcel Dekker Inc, New York, pp 365–391
40. Hewavitharana AK (1996) Method for the extraction of riboflavin for high-performance liquid chromatography and application to casein. Analyst 121(11):1671–1676
41. del Carmen Mondragón-Portocarrero A, Vázquez-Odériz L, Romero-Rodríguez M (2011) Development and validation of an HPLC method for the determination of thiamine and riboflavin in green leafy vegetables using Clara-diastase. J Food Sci 76:C639–C642
42. Valls F, Checa MA, Fernández-Muiño MA, Sancho MT (1999) Determination of thiamin in cooked sausages. J Agric Food Chem 47(1): 170–173
43. Sivaraman T, Kumar TKS, Jayaraman G, Yu C (1997) The mechanism of 2,2,2-trichloroacetic acid-induced protein precipitation. J Protein Chem 16(4):291–297
44. Liddicoat C, Hucker B, Liang H, Vriesekoop F (2015) Thiamin analysis in red wine by fluorescence reverse phase-HPLC. Food Chem 177: 325–329

Chapter 21

Basic Protocols in Biotin Analysis

Linying Liu, Liang Chen, Xinying Zhao, and Hucheng Zhang

Abstract

There are many methods for biotin determination, but the principle, detection time, and suitable applications of these methods are different. In recent years, microbiological assays, high-performance liquid chromatography (HPLC), gas chromatography (GC), and more accurate and efficient analysis methods such as ultra-performance liquid chromatography-mass spectrometry (UPLC-MS) assays for biotin analysis have been reported. However, microbiological assay and HPLC are two methods that are mature and the basis for various applications in biotin analyses. This chapter details the various steps involved in both a microbial and HPLC detection methods for biotin.

Key words Biotin, Analysis, Protocol, Microbiological assay, High-performance liquid chromatography

1 Introduction

Biotin (vitamin B7), which is also known as vitamin H or coenzyme R, belongs to the B vitamin family. Biotin is a coenzyme involved in many enzymes that facilitate carboxylation reactions by incorporating bicarbonate (HCO_3) units. As such biotin is an indispensable substance for the normal synthesis of fatty acids and during gluconeogenesis. Biotin also plays several roles in non-coenzyme-related activities such as gene expression influencing cell growth and development [1]. Therefore, biotin is an essential nutrient to maintain the body's natural growth, development, and maintenance. The chemical structure of biotin molecule consists of two five-membered rings formed by a urea ring and a thiophenol ring with different valeric acid branched chains. In terms of solubility, biotin is more soluble in hot water and slightly alkaline conditions compared to water at room temperature or ethanol.

There are many methods for biotin determination, but the principle, detection time, and suitability of analytical applications of these methods are different. The acid–base titration is the traditional method for biotin analysis. This method is simple and

Lilian Mariutti and Frank Vriesekoop (eds.), *Food Chemistry: Basic Protocols*, Methods and Protocols in Food Science, https://doi.org/10.1007/978-1-0716-4678-6_21,

reliable, and it was once used as a designated method for the determination of biotin used by the United States Pharmacopeia (USP) [2]. However, it is mainly suitable for high-purity biotin as a quality assurance tool, but not for low-purity biotin samples in complex matrices like many biological and food samples. In recent years, microbiological assay, high-performance liquid chromatography (HPLC), and gas chromatography (GC) have been reported for biotin analysis [3, 4]. With HPLC methods having become more commonplace, more accurate and efficient analysis methods have been developed. For example, high-performance liquid chromatography-mass spectrometry (HPLC-MS) method or UPLC-MS have been developed for the detection of a wide range of vitamins including biotin [5, 6]. New rapid analysis methods for biotin are currently being developed [7]. For example, a biosensor-based assay for biotin works on the specific reaction between biotin and the biologically active material in biosensor. The device converts chemical signals into electrical signals for both the qualitative and quantitative analyses of biotin [8–12]. To give another example, the enzyme-linked immunoassay (ELISA) is based on the principle of antigen–antibody reaction, which has also been applied for the quantitative detection method of biotin [13–15]. ELISA-based biotin analysis was found to be suitable for the detection of biotin in samples with complex matrices from which biotin is difficult to extract. All in all, many varied methods for biotin analysis have been developed over time. However, the microbiological and HPLC-based methods are still the two most reliable and commonly used methods [16, 17].

Microbiological Assay for Biotin
Microbiological assay is a relatively simple, but time-consuming method for the detection of low concentrations of biotin in natural organic material. This method is suitable for large-scale determination because of its simple principle, high sensitivity, and low cost. The principle of the microbiological assay is that under certain conditions, the biotin concentration has a linear relationship with the growth of biotin auxotrophic strains such as *Lactobacillus plantarum* [3], and the biotin concentration is directly proportional to the growth rate of the bacteria. The specific sub-methods include plate growth ring method, tube dish method, and a liquid culture tube method among others [4, 18]. This method has long been used by analysts and gradually standardized.

HPLC Assay for Biotin
With the rapid development of analytical techniques and methods, the HPLC method for sample analysis is labor-saving and accurate. Therefore, this method has also become a widely used method for the determination of biotin content [19]. Common detectors used in HPLC analysis are fluorescence detectors and ultraviolet

(UV) detectors. The sensitivity of ultraviolet detection is lower than that of fluorescence detection. However, researchers choose detectors according to the nature of the sample and the anticipated content of biotin in the samples [20, 21]. Since biotin does not naturally fluoresce, using fluorescence derivatization is an effective method to detect trace amounts of biotin by derivatizing biotin with an agent to make its derivatives fluorescent and increasing the detection limit [22–24].

Derivatization of the carboxylic group of biotin by fluorophores has been proposed for the determination of biotin in pharmaceutical preparations containing biotin and other vitamins. For example, fluorescent labeling of biotin with I-pyrenyldiazomethane (PDAM) has been developed [25]. PDAM readily reacted with the carboxylic group of biotin without any catalyst at room temperature, and the ester derivative was separable on a reversed-phase column. The effluent was measured spectrophotometrically at 240 nm or fluorometrically at the excitation and fluorescence wavelengths of 340 and 395 nm. However, the method cannot be recommended in food analysis because the reactions proposed are always performed in an organic medium. The reactions always lead to the formation of numerous interfering products because of the abundance of carboxylic groups in food matrices. On that account, avidins or streptavidin, which are proteins, covalently bound to a fluorescent marker (fluorescein 5-isothiocyanate or FITC) could be used as reagent for a post-column derivatization of biotin which works on the fast complexation of biotin with avidin or streptavidin specifically [26, 27]. The fluorimetric detector operated at an excitation wavelength of 490 nm and at an emission wavelength of 520 nm. The method is proposed to be applied in the determination of biotin in food supplements.

In order to help readers choose suitable detection methods for biotin of various samples, such as food supplements or pharmaceutical preparations containing trace amounts of biotin, we here introduce the detailed steps for the microbiological approach and then the HPLC approach. Since the two methods were the basis methods for the further development of highly efficient automated analysis and multispectral analysis techniques, the protocols (materials and methodology) here we summarized are expected to be reference for current biotin analysis and information for further development of biotin detection methods.

2 Materials

Prepare all solutions using ultrapure water (prepared by purifying deionized water, to attain a sensitivity of 1.0 MΩ·cm at 25 °C). The solution used in the methods refers to the aqueous solution when the solvent is not specified.

2.1 Microbiological Assay Materials

1. *Lactobacillus plantarum* (ATCC 8014).
2. Lactobacillus agar medium and lactobacillus broth medium are purchased from BD company.
3. Free biotin assay medium preparation: 12.0 g casamino acid, 40.0 g glucose, 20.0 g sodium acetate, 1.0 g potassium dihydrogen phosphate, 1.0 g dipotassium hydrogen phosphate, 0.2 g DL-tryptophan, 0.2 g L-cystine hydrochloride, 0.02 g adenine sulfate, 0.02 g guanine hydrochloride, 0.02 g uracil, 0.4 g magnesium sulfate, 0.02 g sodium chloride, 0.02 g ferrous sulfate, 0.02 g manganese sulfate, 0.002 g riboflavin, 0.0002 g p-aminobenzoic acid, 0.004 g pyridoxine hydrochloride, 0.002 g thiamine hydrochloride, 0.002 g niacin, and 0.002 g calcium pantothenate were mixed and then heated and dissolved in 1000 mL water; adjust pH to 6.8 ± 0.2 (25 °C ± 5 °C) and divide into test tubes.
4. Biotin (d-Biotin) standard product ($C_{10}H_{16}N_2O_3S$): purity ≥99%.
5. 50% ethanol solution.
6. 3% sulfuric acid solution.
7. 0.5 mol/L sodium hydroxide solution.

2.2 HPLC Materials

1. Biotin (d-Biotin) standard product ($C_{10}H_{16}N_2O_3S$): purity ≥99%.
2. 0.05% trifluoroacetic acid solution (volume fraction).
3. Acetonitrile (liquid chromatography).
4. Diluent agent: 0.05% trifluoroacetic acid solution:acetonitrile at 75:25.

3 Methods

3.1 Microbiological Assay

3.1.1 *Lactobacillus plantarum* Suspension

1. Transfer the stored *Lactobacillus plantarum* to lactobacillus agar medium, and cultivate it at 36 °C for 18–24 h.
2. Inoculate the above *Lactobacillus plantarum* into lactobacillus broth medium, and culture at 36 °C for 18–24 h.
3. Centrifuge the above lactobacillus broth culture medium at 2000 r/min for 3 min.
4. Discard the supernatant, add 10 mL of 0.85% saline, and mix well.
5. Centrifuge supernatant for 3 min, and repeat the washing step 3 times to remove any biotin carry-over from the initial growth medium.
6. Using 0.85% saline as a blank, measure the light absorbance of the inoculated bacterial suspension at a wavelength of 550 nm.

7. Adjust the light absorbance of the inoculum suspension to between 70% and 80%. The *Lactobacillus plantarum* suspension is suitable and ready for co-incubation in Subheading 3.1.5.

3.1.2 Biotin Standard Solutions

1. Dissolve 10 mg biotin standard in 50 mL of 50% ethanol (alcohol by volume).
2. Draw 1 mL, and dilute the solution to a biotin concentration of 1 μg/mL in 50% ethanol using volumetric flask.
3. Draw 1 mL, and dilute with 50% ethanol solution to a biotin concentration of 10 ng/mL using volumetric flask.
4. Draw 5 mL from the above, and dilute to 250 mL and 500 mL with water respectively as a high-concentration standard solution (0.2 ng/mL biotin) and a low-concentration standard solution (0.1 ng/mL biotin).
5. Prepare the standard solution samples as shown in Table 1.
   5.1 Blank solution tube B1 (0.0 ng biotin): 0.0 mL low-concentration standard solution, 5.0 mL water, that is blank sample for uninoculated.
   5.2 Blank solution tube B2 (0.0 ng biotin): 0.0 mL low-concentration standard solution, 5.0 mL water, that is blank sample for inoculated.
   5.3 Standard solution tube S1 (0.1 ng biotin): 1.00 mL low-concentration standard solution, 4.0 mL water.
   5.4 Standard solution tube S2 (0.2 ng biotin): 2.00 mL low-concentration standard solution, 3.0 mL water.
   5.5 Standard solution tube S3 (0.3 ng biotin): 3.00 mL low-concentration standard solution, 2.0 mL water.
   5.6 Standard solution tube S4 (0.4 ng biotin): 4.00 mL low-concentration standard solution, 1.0 mL water.

Table 1
Standard solution tube preparation

Tube name	B1	B2	S1	S2	S3	S4	S5	S6	S7	S8
Biotin (ng)	0	0	0.1	0.2	0.3	0.4	0.5	0.6	0.7	0.8
Biotin concentration (ng/mL)	0	0	0.01	0.02	0.03	0.04	0.05	0.06	0.07	0.08
Low-concentration standard solution (mL)	0	0	1	2	3	4	5	0	0	0
High-concentration standard solution (mL)	0	0	0	0	0	0	0	3	4	5
Water (mL)	5	5	4	3	2	1	0	2	1	0
Free biotin assay medium (mL)	5	5	5	5	5	5	5	5	5	5

5.7 Standard solution tube S5 (0.5 ng biotin): 5.00 mL low-concentration standard solution.

5.8 Standard solution tube S6 (0.6 ng biotin): 3.00 mL high-concentration standard solution, 2.0 mL water.

5.9 Standard solution tube S7: 4.00 mL high-concentration standard solution, 2.0 mL water (0.8 ng biotin).

5.10 Standard solution tube S8: 5.00 mL high-concentration standard solution, 2.0 mL water (1.0 ng biotin).

5.11 Add 5 mL free biotin assay medium to each tube, and mix. The free biotin assay medium is specific to prevent the interference of biotin-related components in the culture medium to the determination of biotin.

3.1.3 Sample Test Solutions

1. Hydrolyze 3 g sample with 100 mL 3% sulfuric acid solution at 121 °C for 30 min.
2. Rapidly cool in ice.
3. Add 0.5 mol/L sodium hydroxide solution to adjust the pH value of the solution to 4.3–4.7 which is protein isoelectric point to obtain precipitation of proteins in samples.
4. Add sterile water to 250 mL using volumetric flask and mix well.
5. Filter the above with 0.45 μm filter, and discard the first few milliliters of filtrate, draw 5 mL of filtrate, add 20 mL of sterile water, and mix well.
6. Add 0.5 mol/L sodium hydroxide solution, and adjust the pH value to 6.6–7.0 to be suitable for *Lactobacillus plantarum* culture.
7. Dilute it with water to 100 mL using volumetric flask to obtain the sample solution.

3.1.4 Sample Test Tubes

1. Sample solution tube 1: 1.00 mL sample solution, 4.0 mL water.
2. Sample solution tube 2: 2.00 mL sample solution, 3.0 mL water.
3. Sample solution tube 3: 3.00 mL sample solution, 2.0 mL water.
4. Sample solution tube 4: 4.00 mL sample solution, 1.0 mL water.
5. Add 5 mL free biotin assay medium to each tube, and mix.
6. The sample solution tube 1–4 should be done in triplicate.

3.1.5 Co-incubation

1. Sterilize all culture glassware at 121 °C for >5 min.
2. Cool the tubes to room temperature.
3. Aseptically transfer the inoculation solution which is in the biotin standard solution tubes and sample solution tubes into sterile needle tubes. In detail, the biotin standard solutions contain B1, B2, and S1–S8. The sample solutions contain sample solution tube 1–4 with triplicate.
4. Transfer 50 μL *Lactobacillus plantarum* suspension into each sterile needle tubes, except for the standard solution tube B1. The B1 is a control for uninoculated.
5. Incubate at 37 °C ± 1 °C for 19–20 h.
6. Incubate for another 2 h until the light absorbance does not change significantly. The method to detect whether the culture medium reaches the maximum turbidity is to detect that the light absorbance has no obvious change after another 2 h of culture. If the light absorbance of the medium changes within 2 h, continue to culture until it remains unchanged.

3.1.6 Test

1. Shake the cultured tubes using a vortex mixer to obtain an even cell suspension.
2. Using a standard cuvette, measure the light absorbance at 550 nm. Use the blank tube as the control.
3. Determine the absorbance values of the tubes containing the qualitative biotin standards, and create a standard curve of absorbance at 550 nm versus biotin concentration. If there is obvious bacterial growth in the uninoculated blank control tube, it means that there may be mixed bacteria, and the test needs to be repeated.
4. Calculate the average biotin content.
5. Calculate the recovery (%).

3.2 HPLC Assay

3.2.1 Sample Preparation

1. Prepare biotin stock solution: Dissolve the biotin standard product in the diluent agent (0.05% trifluoroacetic acid solution:acetonitrile at 75:25), and ultrasonically make it completely dissolved to prepare a 1 mg/mL biotin stock solution.
2. Prepare biotin standard solution: Dilute the biotin stock solution in the diluent agent using 10 mL volumetric flask to prepare a 0.1 mg/mL biotin standard solution.
3. Prepare sample solutions.
   (a) Take the test sample of 5 mg, and add 40 mL diluent agent and ultrasonic extract in an ultrasonic water bath for 15 min.

(b) Dilute the solution to a concentration of 0.1 mg/mL in the diluent agent using 50 mL volumetric flask.

(c) Cool to room temperature.

(d) Filter the above with 0.45 μm membrane filter to prepare a 0.1 mg/mL sample solution.

3.2.2 Chromatographic Conditions

1. HPLC detector: equipped with ultraviolet detector (detection wavelength is 210 nm).
2. Chromatographic column (Discovery C18): 15 cm × 4.6 mm (inner diameter); filler particle size, 5 μm; ODS column or other equivalent chromatographic column.
3. Eluent: 0.05% trifluoroacetic acid solution:acetonitrile = 75: 25.
4. Flow rate: 1 mL/min.
5. Injection volume: 20 μL.

3.2.3 HPLC Methodology

1. Linearity
   (a) Prepare linearity solutions. Draw 0.1 mg/mL biotin standard solution 0.5 mL, 1.0 mL, 1.5 mL, 2.0 mL, 4.0 mL, and 8.0 mL respectively, dilute, and fix them in a 10 mL volumetric flask with eluent to prepare 5 μg/mL, 10 μg/mL, 15 μg/mL, 20 μg/mL, 40 μg/mL, and 80 μg/mL biotin linearity solution.
   (b) Inject linearity solutions of 20 μL, respectively.
   (c) Record the chromatograms, respectively.
   (d) Construct calibration curve by quadruple injecting six concentration levels of linearity solutions.
   (e) Calculate the equation of regression line, slope, intercept, residual sum of squares, correlation coefficient, limit of detection (LOD), and limit of quantification (LOQ). The LOD and LOQ are calculated as three and ten times a signal to noise.
2. Precision
   (a) Inject sample solution of 20 μL five times.
   (b) Record the peak area of each experiment.
   (c) Calculate the content of biotin in each sample solution.
   (d) Calculate the average content of biotin in sample.
   (e) Calculate the relative standard deviation (RSD).
3. Stability
   (a) Inject sample solution of 20 μL at 0, 2, 4, 8, 12, and 24 h.
   (b) Record the peak area of each experiment.

 (c) Calculate the content of biotin in each sample solution.
 (d) Calculate the average content of biotin in sample.
 (e) Calculate the relative standard deviation (RSD).

4. Recovery
 (a) Add biotin standard product of 4 mg, 5 mg, and 6 mg to the 5 mg test sample, respectively. Every sample is made in triplicate.
 (b) Add 40 mL diluent agent and ultrasonic extract in an ultrasonic water bath for 15 min.
 (c) Dilute the solution to the sample concentration of 0.1 mg/mL in the diluent agent using 50 mL volumetric flask.
 (d) Cool to room temperature.
 (e) Filter the above with 0.45 μm membrane filter to prepare the sample solution.
 (f) Inject sample solutions of 20 μL, respectively.
 (g) Record the peak area of each experiment.
 (h) Calculate the content of biotin in each sample solution.
 (i) Calculate the average content of biotin in sample.
 (j) Calculate the recovery (%).

4 Notes

1. All glassware used in the test needs to be strictly handled, and soak potassium chromate overnight in concentrated sulfuric acid solution. Then put it in the oven after cleaning in the box, and heat at 250 °C for 2 h. This applies to both methods.
2. Before use, the strain should be subcultured three times in succession to enhance the vitality of the strain. Only strong strains of bacteria have better sensitivity to biotin.
3. When measuring the absorbance of the bacterial liquid, you need to wait for the growth and reproduction of the bacterial body in the test tube to completely stop before proceeding.

Acknowledgments

The financial support of major projects of Beijing Polytechnic No. 2021Z009-KXZ is gratefully acknowledged.

References

1. McGuire M, Beerman KA (2011) Water soluble vitamins. In: Nutritional sciences, 2nd edn. Wadsworth, Belmont, pp 458–464
2. Ahemed J, Verma KK (1979) Determination of d-biotin at the microgram level. Talanta 26: 1025–1026
3. Tsuda H, Matsumoto T, Ishimi Y (2011) Biotin, niacin, and pantothenic acid assay using lyophilized lactobacillus plantarum ATCC 8014. J Nutr Sci Vitaminol 57(6):437–440
4. Livaniou E, Costopoulou D, Vassiliadou I, Leondiadis L, Evangelatos GP (2000) Analytical techniques for determining biotin. J Chromatogr A 881(1–2):331–343
5. Zhang H, Chen S, Liao W, Ren Y (2009) Fast simultaneous determination of multiple water-soluble vitamins and vitamin-like compounds in infant formula by UPLC-MS/MS. J Food Agric Environ 7(2):88–93
6. Brouwer VD, Storozhenko S, Stove CP, Daele JV, Straeten DVD, Lambert WE (2010) Ultra-performance liquid chromatography–tandem mass spectrometry (UPLC-MS/MS) for the sensitive determination of folates in rice. J Chromatogr B 878(3–4):509–513
7. Schiewe J, Gobel S, Schwarz M, Neubert R (1996) Application of capillary zone electrophoresis for analyzing biotin in pharmaceutical formulations-a comparative study. J Pharm Biomed Anal 14(4):435
8. Delgado RF, Fernandez RJM, Luque de CMD (2001) Determination of biotin in foodstuffs and pharmaceutical preparations using a biosensing system based on the streptavidin-biotin interaction. Anal Chim Acta 436(1):109–117
9. Indyk HE, Evans EA, Caselunghe MCB, Filonzi EL et al (2000) Determination of biotin and folate in infant formula and milk by optical biosensor-based immunoassay. J AOAC Int 83(5):1141–1148
10. Gao Y, Guo F, Gokavi S, Chow A, Sheng Q, Guo M (2008) Quantification of water-soluble vitamins in milk-based infant formulae using biosensor-based assays. Food Chem 110(3): 769–776
11. Lu B, Iwuohaa EI, Smyth MR, O'Kennedyb R (1997) Development of an "electrically wired" amperometric immunosensor for the determination of biotin based on a non-diffusional redox osmium polymer film containing an antibody to the enzyme label horseradish peroxidase. Anal Chim Acta 345(1–3):59–66
12. Indyk HE, Gill BD, Woollard DC (2014) Biotin content of paediatric formulae, early lactation milk and seasonal bovine milk powders by biosensor immunoassay. Int Dairy J 35(1): 25–31
13. Hong Y (2003) A sensitive competitive ELISA determination of biotin in transformed yeast culture media. J Chin Pharm Sci 12(4): 201–206
14. Chang YS, Wu CH, Chang RJ, Shiuan D (1994) Determination of biotin concentration by a competitive enzyme-linked immunosorbent assay (ELISA) method. J Biochem Biophys Methods 29(3–4):321–329
15. Mishra S, Storer MK, Sherwin CMT, Lewis JG (2005) A simple binding assay for the direct determination of biotin in urine. Clin Chim Acta 360(1–2):60–66
16. Kim JY, Oh CH (2011) Analysis of biotin in Korean representative foods and dietary intake assessment for Korean. Food Sci Biotechnol 20(4):1043
17. Kaczkow G, Czyz E, Anuszewska EL (2001) Elaboration of HPLC method for biotin determination in multiple vitamin drugs and comparison with microbiological method. Acta Pol Pharm 58(2):93–96
18. Fukui T, Iinuma K, Oizumi J, Izumi Y (1994) Agar plate method using Lactobacillus plantarum for biotin determination in serum and urine. J Nutr Sci Vitaminol 40(5):491–498
19. Vinas P, Lopez-Erroz C, Balsalobre N, Hernandez-Cordoba M (2003) Reversed-phase liquid chromatography on an amide stationary phase for the determination of the B group vitamins in baby foods. J Chromatogr A 1007(1–2):77–84
20. Zerzanov A (2007) Using of HPLC coupled with coulometric detector for the determination of biotin in pharmaceuticals. J Pharm Biomed Anal 5(45):730–735
21. Campos GE, Trisconi MJ, Kilinc T, Andrieux P (2010) Optimization and validation of an LC-FLD method for biotin in infant formula, infant cereals, cocoa-malt beverages, and clinical nutrition products. J AOAC Int 93(5): 1494–1501
22. Zerzanova A, Zizkovsky V, Kucera R, Klime J, Jesensky I, Dohnal J, Barron D (2007) Using of HPLC coupled with coulometric detector for the determination of biotin in pharmaceuticals. J Pharm Biomed Anal 45(5):730–735
23. Nojiri S, Kamata K, Nishijima M (1998) Fluorescence detection of biotin using post-column derivatization with OPA in high performance liquid chromatography. J Pharm Biomed Anal 16(8):1357

24. Yomota C, Ohnishi Y (2007) Determination of biotin following derivatization with 2-nitrophenylhydrazine by high-performance liquid chromatography with on-line UV detection and electrospray-ionization mass spectrometry. J Chromatogr A 1142(2):231–235
25. Yoshida T, Rerake A, Nakai C (1988) Liquid chromatographic determination of biotin by using 1-pyrenyldiazomethane as a pre-column florescent labeling reagent. J Chromatogr A 456:421–426
26. Lahely S, Ndaw S, Arella F (1999) Determination of biotin in foods by high-performance liquid chromatography with post-column derivatization and fluorimetric detection. Food Chem 65(2):253–258
27. Staggs CG, Sealey WM, Mccabe BJ, Teague AM, Mock DM (2004) Determination of the biotin content of select foods using accurate and sensitive HPLC/avidin binding. J Food Compos Anal 17(6):767–776

Chapter 22

Analysis of Pantothenate by High-Performance Liquid Chromatography

Iryna Levchuk, Antonina Kapustian, and Larysa Gural

Abstract

Pantothenic acid (vitamin B5) is an essential component of food and a precursor of important biologically active agents in the body. It is involved in energy metabolism and quite common in foods. Vitamin B5 deficiency occurs in certain categories of the population. The production of vitamin B5-enriched dietary supplements, foods, premixes, and feeds is practiced to overcome this problem. Since pantothenic acid is quite labile to the action of high temperatures, for the enrichment of food and feed is usually used its more stable form—calcium pantothenate. The most common and easy to implement method of its identification is high-performance liquid chromatography, which is described in detail in this chapter.

Key words Pantothenic acid, Pantothenate, Calcium-d-pantothenate, Analysis, High-performance liquid chromatography

1 Introduction

Pantothenic acid (vitamin B_5 or pantothenate) is an essential vitamin required for energy metabolism [1, 2]. Pantothenic acid is a precursor of coenzyme A, which is involved in the metabolism of proteins, fats, and carbohydrates. Dimers of pantothenate promote the reduction of total serum cholesterol and low-density lipoprotein and increase the level of high-density lipoprotein [1].

Pantothenic acid is a clear or light yellow viscous substance, soluble in water and ethanol [3], and stable at neutral pH but decomposes when heated [1].

Pantothenic acid consists of pantoic acid linked to β-alanine by an amide bond (Fig. 1a). Pantothenic acid is an optically active compound, but only D-pantothenic acid has biological activity [1, 4, 5, 6–8]. Pantothenic acid exists in two main functional forms: coenzyme A (CoA) (Fig. 1d) and acyl carrier protein (ACP) (Fig. 1c) [1–3, 9–23]. 4-Phosphopantetheine serves as a covalently linked prosthetic group of ACP (Fig. 1b) [1, 5, 9].

Lilian Mariutti and Frank Vriesekoop (eds.), *Food Chemistry: Basic Protocols*, Methods and Protocols in Food Science, https://doi.org/10.1007/978-1-0716-4678-6_22,

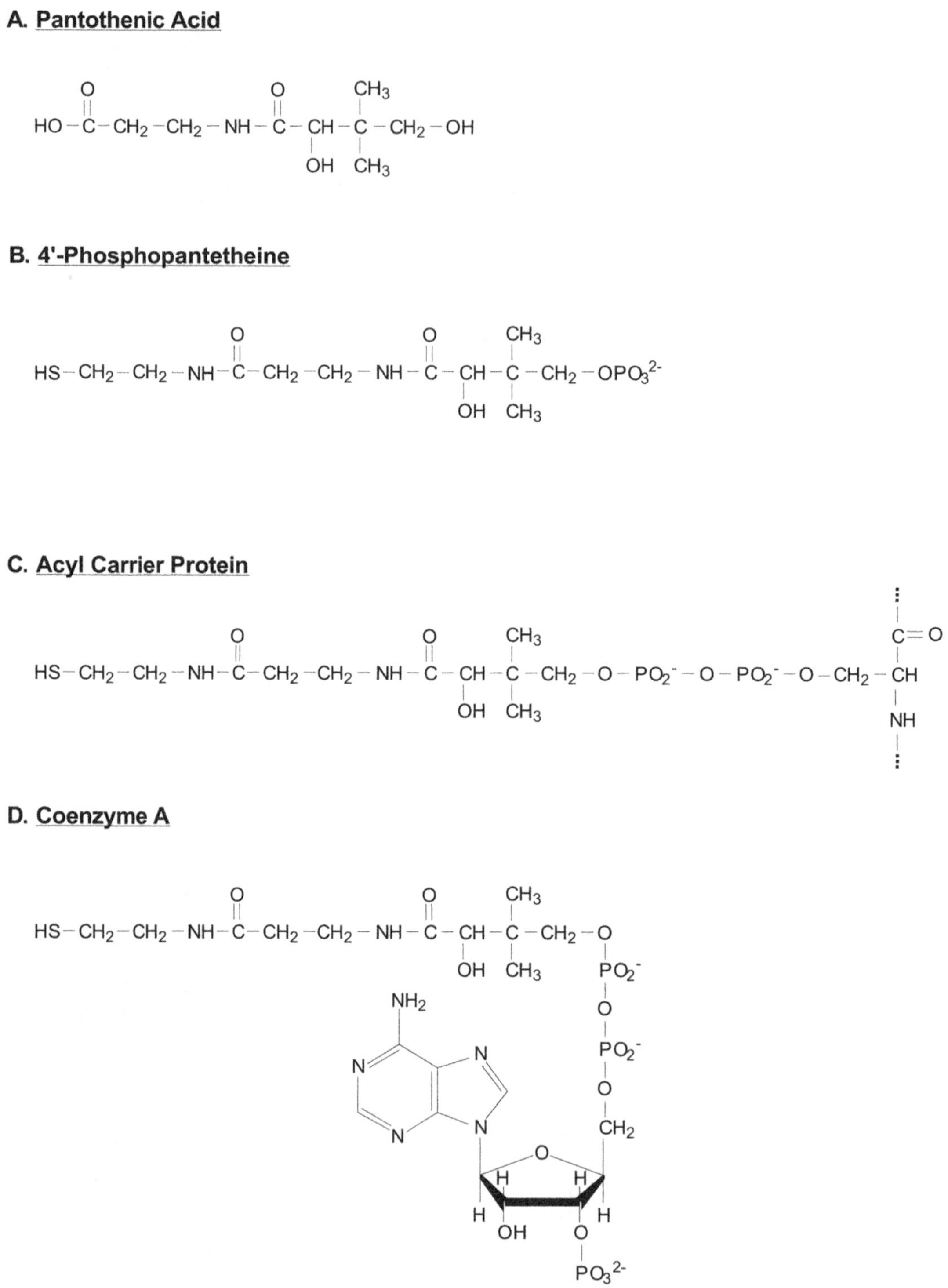

Fig. 1 Chemical structures of (**a**) pantothenic acid, (**b**) 4′-phosphopantetheine, (**c**) acyl carrier protein (ACP), and (**d**) coenzyme A (CoA) [9]

Pantothenic acid from the Greek "pantos" means "everywhere," which indicates its widespread distribution in nature and foods [1, 9–11, 24]. Foods rich in pantothenic acid include yeast, meat, milk, eggs, nuts, avocados, cruciferous vegetables, whole grains, potatoes, and tomato products [9, 13, 21].

Due to its thermal lability and tendency to oxidation, much of the pantothenic acid is lost during food processing. Processing and cleaning of whole grains lead to a loss of pantothenic acid by 37%–74%, while canning of meat, fish, and dairy products leads to losses of 20%–35%. Even greater pantothenic acid losses occur during canning (46%–78%) and freezing (37%–57%) of vegetables [1, 9, 13].

The average intake of pantothenic acid for adults is 5 mg/day [1, 13, 19]. In the general diet, most of the pantothenic acid is approximately 50%–95% conjugated to CoA and ACP and to a lesser extent is present in free form [2, 3, 9, 12, 13]. Bioavailability of native pantothenic acid in foods is 40%–61% [13].

Deficiency of pantothenic acid in humans is extremely rare due to its widespread distribution in products of animal and plant origin [1–3, 9, 11, 13, 19, 21]. People who have a special diet, have malnutrition, or have received a pantothenic acid antagonist may be deficient [12]. Pantothenic acid deficiency can also occur in people who abuse alcohol, women who take oral contraceptives, and people with impaired vitamin absorption due to certain diseases. Signs and symptoms of pantothenic acid deficiency include damage to the skin, liver, adrenal glands, and nervous system [12, 19].

Pharmaceuticals, foods, and feeds are often enriched with D-pantothenic acid [4, 6, 7] to eliminate potential deficiencies in diets. Because D-pantothenic acid is relatively unstable, dietary supplements and other foods are enriched with more stable calcium pantothenate [1].

Therefore, its rapid, reliable, and accurate detection is of great importance for quality and safety control [25]. Various analytical methods are used for this purpose [13].

Methodologies of Analysis

Well-established and classic methods of vitamin B_5 detection are microbiological assays, which explore test organisms that can grow only in a culture medium in the presence of the vitamin. The amount of vitamin is determined by turbidimetric method or by quantitative analysis of the metabolite (e.g., lactic acid) [13, 26].

More selective, quick, and reliable alternatives to microbiological assay are methods of HPLC (high-performance liquid chromatography) coupled with UV and fluorescence detection, LCMSMS (liquid chromatography-mass spectrometry), or immunoassay [10, 13, 25, 27–31].

Enzyme-linked immunosorbent assay (ELISA) with dual antibodies has been used to analyze pantothenic acid in various foods [27, 32]. In a study using surface plasmon resonance (SPR) biosensor assay of monoclonal antibody (MAb) and polyclonal antibody (PAb), a variety of samples (including cereals and baby foods) showed strong correlations with established analytical methods [33]. Pantothenic acid has also been quantified in multivitamin products by stable isotope dilution assays. Detection was then achieved by liquid chromatography-mass spectrometry (LC-MS) [12, 20, 30, 31].

The content of free and bound pantothenic acid in various foods has been determined by reverse-phase liquid chromatography-fluorimetry. This included purification of samples by sequential passage through anion and cation exchange cartridges and derivatization of pantothenic acid for fluorescence analysis [34].

B vitamins in multivitamin supplements are determined by high-performance liquid chromatography with diode-array detection (HPLC-DAD) and mass spectrometry [35]. HPLC-DAD is designed to test stability, determine the shelf life of multivitamin products, as well as routine analysis of all water-soluble vitamins in their most common forms [36].

2 Materials

All equipment and consumables mentioned in the methodology described below can be replaced by similar products from other manufacturers. In this section we describe the method of calcium-D-pantothenate identification, which is based on its extraction from samples with water and subsequent analysis by HPLC-DAD.

2.1 Basic Reagents

Reagents manufactured by Sigma-Aldrich (Regensburg, Germany) with a purity of at least 95% are recommended for this protocol.

Calcium-D-pantothenate is the recommended standard.

The following reagents are required for analysis: sodium heptanesulfonate monohydrate, phosphoric acid, acetonitrile, trichloroacetic acid, and sulfosalicylic acid.

All solutions/standards are to be prepared with 18 MΩ water. It is recommended the use of dark glassware, which should be cleaned by ultrasound and rinsed in 18 MΩ water.

2.2 Instrumentation

The following equipment can be used to prepare samples and their analysis:

1. Water purification system Simplicity UV "Millipore," USA.
2. Electronic scales of the first class "METTLER TOLEDO type AT261 (GmbH, laboratory & Weighing Technologies," Switzerland.

3. Centrifuge Eppendorf (Germany).
4. Vacuum installation for filtration with the mixed cellulose ester (MCE) filter 0.45 μm.

For chromatographic separation and identification, use a high-performance liquid chromatograph Agilent Technologies 1200 (USA) with a diode array detector equipped with:

1. Four-channel eluent supply system and the ability to operate the pump in gradient mode.
2. Degasser.
3. Auto-injector, which allows you to enter different sample volumes.
4. The thermostat of columns.
5. Diode-matrix detector.
6. A computer system with special ChemStation software for automatic integration, as well as the identification of B_5 in the test samples by retention time compared to the B_5 standard and by spectrum.

3 Methods

3.1 Sample Preparation

3.1.1 Extraction of Pantothenate from Liquid Samples

Due to the high solubility of vitamin B_5 in water, its extraction from liquid samples is quite simple. In such matrices, including energy drinks, the samples must be degassed by passing them through an MCE filter 0.45 μm. Other degassing methods may be used, but it is recommended that all samples be passed through a 0.45 μm MCE filter to preserve the HPLC column. Since this method has a high sensitivity, the working concentration of the analyte must be determined experimentally, taking into account the theoretical data on the vitamin content in a particular beverage.

3.1.2 Extraction of Pantothenate from Solid Samples

Vitamin B_5 can be analyzed from solid samples by dissolving them in water. Matrices such as foods and dietary supplements need to be dissolved in water 50–100 mL (hydromodule depends on the concentration of vitamin in the sample). Trichloroacetic acid (TCA) 1–5 mM was added to the extract, which promotes the pH stabilizing and provides the binding proteins removing. This will stabilize the sample pH and helps remove binding proteins [37–38]. Also 5 mg of sulfosalicylic acid (or on the tip of a spatula) is added to the extract (50 mL), stirred, and transferred to a refrigerator for 30 min at 4 °C. Sulfosalicylic acid is commonly used in identification assays as an agent to help release the target compounds from biopolymer supramolecular complexes [39–41]. After stirring, the sample should be centrifuged at 15,000 rpm for 10 min to remove any sample debris, and the

Table 1
Conditions of the gradient mode

Time, min	Solution A, %	Solution B, %
2.5	100	0
10.0	85	15
25.0	30	70
27.0	10	90
34.0	100	0

supernatant is filtered through an 0.45 μm MCE filter and transferred to a clean dark glass vial with a volume of 1 mL.

3.2 Pantothenate Analysis

The chromatographic separation of B_5 is best performed in a gradient mode. In order to increase the efficiency of separation, a specific analytical column Zorbax SB-Aq 4.6 × 150 mm, 5 μm has been used. B_5 detection is best performed at a wavelength of 261 nm ± 4 nm [42].

The mobile phase with 0.0125 M sodium heptanesulfonate monohydrate in 0.1% phosphoric acid, pH 2.5 (solution A), and acetonitrile HPLC (solution B) was used for analysis. The conditions of the gradient mode are shown in Table 1.

The temperature of the column thermostat is 30 °C, the flow rate of the mobile phase on the analytical column is 0.8 mL/min, and the injection volume of the sample extract is 10 μL.

Figure 2 shows the results of elution of standard sample pantothenate (2a), an energy drink extract (2b), and a premix extract (2c). Table 2 shows the accuracy, precision, LOD, and RSD of this analysis. This method also achieved ppm detection limits of the target analytes.

4 Notes

1. Vitamin B5 is not stable in the presence of light. It is recommended to carry out all work away from direct sunlight using dark glass utensils, both for standard and for samples.
2. It is recommended to use pre-column filters to prevent excessive column contamination.
3. The water used for analysis must be free of soluble or colloidal ionic and organic impurities and meet stringent analytical requirements (ISO 3696:1987. Water for analytical laboratory use—Specification and test methods).

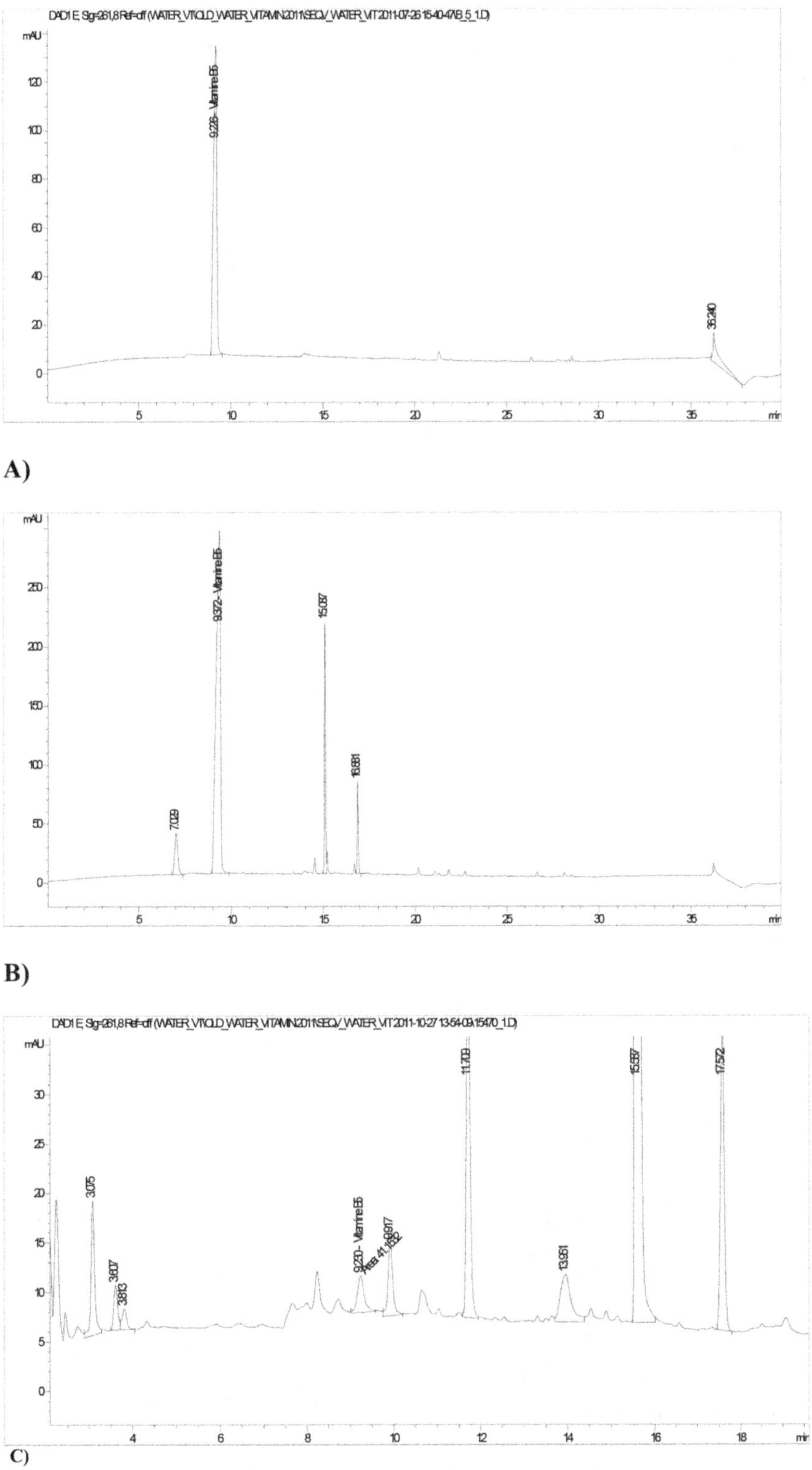

Fig. 2 Chromatographic analysis of vitamin B_5 by HPLC: (**a**) standard solution (calcium-D-pantothenate); (**b**) extract from the energy drink; (**c**) extract from the premix

Table 2
Accuracy, precision, LOQ, LOD, range of linearity, and LOC of described method of pantothenate analysis from liquid samples

Analyte	Accuracy (av. % spike recovery)	Precision (% RSD)	LOQ (ng/mL)	LOD (ng/mL)	Range of linearity (ng/mL)	LOC (R^2 value)
Energy drink	97.9	2.8	10.0	2.5	10–550	0.9999

4. Since this method has a high sensitivity, the working concentration of the analyte must be determined experimentally, taking into account the theoretical data on the vitamin content in samples.

References

1. Sampedro A, Rodriguez-Granger J, Ceballos J, Aliaga L (2015) Pantothenic acid: an overview focused on medical aspects. Eur Sci J 11(21): 1–18
2. Nohr D, Biesalski HK, Back EI (2011) Vitamins. In: Pantothenic acid. Encyclopedia of dairy sciences, 2nd edn, pp 694–696. https://doi.org/10.1016/B978-0-12-374407-4.00488-X
3. Nohr D, Biesalski HK (2016) Encyclopedia. In: Vitamin B5. Reference module in food. https://doi.org/10.1016/B978-0-08-100596-5.01071-4
4. Zhang B, Chen L, Jin J-Y, Zhong N, Cai X, Zou S-P, Zhou H-Y, Liu Z-Q, Zheng Y-G (2021) Strengthening the (R)-pantoate pathway to produce D-pantothenic acid based on systematic metabolic analysis. Food Biosci 43: 101283. https://doi.org/10.1016/j.fbio.2021.101283
5. Coxon KM, Chakauya E, Ottenhof HH, Whitney HM, Blundell TL, Abell C, Smith AG (2005) Pantothenate biosynthesis in higher plants. Biochem Soc Trans 33(4):743–746. https://doi.org/10.1042/BST0330743
6. Zou S-P, Wang Z-J, Zhao K, Zhang B, Niu K, Liu Z-Q, Zheng Y-G (2021) High-level production of D-pantothenic acid from glucose by fed-batch cultivation of Escherichia coli. Biotechnol Appl Biochem 68(6):1227–1235
7. Zou S-P, Zhao K, Wang Z-J, Zhang B, Liu Z-Q, Zheng Y-G (2021) Overproduction of D-pantothenic acid via fermentation conditions optimization and isoleucine feeding from recombinant Escherichia coli W3110. 3 Biotech 11:295. https://doi.org/10.1007/s13205-021-02773-0
8. Zou S, Zhao K, Tang H, Zhang Z, Zhang B, Liu Z, Zheng Y (2021) Improved production of D-pantothenic acid in Escherichia coli by integrated strain engineering and fermentation strategies. J Biotechnol 339(20):65–72. https://doi.org/10.1016/j.jbiotec.2021.07.014
9. Miller JW, Rucker RB (2012) Pantothenic acid. AG-PPKN-0909 R1 CH25 04-04-06 07:27:58
10. Huisjes R, Card DJ (2019) Chapter 8 – Methods for assessment of pantothenic acid (vitamin B5). In: Laboratory assessment of vitamin status, vol 8, pp 173–179. https://doi.org/10.1016/B978-0-12-813050-6.00008-5
11. Shetty SA, Young MF, Taneja S, Rangiah K (2020) Quantification of B-vitamins from different fresh milk samples using ultra-high performance liquid chromatography mass spectrometry/selected reaction monitoring methods. J Chromatogr A 1609:460452. https://doi.org/10.1016/j.chroma.2019.460452
12. Xu J, Patassini S, Begley P, Church S, Waldvogel HJ, Faull RLM, Unwin RD, Cooper GJS (2020) Cerebral deficiency of vitamin B5 (d-pantothenic acid; pantothenate) as a potentially-reversible cause of neurodegeneration and dementia in sporadic Alzheimer's disease. Biochem Biophys Res Commun 527(3): 676–681. https://doi.org/10.1016/j.bbrc.2020.05.015
13. Dietary Reference Intakes for Thiamin, Riboflavin, Niacin, Vitamin B6, Folate, Vitamin B12, Pantothenic Acid, Biotin, and Choline (1998) Institute of Medicine (US) Standing Committee on the Scientific Evaluation of

Dietary Reference Intakes and its Panel on Folate, other B vitamins, and choline, vol 10. National Academies Press (US), Washington, pp 357–373

14. Tsuchiya Y, Pham U, Gout I (2014) Methods for measuring CoA and CoA derivatives in biological samples. Biochem Soc Trans 42: 1107–1111. https://doi.org/10.1042/BST20140123
15. Ottenhof HH, Ashurst JL, Whitney HM, Adrian Saldanha S, Schmitzberger F, Gweon HS, Blundell TL, Abell C, Smith AG (2004) Organisation of the pantothenate (vitamin B5) biosynthesis pathway in higher plants. Plant J 37(1):61–72. https://doi.org/10.1046/j.1365-313X.2003.01940.x
16. Jung S, Kim MK, Choi BY (2017) The long-term relationship between dietary pantothenic acid (vitamin B5) intake and C-reactive protein concentration in adults aged 40 years and older. Nutr Metab Cardiovasc Dis 27(9): 806–816. https://doi.org/10.1016/j.numecd.2017.05.008
17. Leonardi R, Jackowski S (2007) Biosynthesis of pantothenic acid and coenzyme A. EcoSal Plus 2(2):10. https://doi.org/10.1128/ecosalplus.3.6.3.4
18. Kobayashi D, Kusama M, Ma O, Nakahata N (2011) The effect of pantothenic acid deficiency on keratinocyte proliferation and the synthesis of keratinocyte growth factor and collagen in fibroblasts. J Pharmacol Sci 115(2): 230–234. https://doi.org/10.1254/jphs.10224SC
19. Combs GF Jr, McClung JP (2022) Chapter 15 – Pantothenic acid. The vitamins. In: Fundamental aspects in nutrition and health, 6th edn, pp 435–452. https://doi.org/10.1016/B978-0-323-90473-5.00009-4
20. Ismail N, Kureishy N, Church SJ, Scholefield M, Unwin RD, Xu J, Patassini S, Cooper GJS (2020) Vitamin B5 (D-pantothenic acid) localizes in myelinated structures of the rat brain: potential role for cerebral vitamin B5 stores in local myelin homeostasis. Biochem Biophys Res Commun 522(1):220–225. https://doi.org/10.1016/j.bbrc.2019.11.052
21. Miller JW, Rucker RB (2020) Chapter 16 – Pantothenic acid. Present knowledge in nutrition. In: Basic nutrition and metabolism, 11th edn, pp 273–287. https://doi.org/10.1016/B978-0-323-66162-1.00016-0
22. Velazquez-Arellano A, Hernandez-Vazquez Ade J (2020) Chapter 35: Vitamins as cofactors for energy homeostasis and their genomic control, with special reference to biotin, thiamine, and pantothenic acid. In: Principles of nutrigenetics and nutrigenomics fundamentals of individualized nutrition, pp 271–277. https://doi.org/10.1016/B978-0-12-804572-5.00035-5
23. Shibata K, Nakai T, Fukuwatari T (2012) Simultaneous high-performance liquid chromatography determination of coenzyme A, dephospho-coenzyme A, and acetyl-coenzyme A in normal and pantothenic acid-deficient rats. Anal Biochem 430(2):151–155. https://doi.org/10.1016/j.ab.2012.08.010
24. Ball GFM (1998) Bioavailability and analysis of vitamins in foods. In: Pantothenic acid, pp 409–422. https://doi.org/10.1007/978-1-4899-3414-7
25. Zeng L, Guo L, Wang Z, Xu X, Song S, Xu L, Kuang H, Li A, Xu C (2021) Immunoassays for the rapid detection of pantothenic acid in pharmaceutical and food products. Food Chem 348(30):129114. https://doi.org/10.1016/j.foodchem.2021.129114
26. Solberg O, Hegna IK (1979) Microbiological assay of pantothenic acid. Methods Enzymol 62:201–204. https://doi.org/10.1016/0076-6879(79)62218-8
27. Gonthier A, Boullanger P, Fayol V, Hartmann D (1998) Development of an ELISA for pantothenic acid (vitamin B5) for application in the nutrition and biological fields. J Immunoass 19(2–3):167–194. https://doi.org/10.1080/01971529808005479
28. Andrieux P, Fontannaz P, Kilinc T, Giménez EC (2012) Pantothenic acid (vitamin B5) in fortified foods: comparison of a novel ultra-performance liquid chromatography-tandem mass spectrometry method and a microbiological assay (AOAC official method (SM) 992.07). J AOAC Int 95(1):143–148. https://doi.org/10.5740/jaoacint.10-333
29. Islam MA, Park E, Jeong B, Gwak Y-J, Kim J, Hong W-H, Sujin P, Jung J, Na-Y Y, Kim Y-K, Chun J (2022) Validation of vitamin B5 (pantothenic acid) and B6 (pyridoxine, pyridoxal, and pyridoxamine) analyses in seafood. J Food Compos Anal 109:104518. https://doi.org/10.1016/j.jfca.2022.104518
30. Rychlik M (2003) Simultaneous analysis of folic acid and pantothenic acid in foods enriched with vitamins by stable isotope dilution assays. Anal Chim Acta 495(1–2): 133–141. https://doi.org/10.1016/j.aca.2003.08.020
31. Mittermayr R, Kalman A, Trisconi M-J, Heudi O (2004) Determination of vitamin B5 in a range of fortified food products by reversed-phase liquid chromatography-mass spectrometry with electrospray ionisation. J Chromatogr

A 1032(1–2):1–6. https://doi.org/10.1016/j.chroma.2003.11.062

32. Finglas PM, Faulks RM, Morris HC et al (1988) The development of an enzyme-linked immunosorbent assay (ELISA) for the analysis of pantothenic acid and analogues, part II-determination of pantothenic acid in foods. J Micronutr 4:47–59
33. Haughey SA, Elliott CT, Oplatowska M, Stewart LD, Frizzell C, Connolly L (2012) Production of a monoclonal antibody and its application in an optical biosensor based assay for the quantitative measurement of pantothenic acid (vitamin B5) in foodstuffs. Food Chem 134(1):540–545. https://doi.org/10.1016/j.foodchem.2012.02.116
34. Pakin C, Bergaentzle M, Hubscher V, Aoude-Werner D, Hasselmann C (2004) Fluorimetric determination of pantothenic acid in foods by liquid chromatography with post-column derivatization. J Chromatogr A 1035(1):87–95. https://doi.org/10.1016/j.chroma.2004.02.042
35. Sim H-J, Kim B, Lee J (2016) A systematic approach for the determination of B-group vitamins in multivitamin dietary supplements by high-performance liquid chromatography with diode-array detection and mass spectrometry. J AOAC nt 99(5):1223–1232. https://doi.org/10.5740/jaoacint.16-0093
36. Rakuša ŽT, Grobin A, Roškar R (2021) A comprehensive approach for the simultaneous analysis of all main water-soluble vitamins in multivitamin preparations by a stability-indicating HPLC-DAD method. Food Chem 337(1):127768. https://doi.org/10.1016/j.foodchem.2020.127768
37. Greenberg NA (2006) Comparison of the abilities of trichloroacetic, picric, sulfosalicylic, and tungstic acids to precipitate protein hydrolysates and proteins. J Food Sci 44(3):735–737. https://doi.org/10.1111/j.1365-2621.1979.tb08487.x
38. Nils AR et al (2021) Improved estimation of in vitro protein digestibility of different foods using size exclusion chromatography. Food Chem 358(1):129830
39. Huang Z et al (2022) The mechanisms of sodium chloride stress mitigation by salt-tolerant plant growth promoting Rhizobacteria in wheat. Agronomy 12(3):543. https://doi.org/10.3390/agronomy12030543
40. Silventoinen P, Rommi K, Holopainen-Mantila U et al (2019) Biochemical and techno-functional properties of protein- and fibre-rich hybrid ingredients produced by dry fractionation from rice bran. Food Bioprocess Technol 12:1487–1499. https://doi.org/10.1007/s11947-019-02307-w
41. Zhao C, Guo H, Wang J, Wang Y, Zhang R (2021) Melatonin enhances drought tolerance by regulating leaf stomatal behavior, carbon and nitrogen metabolism, and related gene expression in maize plants. Front Plant Sci 12:779382. https://doi.org/10.3389/fpls.2021.779382
42. Hudson TS, Subramanian S, Allen RJ (1984) Determination of pantothenic acid, biotin, and vitamin B12 in nutritional products. J AOAC 67(5):994–998. https://doi.org/10.1093/jaoac/67.5.994

Chapter 23

Analysis of Pyridoxine and Its Respective Vitamers by High-Performance Liquid Chromatography

Ran Yu

Abstract

Pyridoxine (vitamin B_6) is an essential vitamin which plays a crucial role in the metabolism of amino acids and lipids, but also involved in immune responses and alleviating oxidative stress. There are two other principle forms of vitamin B_6, namely, pyridoxal and pyridoxamine. Intracellularly vitamin B_6 is converted to its bioactive 5′-phosphate form, pyridoxal 5′-phosphate (PLP), which is essential for many organisms and serves as versatile coenzymes in more than 100 enzymatic reactions. Deficiencies in this vitamin generally lead to the development of inflammatory reaction and major chronic diseases. The following chapter describes a simple and reproducible method for the analysis of pyridoxine via high-performance liquid chromatography utilizing fluorescence detection.

Key words Pyridoxine, Vitamers, High-performance liquid chromatography, Fluorescence

1 Introduction

In its pure form, pyridoxine is white crystalline powder that was first found in yeast extract and described as vitamin B_6 [1–4]. As such pyridoxine is better known as vitamin B_6; however, the term defines a series of compounds with a basic structure of 2-methyl-3-hydroypyridine, including pyridoxine, pyridoxal, pyridoxamine, pyridoxine-5′-phosphate (PNP), pyridoxal-5′-phosphate (PLP), and pyridoxamine-5′-phosphate (PMP) (Fig. 1). The majority of these vitamers are converted into PLP through metabolism, which is the typical active form of vitamin B_6.

Pyridoxine is stable under most conditions other than heat and light. Significant pyridoxine losses occur upon heating of low-moisture food products [5]. The heterocyclic ring structure of vitamin B_6 contains three or two –OH groups that are easily oxidizable with a lone pair of electrons [6]. Under both acidic and alkaline conditions, the major oxidation product of vitamin B6 is

Lilian Mariutti and Frank Vriesekoop (eds.), *Food Chemistry: Basic Protocols*, Methods and Protocols in Food Science, https://doi.org/10.1007/978-1-0716-4678-6_23,

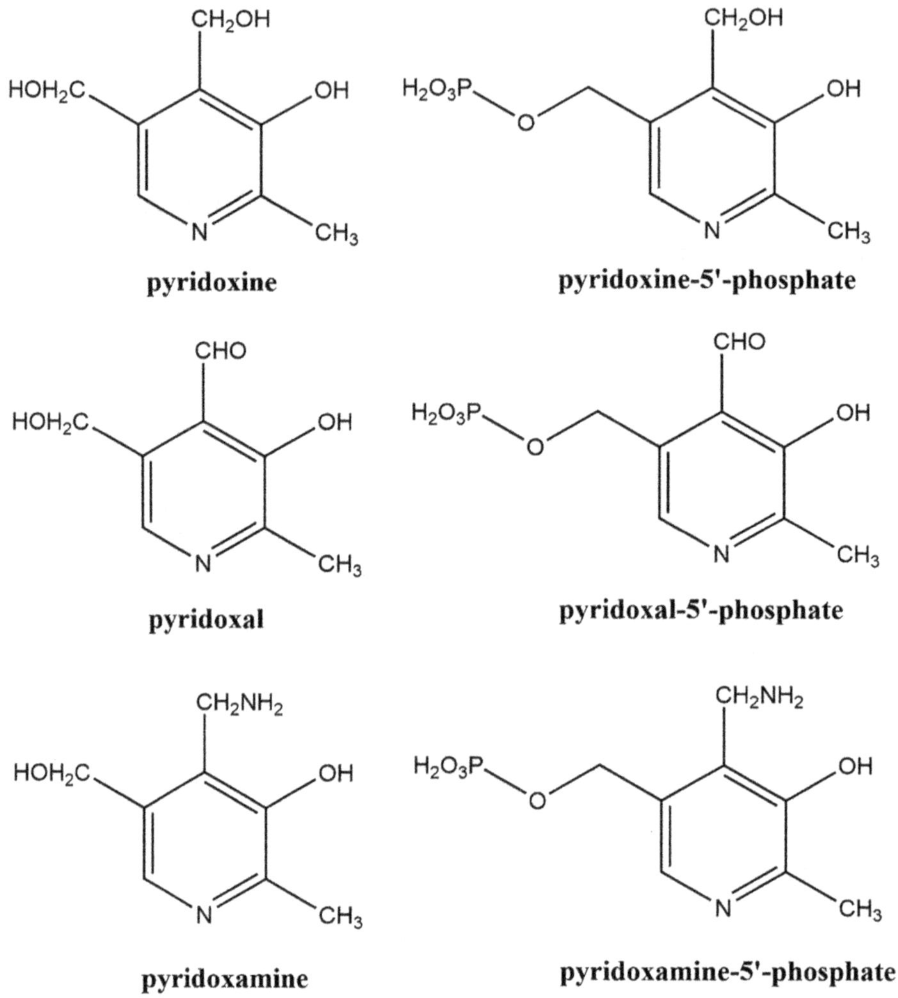

Fig. 1 Structure of the pyridoxine vitamers

pyridoxal [7] with the oxidative decomposition being greater under alkaline conditions [8].

Pyridoxine is present in many natural unprocessed foods and cannot be produced in the human body; therefore it must be absorbed from food sources like various vegetables and fruits [9]. PLP, the active coenzyme form of vitamin B_6, is involved in over 140 enzymatic activities [10] and is of great metabolic relevance for amino acid biosynthesis [11, 12]. Vitamin B_6 deficiency is related to inflammations, nervous system disorder [13], gastrointestinal issues, and chronic disease risks [14]. PLP also has the ability to act as a scavenger of reactive oxygen species [15].

Methodologies of Analysis

Pyridoxine is traditionally easy to analyze. Typically, HPLC methods coupled with UV-vis, fluorescence, or mass spectrometry are used to determine the pyridoxine content from various food matrices [16–18]. Due to the natural fluorescence of pyridoxine, fluorescence detection is usually chosen as the most commonly used method to analyze this vitamin and its vitamers. The detection limits of fluorescence methods are in the ug/L range with high specificity and sensitivity [19]. Specific excitation and emission wavelengths are typically adopted during the analysis of pyridoxine to alleviate the potential interference of matrix-dependent impurities. Being highly water soluble, extractions in water are recommended, with the use of inorganic acids being employed for improved total recovery [20]. Pyridoxal, pyridoxamine, and pyridoxine are extracted from food by acid hydrolysis of the food matrix and often dephosphorylated enzymatically using acid phosphatase. By reaction with glyoxylic acid in the presence of Fe^{2+} as a catalyst, pyridoxamine can be transformed into pyridoxal, which is then reduced to pyridoxine by the action of sodium borohydride in alkaline medium. Total pyridoxine is then quantified in the same solution by HPLC by means of fluorometric detection [16]. When analyzing for pyridoxine or its respective vitamers, it is crucial that all manipulations are performed in amber glassware. Failure to do so may lead to an underestimate in the results due to pyridoxine's sensitivity to light.

2 Materials

All equipment and consumables mentioned in this chapter can be substituted by alternative products from other manufacturers. In this chapter we describe a tried set of methodologies based on the equipment and consumables mentioned.

2.1 Basic Reagents

All reagents/standards referred to in this chapter are of 95% purity or greater. All solutions/standard are made with 18 MΩ water (Sartorius arium 611VF water purification system—Sartorius AG, Göttingen, Germany). It is recommended that all glassware used in this analysis is cleaned via sonication and rinsed in 18 MΩ water. A gradient buffered methanol mobile phase is utilized for the chromatographic separation.

2.2 Instrumentation

The example HPLC system consists of an Agilent 1200 HPLC system (Agilent Technologies, Santa Clara, CA, USA), a column oven (35 °C), and an autosampler (20 μL injection) coupled with a fluorescence detector set at 290 nm excitation and 396 nm emission wavelengths, respectively. Chromatographic separation can be performed on an Imtakt Scherzo SW-C18 column (4.6 mm × 150 mm, 3 μm, Shiseido) at a flow rate of 0.7 mL/min. The mobile phase gradient employed was as follows: solvent A (ammonium formate (0.1% formic acid) in methanol); solvent B (100% methanol); using the following gradient elution profile, 0–14 min 100:0 A:B; 14–20 min 30:70 A:B; 20–21 min 100:0 A:B; 21–30 min 100:0 A:B; at a total run time of 30 min [21].

3 Methods

3.1 Sample Preparation (see Note 1)

The extraction procedure includes a combination of acid digestion followed by an enzymatic hydrolysis. The acid hydrolysis breaks protein complexes and effectively liberates the pyridoxin vitamers from sample matrix. Phosphorylated vitamin esters are typically converted to their free vitamer forms by the use of phosphatase enzymes. The extraction of vitamin B_6 by enzymatic mixture has been recommended for hydrolyze of the analyte prior to its quantification. If analyzing for total pyridoxine, it is recommended that taka-diastase or acid phosphatase is used to extract pyridoxine from samples. These enzymes can cleave the phosphate bonds in the vitamers into the free forms to obtain good dephosphorylation [20, 22, 23], but for full vitamer speciation, it is recommended that these enzymes are not used [24].

All samples were homogenized properly by using a professional blender (Hanil, HMF-3500TG) and divided into amber glassware and stored at −70 °C [21]. It is also recommended to prevent bacterial growth in samples that they are only prepared prior to analysis but can be stored for up to 72 h under refrigerated conditions.

3.1.1 Pyridoxine Extraction from Liquid Samples

Due to the high solubility of the pyridoxine vitamers in water, liquid sample analysis is reasonably straightforward. Unless sample cleanup is required due to matrix interferences, the only sample preparation required is particulate removal, which can be facilitated by passing the sample through a 0.45 μm RC filter. To preserve the HPLC column, it is best to use a guard-column.

Fifty milliliters of 0.1 M HCl is added to 10 mL sample which is left for 30 min in a water bath at 100 °C. After cooling to ambient temperature, the pH must be adjusted to 4.5 with 2.5 M sodium acetate buffer in order to facilitate the enzymatic hydrolysis reaction. For the enzymatic hydrolysis, follow the enzyme manufacturer's instructions, since these vary from manufacturer to manufacturer. After enzyme incubation, samples should be heated to 100 °C for 5 min to inactivate the enzyme. All samples and standards should be prepared under dim light and stored in amber flask [6] (*see* **Note 2**). Before HPLC analysis, all samples should be filtered through a 0.45 μm RC filter.

3.1.2 Pyridoxine Extraction from Solid Samples

Free pyridoxine vitamers can easily be analyzed from solid samples by dissolving them in water. Do ensure stability of the phosphorylated form; it is recommended that a 10 mM ammonium formate (0.1% formic acid) is used [21]. This will help to stabilize the pH during extraction and make it easier for later filtration and injection in to the HPLC system. For total pyridoxine vitamer testing, using hydrochloric acid can aid in the breakdown on binding proteins releasing the vitamers. The pH of the sample solution is adjusted to 4.5 and then injected into the HPLC system. In matrices such as cereal-based food, samples must be processed into a fine powder and homogenized. Longer extraction times can lead to a reduction in recovery, and the mixing should be always shielded from light.

For starchy food, after acid hydrolysis, taka-diastase, acid phosphatase, and β-glucosidase enzymes may be added to the sample before which is incubated at 37 °C to enhance extraction efficiencies. All samples should be filtered through a 0.45 μm RC filter before the transfer into the HPLC device [21].

3.2 Separation and Analysis of Pyridoxine Vitamers (see Note 2)

The elution of vitamers occurs on a buffer to methanol gradient with a final composition described in Subheading 2.2.

Figure 2 depicts the elution order of the described vitamers. Table 1 shows the precision, LOD, and RSD of this analysis. This method also achieved ppm detection limits of the target analytes.

4 Notes

1. Pyridoxine, pyridoxal, and pyridoxamine are not stable in the presence of light. It is recommended that all work is carried out away from direct sunlight utilizing amber or covered glassware for both standards and samples.
2. Precolumn filters are recommended for all samples to prevent excessive column fouling.

Due to the high specificity of the fluorescence detection, it is critical that all glassware used is thoroughly cleaned using sonication and washed in 18 MΩ water.

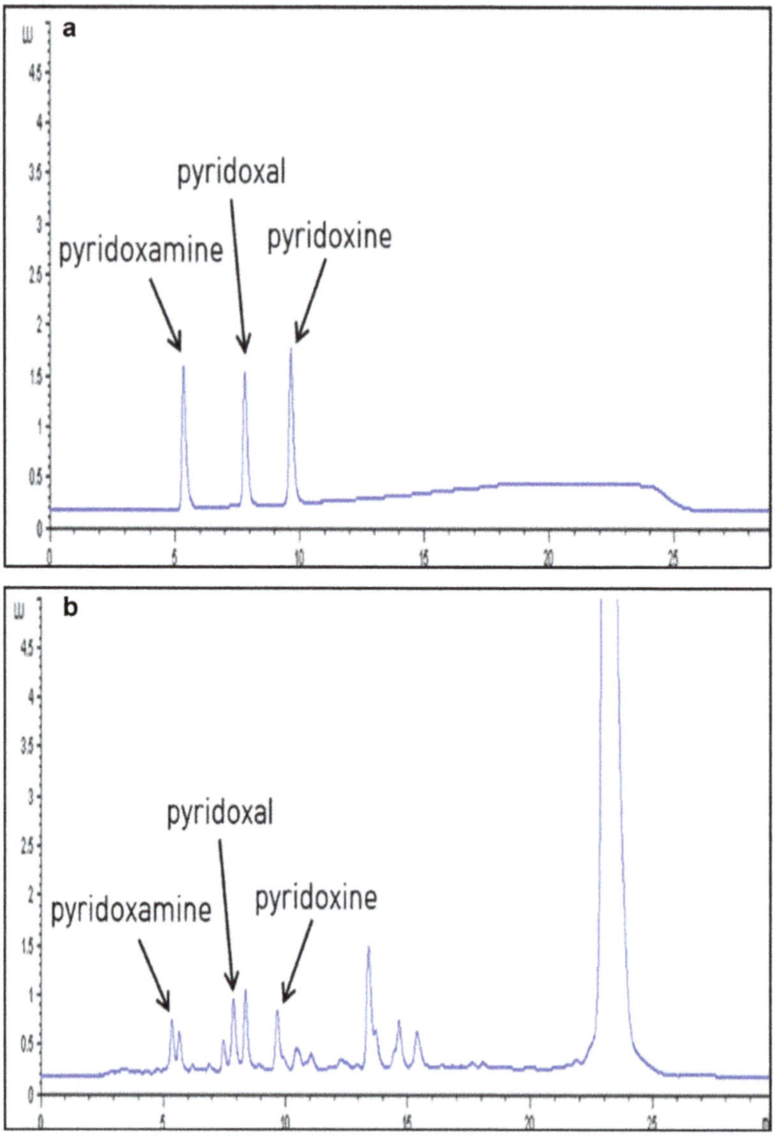

Fig. 2 Chromatographic analysis of pyridoxine vitamers by HPLC: (**a**) standard solution of pyridoxine vitamers; (**b**) sample (salted oyster) containing pyridoxine vitamers [21]

Table 1
Accuracy, precision, limit of quantitation and, linearity of calibration for sample analysis [21]

Analyte	Accuracy (av. % spike recovery)	Precision (% RSD)	LOQ (ng/mL)	LOD (ng/mL)	Range of linearity (ng/mL)	LOC (R^2 value)
Pyridoxine	99.3	3.7	0.83	0.27	20–1259	0.9998
Pyridoxal	98.1	2.5	0.92	0.29	20–1253	0.9999
Pyridoxamine	99.2	5.2	0.64	0.19	20–1251	1.0000

References

1. Calori IR, Gusmão LA, Tedesco AC (2021) B6 vitamers as generators and scavengers of reactive oxygen species. J Photochem Photobiol 7: 100041
2. GyÜRgy P, Eckardt RE (1939) Vitamin B6 and skin lesions in rats. Nature 144(3646): 512–512
3. GyÖRgy P (1934) Vitamin B2 and the pellagra-like dermatitis in rats. Nature 133(3361):498–499
4. Van den Eynde MDG et al (2021) Quantification of the B6 vitamers in human plasma and urine in a study with pyridoxamine as an oral supplement; pyridoxamine as an alternative for pyridoxine. Clin Nutr 40(7):4624–4632
5. Bui LTT, Small DM (2012) The stability of pyridoxine hydrochloride used as a fortificant in Asian wheat flour noodles. Food Chem 130(4):841–846
6. Lebiedzińska A et al (2007) Reversed-phase high-performance liquid chromatography method with coulometric electrochemical and ultraviolet detection for the quantification of vitamins B1 (thiamine), B6 (pyridoxamine, pyridoxal and pyridoxine) and B12 in animal and plant foods. J Chromatogr A 1173(1): 71–80
7. Nipane SV, Gurame VM, Gokavi GS (2011) Kinetics and mechanism of oxidation of pyridoxine by Enneamolybdomanganate(IV). Inorg Chem Commun 14(7):1102–1106
8. Albarrán G, Ramírez-Cahero F, Aliev R (2008) Radiolysis of pyridoxine (vitamin B6) in aqueous solution under different conditions. Radiat Phys Chem 77(5):605–611
9. Arvapally M et al (2021) Development of an analytical method for the quantitative determination of multi-class nutrients in different food matrices by solid-phase extraction and liquid chromatography-tandem mass spectrometry using design of experiments. Food Chem 341:128173
10. Ulvik A et al (2016) Vitamin B-6 catabolism and long-term mortality risk in patients with coronary artery disease. Am J Clin Nutr 103(6):1417–1425
11. Ueland PM et al (2015) Direct and functional biomarkers of vitamin B6 status. Annu Rev Nutr 35(1):33–70
12. Vrolijk MF et al (2020) Inter-individual differences in pharmacokinetics of vitamin B6: a possible explanation of different sensitivity to its neuropathic effects. PharmaNutrition 12: 100188
13. Ahmed S et al (2020) Vitamin B6-dependent epilepsy due to pyridoxal phosphate-binding protein (PLPBP) defect – first case report from Pakistan and review of literature. Ann Med Surg 60:721–727
14. Schorgg P et al (2022) Increased vitamin B6 turnover is associated with greater mortality risk in the general US population: a prospective biomarker study. Clin Nutr 41(6):1343–1356
15. Zhang X, Tang X, Daly TM (2020) A one-step NIST traceable HPLC method for quantitation of vitamin B6 and 4-pyridoxic acid in human plasma. Pract Lab Med 21:e00160
16. Bergaentzlé M et al (1995) Determination of vitamin B6 in foods by HPLC — a collaborative study. Food Chem 52(1):81–86
17. Roelofsen-de Beer RJAC et al (2017) Simultaneous measurement of whole blood vitamin B1 and vitamin B6 using LC-ESI–MS/MS. J Chromatogr B 1063:67–73
18. Cellar NA et al (2016) A new sample preparation and separation combination for precise, accurate, rapid, and simultaneous determination of vitamins B1, B2, B3, B5, B6, B7, and B9 in infant formula and related nutritionals by LC-MS/MS. Anal Chim Acta 934:180–185
19. Llorent-Martínez EJ et al (2006) A multicommuted fluorescence-based sensing system for simultaneous determination of vitamins B2 and B6. Anal Chim Acta 555(1):128–133
20. Kall MA (2003) Determination of total vitamin B6 in foods by isocratic HPLC: a comparison with microbiological analysis. Food Chem 82(2):315–327
21. Islam MA et al (2022) Validation of vitamin B5 (pantothenic acid) and B6 (pyridoxine, pyridoxal, and pyridoxamine) analyses in seafood. J Food Compos Anal 109:104518
22. Reitzer-Bergaentzle M, Marchioni E, Hasselmann C (1993) HPLC determination of vitamin B6 in foods after pre-column derivatization of free and phosphorylated vitamers into pyridoxol. Food Chem 48(3): 321–324
23. Bitsch R, Möller J (1989) Analysis of B6 vitamers in foods using a modified high-performance liquid chromatographic method. J Chromatogr A 463:207–211
24. Yaman M, Mızrak ÖF (2019) Determination and evaluation of in vitro bioaccessibility of the pyridoxal, pyridoxine, and pyridoxamine forms of vitamin B6 in cereal-based baby foods. Food Chem 298:125042

Part VII

Pigments

Chapter 24

Guidelines for the Carotenoid Analysis in Foods

Adriana Z. Mercadante

Abstract

The carotenoid analysis in foods includes several steps: sampling, subsampling, and size reduction to obtain a homogenized size-laboratory sample, followed by exhaustive extraction usually with acetone and liquid-liquid partition to ether/diethyl ether. At this point, the carotenoid extract is optionally saponified with 10% KOH in methanol (MeOH) when carotenoid esters are not the target, followed by alkali washout, complete solvent evaporation, and storage under low temperature. Before chromatographic separation, the extract is dissolved in methyl *t*-butyl ether (MTBE)/MeOH and analyzed by high-performance liquid chromatography (LC) on a C30 column with gradient elution of MeOH/MTBE for saponified samples and water/MeOH/MTBE for non-saponified ones. Detection is achieved by a diode array detector followed by a mass spectrometer with atmospheric pressure chemical ionization at positive mode. Identification is based on the combined information provided by elution order on the C30 column, UV-visible spectrum features (maximum wavelengths, spectral fine structure, and *cis*-peak intensity), and mass spectrum (protonated molecule and tandem mass fragments from both polyene chain and end-groups). Quantification is carried out with external calibration curves. Finally, selectivity, recovery, repeatability, and limits of quantification and detection are determined to evaluate the method suitability and result reliability.

Key words Total carotenoids, Carotenoid esters, Carotenoid quantification, Carotenoid identification, Method validation, LC-DAD, LC-MS, UV-vis, Calibration curve

1 Introduction

Carotenoids are natural pigments found in foods, animals, and microorganisms. These molecules are chemically grouped in carotenes, with only carbon and hydrogen, and xanthophylls, which show additional oxygen in the structure. In foods, most carotenoids have 40 carbons and often found acylated to fatty acids in fruits. The most characteristic feature of the carotenoid structure is the extensive system of conjugated double bonds (cdb) in the chain, which is responsible for the light absorption and solubility properties. Figure 1 shows some carotenoid structures.

Lilian Mariutti and Frank Vriesekoop (eds.), *Food Chemistry: Basic Protocols*, Methods and Protocols in Food Science, https://doi.org/10.1007/978-1-0716-4678-6_24,

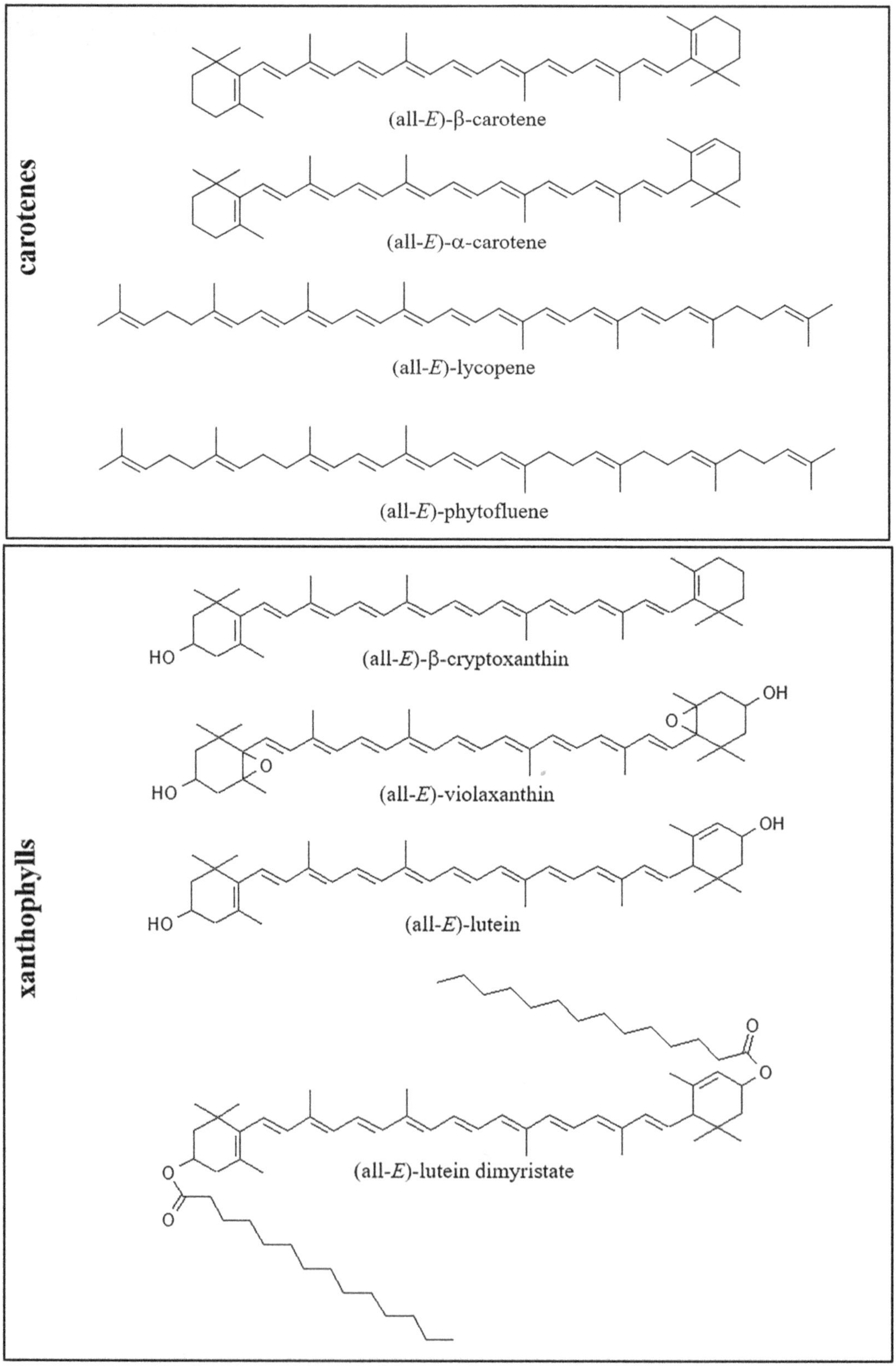

Fig. 1 Structure of (all-*E*)-carotenes and both free and esterified (all-*E*)-xanthophylls

Although carotenoids are naturally stabilized in the matrix, the cut or disruption of the tissues favors their exposure to oxygen, light, endogenous acids, and oxidative enzymes, catalyzing their *trans* (*E*), the most predominant geometric isomer in nature, to *cis* (*Z*) isomerization (Fig. 2), oxidation, and rearrangement of 5,6-epoxide to 5,8-furanoid group (Fig. 3). Spontaneous isomerization also occurs in solution, although at slow rates at room temperature [1]. In addition, traces of acetone in the carotenoid extract can induce aldol condensation during the saponification step, with artifact formation. As a consequence, acetone should be removed from extracts, and all analysis steps carried out in acid-free environments, under dim light and extracts/standards, should be kept dried, under low temperature and inert atmosphere [2].

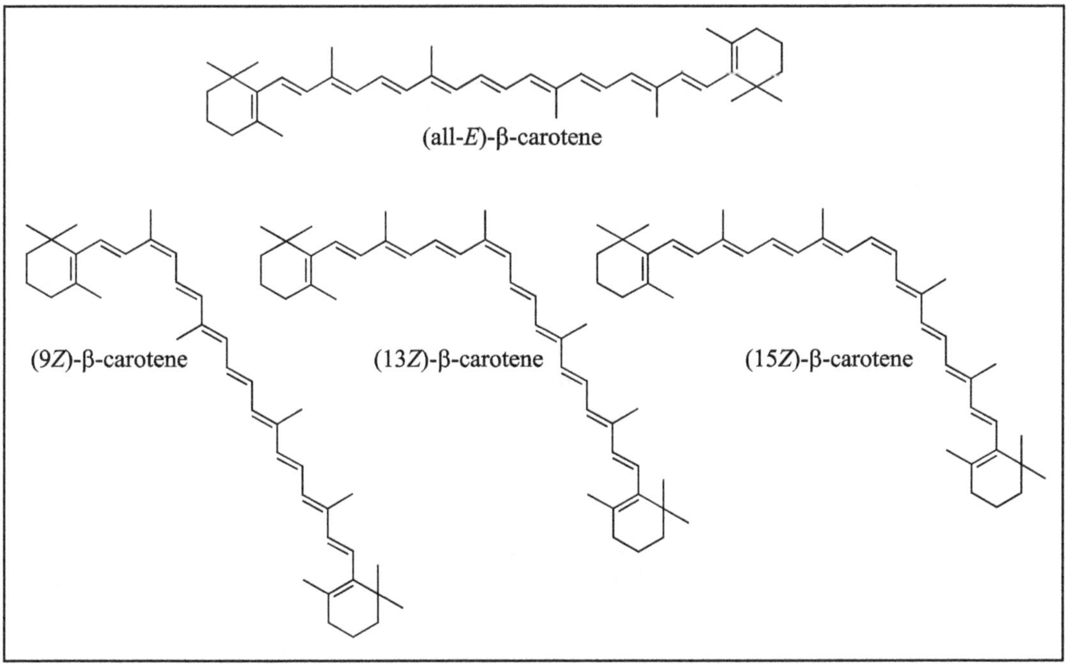

Fig. 2 Structure of the β-carotene isomers most found in nature

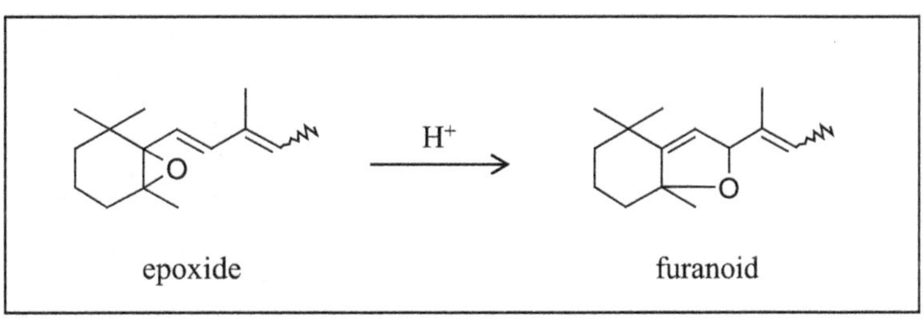

Fig. 3 Epoxide to furanoid rearrangement, involving a hypsochromic shift of 20 nm

The carotenoid analysis in foods usually embraces many steps: sampling, subsampling and sample preparation, extraction, liquid-liquid partition, saponification (not applied when esters are the target), solvent evaporation, chromatographic separation on a reversed-phase column, identification, and quantification. In fact, whenever possible saponification should be avoided due to lower both carotenoid recovery and analysis precision [3, 4]. Validation is used to evaluate method suitability for its intended goal and reliability of the results, and for this purpose, many protocols and guidelines are available [5–9]. Each laboratory has to decide which performance characteristics need to be investigated to validate the method for the desired purpose [7, 9].

Due to carotenoid physical–chemical characteristics, the most used chromatographic system is liquid chromatography (LC), either high or ultra-high performance, coupled to diode array detector (DAD) and mass spectrometry (MS). Moreover, the polymeric reversed-phase C30 column, specially developed for the separation of geometrical isomers of asymmetrical carotenoids, is the most used nowadays [2, 10, 11]. Regarding MS, among the ionization techniques compatible with LC-MS, the most used one for carotenoid identification has been atmospheric pressure chemical ionization (APCI), which is able to form abundant positively or negatively charged molecular ions and fragments, along with protonated or deprotonated molecules of both carotenes and xanthophylls [11–13]. Although ionization in the positive mode is the most used for carotenoids, the negative mode can complement the spectrum information obtained in the positive mode [11]. For example, in positive mode, an extremely high background noise, probably originating from the lipids present in the sample, hindered the identification of the protonated molecules of starfish carotenoids [13]. In the negative mode conditions, the negatively charged molecular ions of free carotenoids were preferentially formed, and the lipids were not ionized [13]. However, according to our experience, carotenoid esters show very low [11], or none at all, molecular ions or deprotonated molecules by APCI in negative mode.

The combined information provided by elution order on C30 column, co-chromatography with authentic standards, UV-visible (UV-vis) spectrum (maximum wavelengths (λ_{max}), spectral fine structure (% III/II), and *cis*-peak intensity (% A_B/A_{II})), and mass spectrum (molecular weight and characteristic fragment ions) are the minimum requirement to identify carotenoids in foods [14]. In fact, without an MS detector, carotenoid esters cannot be identified since acylation does not affect the chromophore and consequently the UV-vis spectra of these compounds.

2 Material

Use analytical grade solvents for extraction, saponification, and dilutions for spectrophotometric quantification. Use HPLC grade solvents and ultrapure water for all the steps involving LC analysis. Prepare the methanolic alkaline solution on the same day as the use.

Filter solvents and sample extracts before LC analysis using, respectively, PTFE membrane filters of 0.22 and 0.45 μm. Sonicate the mobile phases for 10 min before LC analysis when equipment does not have a degasser.

When applying MS, rinse the glassware with ultrapure water and MeOH HPLC grade, avoid blender, and use only high-quality paper and plastic materials to avoid contamination by residues, e.g., from detergent, plastic, oil and paper, and further MS detection.

3 Methods

Avoid exposure to light, oxygen, acid, and temperature higher than 35 °C during analysis. All methods are carried out at room temperature, and extracts/standards stored under nitrogen atmosphere at least at −18 °C, preferably at −25 or − 70 °C.

A full description of the food, both common and scientific names, cultivar/variety/race, stage of maturity, and other important information for food characterization and experiment reproduction, e.g., part of the plant analyzed, should be reported. Note that the term "replicate" gives the total method variation, i.e., entire procedure repetition using the composite sample (from Subheading 3.2, 3.3, 3.4, 3.5, and 3.6), and "duplicate/triplicate" provides instrument precision, e.g., injection of the same extract twice.

3.1 Sampling and Sample Preparation

Sampling is a combination of procedures/steps to obtain the laboratory sample, subsampling, and its preparation to get a representative and homogeneous sample of the initial material [15]. There are many terms used to describe a representative sample, e.g., homogeneous [16], mixed [17], pooled [18], milled [19, 20], minced [21], and composite [22].

Considering that no general procedure or protocol for sampling, subsampling, sample preparation, and storage applied to carotenoid analysis is available, we describe here the strategies used in our laboratory [15, 16, 23–25].

1. Randomly choose about 2 kg of fresh food or three to five packs of a single lot of processed food (*see* **Note 1**).

2. When samples arrive at the laboratory, discard those fruits with apparent physical injuries, and remove inedible parts. Once you start, subsamples must be prepared on the same day.
3. For the subsampling of big fruits and vegetables, e.g., papaya, longitudinally quarter and separate opposite sections. For subsampling of small size foods, e.g., blueberries, randomly separate about 10 units.
4. Either dice into small pieces, or homogenize the subsamples to obtain a representative sample (*see* **Note 2**).
5. After size reduction, immediately freeze by liquid nitrogen immersion, separate in small portions, and pack under vacuum in polyethylene bags for storage at low temperature ($T < -26$ °C). Alternatively, freeze-dry the samples until constant weight, vacuum-pack, and store at low temperature ($T < -26$ °C) [22, 26–28] (*see* **Note 3**).

3.2 Carotenoid Extract Preparation

1. Weigh the composite sample in an analytical or semi-analytical balance. The amount of sample depends on its carotenoid concentration and amount of water (fresh/frozen or dried) [16, 22, 24, 28–34] (*see* **Note 4**). Optionally, add about the same weight as the sample of Na_2CO_3 to neutralize naturally present organic acids in acidic fruits, such as oranges [35].
2. Exhaustively extract fresh/frozen samples with acetone. Alternatively, freeze-dried samples can also be exhaustively extracted with ethyl acetate and methanol (*see* **Note 5**). For samples rich in protein, e.g., yogurt and milk-based beverages, extract with 20 mL of tetrahydrofuran (THF) by vortexing for 1 min, followed by 5 min centrifugation at 3500 *g* at 10 °C, repeating the process until the complete absence of sample color [36]. Afterward, follow step 4 and so on.
3. Vacuum-filter through a paper on a Buchner funnel or sintered glass filter. Transfer the retained solid residue, and repeat step 2 until the residue becomes colorless (*see* **Note 6**). Combine all the filtrates. Additional steps are necessary in the case of a sample with high oil content (*see* **Note 7**).
4. For liquid–liquid partition, add about 100 mL petroleum ether/diethyl ether (1:1) in a separatory funnel, and gradually transfer the combined carotenoid extract by washing it with distilled water after each time. Add water slowly to allow its flow along the funnel walls to avoid emulsion formation. Leave the two layers to separate without agitation, and discard the lower aqueous phase (with acetone or THF). After the addition of the entire extract, wash the ethereal phase 4–5 times with water to remove the residual acetone or THF until the ether aspect changes from cloudy to translucent.

5. Alternatively, the liquid–liquid partition can also be carried out by centrifugation at 16900 *g* for 10 min at 20 °C. Use 10% NaCl aqueous solution instead of distilled water to avoid emulsion formation [35]. Collect the upper phase containing carotenoids.
6. Add an equal volume of 10% KOH in methanol to the ethereal carotenoid extract, and leave the mixture for overnight saponification at room temperature in the dark. Shaking is not necessary. Remove the alkali by carefully washing several times with distilled water in the same way as in liquid–liquid partition, until neutral pH. Collect the carotenoid upper phase. Do not carry out this step when carotenoid esters are the target of the analysis (*see* **Note 8**).
7. Add about one coffee spoon of anhydrous Na_2SO_4 to remove traces of water, transfer to a round-bottom flask, and evaporate the solvent to a small volume (1–2 mL) in a rotary evaporator (T < 40 °C).
8. Quantitatively transfer the extract to a small flask by rinsing the round-bottom flask with a small volume of ethyl acetate or petroleum ether/diethyl ether or diethyl ether, dry completely under N_2, and store at low temperature (T < −25 °C) until analysis. Visually verify if the extract is completely dried, and in case of oil presence, *see* **Note** 7.
9. Repeat all steps 1–8 to obtain replicates of the carotenoid extracts, usually three.

3.3 Separation by Liquid Chromatography (LC)

1. Condition the C30 column for 50–60 min under the initial mobile phase conditions. The most common solvents used in this column type are MeOH, MTBE, and water [23, 26–28, 30, 35] (e.g., *see* **Notes 9–11**).
2. Dissolve the dried carotenoid extract in a known small volume (≤1 ml) of methyl *tert*-butyl ether (MTBE)/MeOH (1:1). Firstly, add MTBE, and apply 1 min sonication if the extract is not completely dissolved. Afterward, add MeOH, and apply again 1 min-ultrasound bath to assure homogenization.
3. Inject the extract into the equipment. The response of the major peak should not exceed 0.8 AU (*see* Subheading 3.6).
4. Process the data to obtain the chromatogram at 450 nm and also at different wavelengths when the sample also contains lycopene (470 nm), ζ-carotene (400 nm), phytofluene (348 nm), and phytoene (285 nm).

The elution order mostly depends on the type of column stationary phase, which drives the mechanism of separation. Some elution patterns on the C30 column are shown below and observed in Figs. 4 and 5, providing the first insight to support the

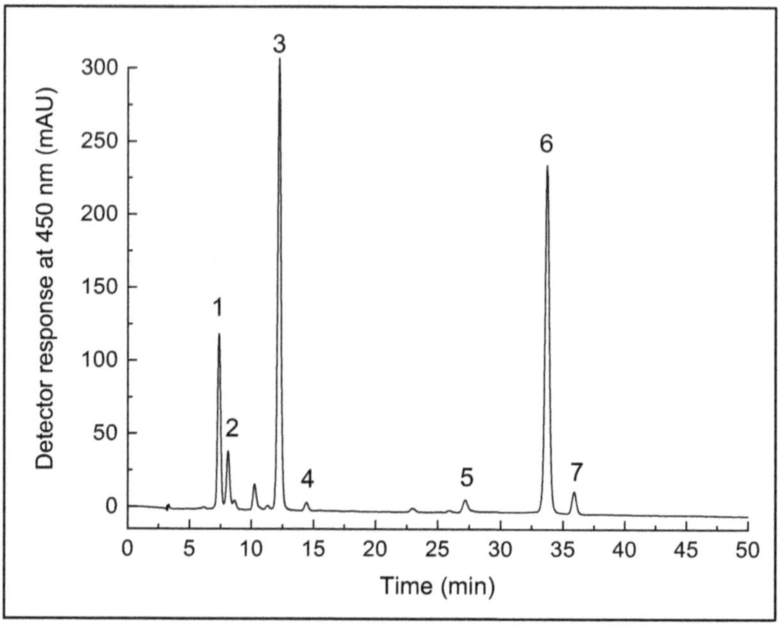

Fig. 4 Chromatogram obtained by LC-DAD of saponified carotenoid extract from kale. Peak identification: 1- (all-*E*)-neoxanthin + (all-*E*)-violaxanthin, 2- (9*Z*)-neoxanthin, 3- (all-*E*)-lutein, 4- (all-*E*)-zeaxanthin, 5- (13*Z*)-β-carotene, 6-(all-*E*)-β-carotene, 7- (9*Z*)-β-carotene. Chromatographic conditions: C30 YMC column (5 μm, 250 mm × 4.6 mm) with temperature set at 29 °C, using as mobile phase a linear gradient of MeOH/MTBE from 95:5 to 70:30 in 30 min, followed by 50:50 in 20 min at 0.9 mL/min

identification process. Tables with elution order and retention time values on C30 columns are found in the literature [12, 23, 26, 37–39].

3.4 Identification by LC-DAD-MS

3.4.1 Elution Order on the C30 Column

1. Identify firstly the most common carotenoids with the standards available, e.g., (all-*E*)-β-carotene and (all-*E*)-lutein (*see* **Note 12**).
2. The most polar free C40-xanthophylls elutes earlier than C40-carotenes, e.g., zeaxanthin < β-cryptoxanthin < β-carotene.
3. Retention time increases as the number of conjugated double bonds also increases. For example, phytoene < phytofluene < ζ-carotene < α-carotene < β-carotene < (9*Z*)-β-carotene ≅ β-zeacarotene < δ-carotene < γ-carotene < lycopene; lutein < zeaxanthin.
4. The (9*Z*)-isomer elutes after, while (15*Z*)- and (13*Z*)-isomers elute before the corresponding (all-*E*)-isomer; in other words: 15*Z* < 13*Z* < 13'*Z* < all-*E* < 9*Z* < 9'*Z*.
5. Carotenoids possessing a 5,8-furanoid ring elute after their corresponding 5,6-epoxycarotenoid.

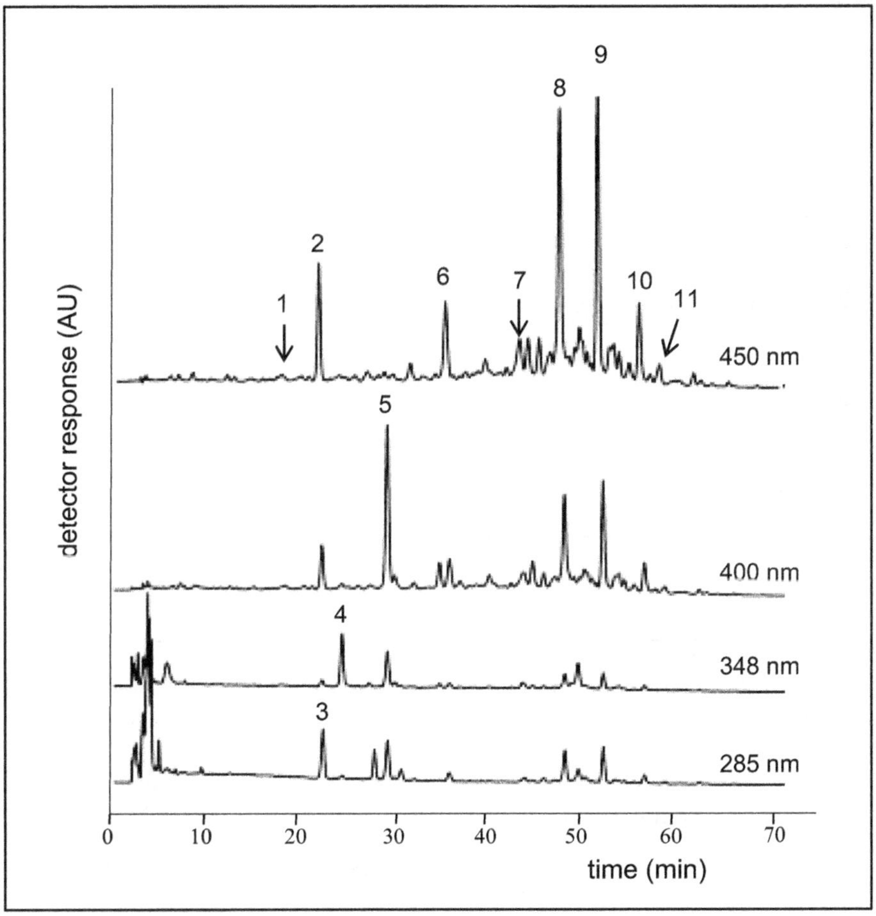

Fig. 5 Chromatogram obtained by LC-DAD of non-saponified carotenoid extract from mandarin. Peak identification: 1- (13*Z*)- or (15*Z*)-β-cryptoxanthin, 2- (all-*E*)-β-cryptoxanthin, 3- phytoene, 4- phytofluene, 5- ζ-carotene, 6- (all-*E*)-β-carotene, 7- (13*Z*)- or (15*Z*)-β-cryptoxanthin laurate, 8- (all-*E*)-β-cryptoxanthin laurate, 9- (all-*E*)-β-cryptoxanthin myristate, 10- (all-*E*)-β-cryptoxanthin palmitate, 11- (9*Z*)-β-cryptoxanthin palmitate. Chromatographic conditions: C30 YMC column (5 μm, 250 mm × 4.6 mm) with temperature set at 35 °C, using as mobile phase a linear gradient of A: MeOH/MTBE/H_2O (81:15:4) and B: MeOH/MTBE/H_2O (16: 80.4:3.6) from 99 to 66% A in 30 min, maintaining this condition for 5 min, changing from 66 to 44% A in 15 min, keeping this condition for 5 min, changing from 44 to 22% A in 15 min and from 22 to 0% A in 5 min, returning to the initial condition (99% A) in 5 min, and keeping this condition for 5 min at 0.9 mL/min

6. The retention time of esters from the same carotenoid increases as the molecular weight (MW) of the fatty acid (FA) also increases, except for those acylated with unsaturated fatty acids that elute earlier than the corresponding saturated FA.
7. Homodiesters elute before heterodiesters with the same MW.

3.4.2 UV-Visible Spectrum

The carotenoid spectrum usually shows three maximum wavelength peaks (λ_{max}) in the visible region and one peak in the UV region for (*Z*)-isomers, as shown in Fig. 6. Both the position of the λ_{max} and the shape (spectral fine structure) of the spectrum reflect

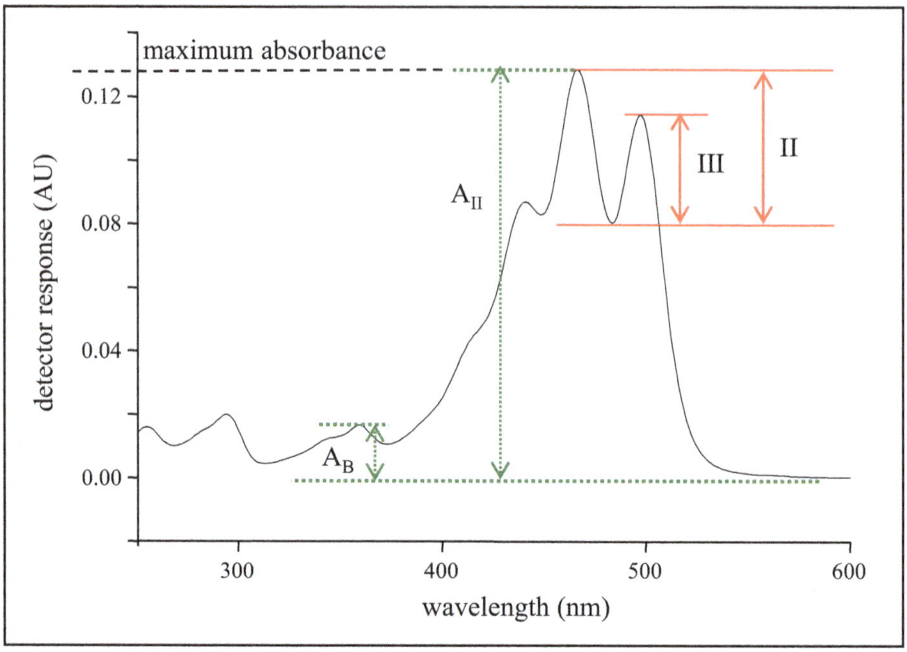

Fig. 6 UV-visible spectrum of lycopene, obtained by LC-DAD in MTBE/MeOH, indicating the calculation of spectral fine structure (% III/II = III/II × 100) and *cis*-peak intensity (% $A_B/A_{II} = A_B/A_{II} \times 100$)

the chromophore. The rationale for spectrum interpretation and data compilation are found in the literature [2, 12, 40–43].

1. Acquire and process the UV-vis spectrum between 220 and 700 nm.
2. Tabulate the three maximum wavelengths of each peak, along with the calculation of spectral fine structure (%III/II) and *cis*-peak intensity (%A_B/A_{II}) values as shown in Fig. 6. Use a ruler to measure the distances (*see* **Note 13**).
3. Consider that as the number of cdb in the carotenoid structure increases, the λ_{max} values also increase. See examples in Table 1.
4. Shoulder instead of peak is observed at the lowest wavelength when a β-ring end-group is allylic to the polyene chain, along with a low %III/II value (Fig. 7).
5. Functional groups that do not change the chromophore cause no modifications on the spectrum, e.g., the spectra of (all-*E*)-zeaxanthin and (all-*E*)-β-carotene are identical and so are those of (all-*E*)-lutein, (all-*E*)-α-carotene, and (all-*E*)-lutein dimyristate (Table 1).
6. Consider that (*Z*)-isomers show decreased %III/II compared to that of the corresponding (all-*E*)-one and increased %A_B/A_{II} (~10% = 9*Z*, ~45% = 13*Z* and ~ 56% = 15*Z*) values, as shown in Fig. 7. Use this information altogether to differentiate between (*Z*)-isomers of β-carotene and (all-*E*)-α-carotene. See value examples in Table 1.

Table 1
Characteristics of the UV-visible spectrum of selected carotenoids, obtained by LC-DAD in MeOH/ MTBE

Carotenoid (number of total cdb and of β-ring end-group)[a]	λ_{max} (nm)	% III/II	% A_B/II
(13*Z*)-lutein (10 and 1β)	330, 405, 436, 465	22	44
(13'*Z*)-lutein (10 and 1β)	30, 410, 438, 465	29	43
(all-*E*)-lutein (10 and1β)	420, 445, 473	65	0
(all-*E*)-zeaxanthin (11 and 2β)	(421), 450, 477	20	0
(9*Z*)-lutein (10 and 1β)	330, 417, 439, 467	67	8
(9'*Z*)-lutein (10 and 1β)	330, 419, 440, 468	63	8
(15*Z*)-β-carotene (11 and 2β)	337, (420), 449, 472	10	61
(13*Z*)-β-carotene (11 and 2β)	338, (420), 444, 470	12	47
(all-*E*)-α-carotene (10 and 1β)	420, 445, 473	66	0
(9*Z*)-α-carotene (10 and 1β)	330, 420, 444, 472	60	9
(all-*E*)-β-carotene (11 and 2β)	(421), 452, 478	25	0
(9*Z*)-β-carotene (11 and 2β)	338, (420), 447, 472	20	18
(all-*E*)-lutein dimyristate (10 and 1β)	423, 446, 474	65	0

Sources: De Rosso & Mercadante [23], Mariutti et al. [22], Rodrigues et al. [27]
[a]Compounds are displayed in order from the shortest to the longest elution time on the C30 column. cdb, conjugated double bonds. Parenthesis indicates shoulder instead of a peak, as can be seen in Fig. 7

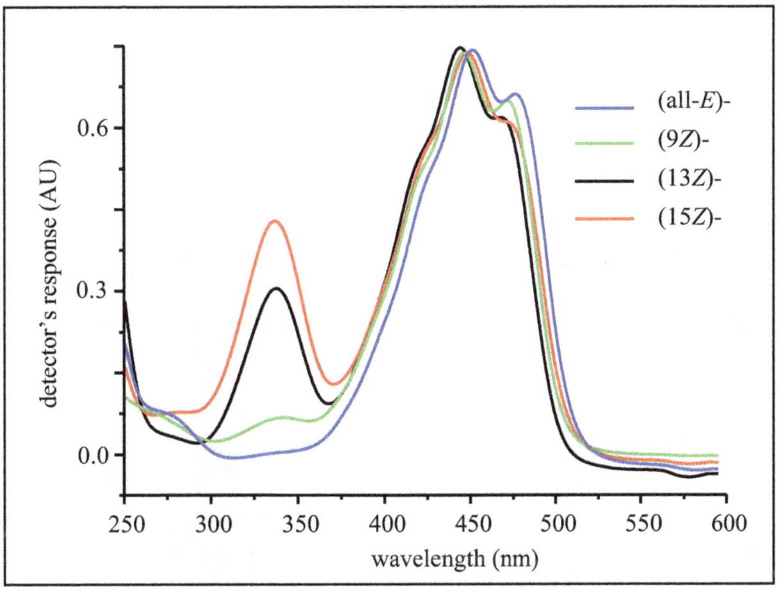

Fig. 7 UV-visible spectra of β-carotene isomers, obtained by LC-DAD in MTBE/ MeOH. Note the shoulder instead of the peak at 420–41 nm

7. Compare the tabulated information with data obtained with standards and literature, considering the carotenoid elution order from the column (3.4.1) and mass spectrum (3.4.3).

3.4.3 Mass Spectrometry Spectrum

The detection of the molecular mass and fragments depends not only on the carotenoid structure but also on the type and conditions of the ionization source [11, 44]. The rationale for spectrum interpretation, characteristic mass fragments, and data compilation is found in the literature [11, 12, 23, 26, 41].

1. Acquire the mass spectrum with APCI at positive mode setting parameters that allow the formation of the in-source protonated molecule ($[M + H]^+$) as base peak (*see* **Note 14**).
2. Locate the protonated molecule ($[M + H]^+$) with an odd number. It is usually detected, except for lutein (both free and esterified) (*see* **Notes 14–16**).
3. Locate the loss of C_7H_8 ($[M + H\text{-}92]^+$ and/or ($[M\text{-}92]^+$), characteristic of toluene elimination from the polyene chain, to confirm the protonated molecule assignment, especially by MS/MS (*see* **Note 17**).
4. Locate MS/MS diagnostic ions indicative of end-groups and/or functional groups which are shown in Table 2.
5. Elimination of the FA moiety, $[M + H\text{-FA}]^+$ in monoesters and $[M + H\text{-FA1-FA2}]^+$ in diesters, is always observed. Thus, the fragment ion representing the carotenoid backbone is always detected.
6. Combined losses are also expected, e.g., $[M + H\text{-}92\text{-}18]^+$ and $[M + H\text{-}18\text{-}18]^+$ in violaxanthin, $[M + H\text{-FA-}18]^+$ or $[M + H\text{-FA1-FA2-}18\text{-}18]^+$ giving rise to a fragment at *m/z* 531 for C40 xanthophyll esters.

Table 2
Fragments obtained by APCI(+) from tandem mass spectra of carotenoids

End-groups and chemical functions	Characteristic fragments (MS and/or MS/MS)
β-ring	$[M + H\text{-}137]^+$, $[M\text{-}137]^+$
ε-ring	$[M + H\text{-}56]^+$, 123
ψ-group	$[M + H\text{-}69]^+$, $[M\text{-}69]^+$
hydroxyl	$[M + H\text{-}18]^+$
aldehyde	$[M + H\text{-}18]^+$, $[M + H\text{-}28]^+$, $[M + H\text{-}56]^+$
ketone	$[M + H\text{-}18]^+$
5,6-epoxide or 5,8-furanoid	$[M + H\text{-}18]^+$, 205 (in β-ring), 221 (in hydroxylated β-ring)

Source: De Rosso & Mercadante [23], Mariutti et al. [13], Van Breemen et al. [11]

7. Regioisomeric forms of lutein esters are distinguished, considering that the most abundant in-source fragment ion corresponds to the loss of the hydroxyl group or FA at the C-3′. The assignment of the position of the FA by MS and MS/MS spectra in esters of other asymmetric xanthophylls is not possible, e.g., the fragmentation pattern of esters of β-cryptoxanthin and zeinoxanthin is essentially the same.

3.4.4 Combined Approach for Carotenoid Identification

The carotenoid analysis is recognized as inherently difficult, and errors can be introduced at all the steps. The key combined information obtained from elution order, UV-visible, and MS spectra must perfectly fit all the carotenoid structure characteristics, e.g., chromophore, molecular weight, and polarity, as in a puzzle. In case one piece of information does not combine with the rest of them, do not assign a structure. Moreover, misidentification is common to occur when the recommended minimum criteria for carotenoid identification are not fulfilled [14].

It is worth highlighting that the structure of a carotenoid is only unequivocally established through nuclear magnetic resonance (NMR) experiments along with circular dichroism for carotenoids with chiral centers or atoms. The exact location assignment of the (*Z*)-double bond is only possible by NMR [14].

3.5 Quantification by UV-vis Spectrophotometry

The spectrophotometric equipment is used for total carotenoid determination of an extract and also for the quantification of a specific carotenoid standard in a stock solution for further quantification by LC-DAD. High purity authentic standards of carotenoids are either commercially obtained (e.g., Sigma, CaroteNature®, Extrasynthèse) or isolated from natural sources and in some cases synthesized in the laboratory.

1. Choose a solvent that is both able to completely dissolve the dried extract/standard and has a known specific absorption coefficient ($A_{1cm}^{1\%}$) for the desired carotenoid (*see* **Note 18**).
2. Completely dissolve the extract in a known volume of solvent; the solution must be translucent, with no visible particles. Use sonication for 5 min to assure complete carotenoid solubilization.
3. Measure the spectrum from 220 to 600 nm, and use the maximum absorbance indicated in Fig. 6 for quantification. The maximum absorbance must be between 0.2 and 0.8.
4. Determine the concentration using the Beer–Lambert law, according to Eq. 1. The absorption coefficient ($A_{1cm}^{1\%}$) values in different solvents are found in the literature [40] (*see* **Note 19**).

$$\mu g/ml = \frac{abs \times 10^4}{A_{1cm}^{1\%}} \quad (1)$$

where *abs* is the maximum absorbance and $A_{1cm}^{1\%}$ is the absorption coefficient of the specific carotenoid in the solvent used.

3.6 Quantification of Carotenoids by LC-DAD

3.6.1 External Calibration Curve

1. Determine spectrophotometrically the concentration of the standard stock solution, according to Subheading 3.5.
2. Dilute the stock solution into work solutions with 5–8 different concentration levels about equally spaced over the concentration range of interest. Dry the solvent completely, and redissolve it in the same injection mixture solvent as the sample (Subheading 3.3). Repeat this process three times to obtain replicate solutions of each concentration.
3. Inject each work solution into the LC, under the same conditions as the samples (*see* **Note 20**). Process the chromatogram to obtain the peak area (area units, AU).
4. Plot the response in AU (y-axis) against the known concentration of the work solutions (x-axis), and calculate the linear regression equation of the calibration curve, according to Eq. 2 (*see* **Notes 21–23**).

$$y = a + bx \quad (2)$$

where a is the intercept, b is the slope, x is the carotenoid concentration, and y is the measurement response (area units).

3.6.2 Sample Carotenoid Extract

1. Inject the extracts according to Subheading 3.3, and process the chromatogram to obtain the area unit of each peak (*see* **Note 24**). Carotenoids already identified according to Subheading 3.4.
2. Calculate the concentration of the carotenoid(s) in the sample using the corresponding calibration curve (*see* **Notes 25** and **26**). Remember that carotenoid concentration in the sample must fit into those of the calibration curves.
3. Calculate the vitamin A value of the sample using the NAS-IOM [45] conversion factors (*see* **Note 27**).

3.7 Evaluation of Method Performance Characteristics

The typical performance characteristics of an LC-DAD method that should at least be evaluated are specificity, working concentration range and linearity, accuracy or recovery, precision, limits of detection (LOD), and limits of quantification (LOQ) [6, 15]. In fact, the purpose of the analysis determines which characteristics are the most important and which may not be essential. To minimize the number of required experiments, Fig. 8 shows a flowchart for simultaneously evaluating those mentioned performance characteristics in the order recommended by Bratinova et al. [9].

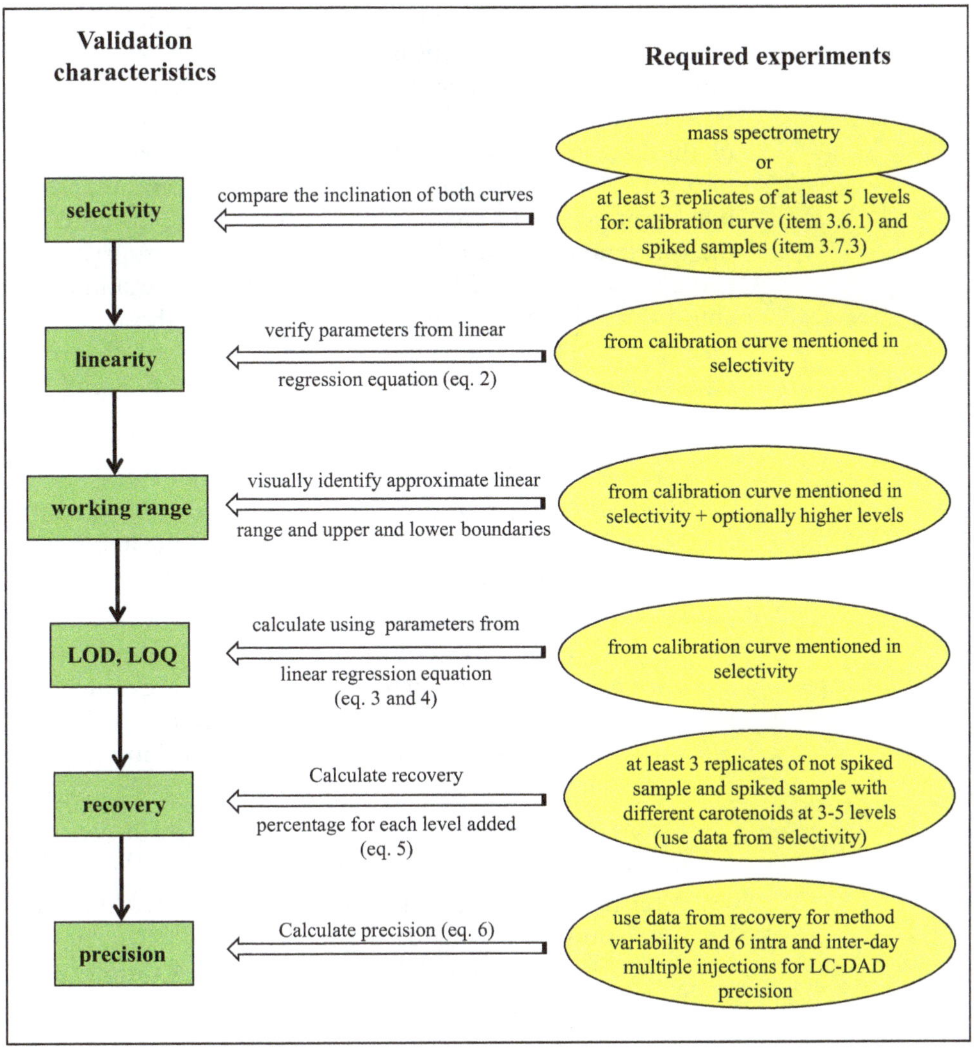

Fig. 8 Flowchart for the simultaneous evaluation of different method performance characteristics for carotenoid analysis by LC-DAD

3.7.1 Selectivity

Both terms, selectivity [8] and specificity [6, 9], mean the ability to unequivocally assess the carotenoid in the presence of other interfering compounds extracted along with the target compounds from the food. Selectivity is evaluated by different means, two of which are described below [15].

1. MS information indicates whether the quantified peak is pure or not. Requires no further action.
2. Compare an analytical curve with the addition of the carotenoid to the sample (Subheading 3.7.3) to that of the standard (calibration curve, Subheading 3.6.1). If both curves are parallel, there is no matrix effect, and the method is considered quantitatively selective.

3.7.2 Linearity, Working Range, Limits of Detection (LOD), and Limits of Quantification (LOQ)

Linearity expresses the ability to obtain a signal directly or through a well-defined mathematical transformation, proportional to the carotenoid concentration in the sample, within a working concentration range. The working range is the interval over which the method provides results with an acceptable uncertainty [7]. LOD means the lowest amount of carotenoid that can be detected but not quantified as an accurate value. LOQ is the lowest amount of the carotenoid that can be determined with appropriate precision and accuracy [5, 7, 15]. Synonyms of LOD are minimum detectable value, detection limit, and CCβ, while those of LOQ are quantification limit, quantitation limit, limit of determination, reporting limit, limit of reporting, and application limit [5–7, 15].

1. Use the data of the calibration curves obtained in Subheading 3.6.1 and the equation generated.
2. Verify the values of the coefficient of determination (r^2) (*see* **Note 28**). Evaluate also the significance of the linear regression and if the residual plots show a normal distribution to verify the curve linearity.
3. Include higher concentrations than the highest level of the calibration curve to determine the working range. Visually examine the responses to identify approximate linear range and upper and lower boundaries of the working range for the instrument.
4. Determine LOD and LOQ using parameters of the calibration curve, according to Eqs. 3 and 4 (*see* **Notes 29** and **30**).
5. Alternatively, dilute and inject into the LC-DAD known concentrations of the standard until achieving signal/noise (S/N) values 3:1–2:1 for LOD and 10:1 for LOQ to estimate such limits.

$$\mathrm{LOD} = \frac{3.3 \times SD}{S} \times 100 \tag{3}$$

$$\mathrm{LOD} = \frac{10 \times SD}{S} \times 100 \tag{4}$$

where *SD* is obtained from the residue of a regression line or of the *y*-intercept of the regression line (*b* value from Eq. 2) and *S* is the slope of the calibration curve.

3.7.3 Accuracy

Accuracy expresses how close a single result is to the concentration that is accepted as the true or reference value for a compound [7]. Accuracy evaluation is preferably carried out by the analysis of a certified reference material (CRM); however, only a few certified values are found for some carotenes and free xanthophylls (BCR-485 mixed vegetables from the European Commission and SRM2383a baby food composite from NIST). There are no certified values for carotenoid esters. As a consequence, the most applied approach for carotenoids is an indirect estimation of accuracy by quantifying the amounts of spiked standard recovered from the sample, described herein [15, 35, 36, 46, 47].

1. Spike known amount of carotenoid standards with different polarities, e.g., (all-*E*)-lutein, (all-*E*)-β-cryptoxanthin, and (all-*E*)-β-carotene, into the sample in 2–5 addition levels for each carotenoid (*see* **Notes 31** and **32**). Choose concentrations corresponding to the midpoint of the interval and both at the lowest and the highest concentration range of quantification.
2. Extract the spiked samples and the original one (not spiked), 3–6 replicates for each level, according to Subheading 3.2 (*see* **Note 33**).
3. Analyze by LC-DAD and calculate the concentration of the target carotenoids as described before.
4. Calculate the recovery percentage for each added level, according to Eq. 5 (*see* **Notes 34** and **35**).

$$\text{Recovery}(\%) = \frac{C_S - C_O}{C_A} \times 100 \tag{5}$$

where C_S is the concentration average of the carotenoid measured in the spiked sample, C_O is the concentration average measured in the original sample (not spiked), and C_A is the carotenoid concentration added to the spiked sample.

3.7.4 Precision

Precision shows how close results are to one another in a set of measurements for the same sample, expressed as relative standard deviation (RSD). Repeatability is the term used when results are obtained intra-lab by the same analyst, using the same equipment and reagents within a short period, while reproducibility is used for precision results between laboratories [7, 15].

1. Use the results of the accuracy experiments (Subheading 3.7.3) to determine method variability among at least three replicate analyses, and calculate the method repeatability according to Eq. 6 (*see* **Notes 36–38**).
2. Carry out intra- and inter-day multiple injections, e.g., 6 ones in 2 different days, of a solution containing the extract or standards to assess the precision of the chromatographic system. Calculate according to Eq. 6 (*see* **Note 39**).

$$\text{Precision}(\%) = \frac{SD}{\text{average}} \times 100 \tag{6}$$

where *SD* is the standard deviation.

4 Notes

1. In case of evaluation of a specific effect on the carotenoid composition, e.g., difference between cultivars, a strict sampling is necessary, such as distribution of the plants in the orchard/plantation, hour and day of harvest, and vegetable maturity [24, 30].
2. Depending on the sample size, firstly freeze-dry, and afterward ground into powder, homogenize, vacuum pack, and store at T < −26 °C [35].
3. Report all the information of the sample, sampling protocol, and storage conditions.
4. The amount of fresh foods usually varies from 5 to 75 g, e.g., 5 g of milk-based beverage [16], 10–15 g of loquat [30], 15 to 20 g of banana [29], 7–10 g of jackfruit [31], 30–75 g of cashew-apple [32], 10 g arugula [34], and 15 g papaya [33]. For freeze-dried samples, 1 to 5 g is usually used, e.g., 5 g of mana-cubiu [48], murici [22], *Caryocar villosum* pulp [28], 1 g of orange hybrids [24].
5. Extraction can be carried out by maceration using a mortar and pestle (celite or Hyflosuercel is usually added), vortexing, shaking, electric blending, magnetic stirring, or high-performance homogenizer.
6. To confirm that extraction was exhaustive, extract the sample once again, filter, and verify the absence of carotenoids in the last extract, either visually or at 450 nm.
7. Physically remove the oil from the sample extract, as follows: before ether transference, keep the carotenoid extract in the freezer (temperature < −18 °C) for at least 2 h, followed by filtration using cold glassware and washing with cold acetone [23].
8. Although saponification—step 6—is usually carried out to facilitate carotenoid analysis, it is not applicable when carotenoid esters are the target. This step is also generally not used when samples do not contain carotenoid esters nor high amounts of lipids, such as green leaves and few fruits, e.g., kiwi. In fact, whenever possible saponification should be avoided due to lower carotenoid recovery and analysis precision [3, 4].
9. With the C30 column (5 μm, 250 mm × 4,6 mm), use a linear gradient of MeOH and MTBE, from 95:5 to 70:30 in 30 min followed by 50:50 in 20 min with flow rate at 0.9 mL min^{-1} and column temperature at 29 °C, which is able to separate carotenoids from different matrices [23, 28–31, 49].

10. For separation of carotenoid esters on the C30 column, use mixtures of methanol, MTBE, and water as follows: A (81: 15:4, v/v/v) and B (16:80.4:3.6, v/v/v) in linear gradient mode; for most fruits apply at 1.0 mL/min from 99% to 44% solvent A in 39 min, 44% to 0% solvent A in 6 min, returning to the initial condition (99% solvent A) in 5 min and remaining in this condition for 5 min, totaling a 55 min run [27, 38]; for citrus fruits, set the flow rate at 0.9 mL/min and column temperature at 35 °C, changing from 99 to 66% A in 30 min maintaining this condition for 5 min, from 66 to 44% A in 15 min keeping this condition for 5 min, from 44 to 22% A in 15 min and from 22 to 0% A in 5 min, returning to the initial conditions (99% A) in 5 min and keeping this condition for 5 min [26].
11. The addition of 0.1% triethylamine (TEA) to the mobile phase enhances recovery on the C_{30} column; however, it must be excluded when the MS detector is used because TEA is easily ionized in the APCI source, and as a result, the carotenoid ion signals decrease [23].
12. Use extracts with known composition in case standards are not available in your laboratory. For example, the qualitative composition of major carotenoids in green leafy vegetables is usually the same (*see* Fig. 4).
13. The presence of a conjugated aldehyde group may induce complete loss of fine structure, depending on the polyene chain length.
14. For lutein, the protonated molecule ($[M + H]^+$), when detected, is much smaller than the fragment ion $[M + H\text{-}18]^+$. Thus, locate the loss of water ($[M + H\text{-}18])^+$ in the MS of lutein and its derivatives. Similar behavior is also observed for α-cryptoxanthin.
15. As expected, all (*Z*)-isomers present identical MS fragmentation patterns to those of the (all-*E*)-counterpart. Thus, MS does not differentiate geometrical isomers.
16. The MS and MS/MS spectra of structural carotenoid isomers with epoxy and/or furanoid groups are virtually the same.
17. The fragment ions $[M\text{-}106]^+$ and $[M + H\text{-}106]^+$ are also very characteristic of the carotenoid polyene chain, but not frequently detected by APCI(+).
18. Solvents most used are ethanol for samples containing high-lutein contents and petroleum ether for the other samples, e.g., rich in β-carotene.
19. The concentration of the stock solution estimated spectrophotometrically tends to be more accurate than weighing, considering that a very low amount of carotenoids is usually used.

20. Mixture known concentrations of different carotenoids to obtain signals of different peaks with only one injection.
21. Optionally correct the concentration by the standard purity in case it is lower than 95% [5].
22. The model of linear regression is the classical one used, which assumes a linear correlation between measured area units (yi) and corresponding concentrations (xi). This model requires response normal distribution and homogeneity of variances over the working range, which is normally observed in a working range of a two-order maximum of concentration magnitude. However, different mathematical or statistical tools can be used in case those requirements are not fulfilled.
23. High correlation coefficient values (e.g., $r > 0.99$) are often reported.
24. Make an adequate dilution when any carotenoid concentration of the sample solution overcame the highest point of the calibration curve.
25. When standards of all carotenoids are not available, use the calibration curve of an available standard to quantify other peaks eluting close to it, e.g., neoxanthin, violaxanthin, and luteoxanthin quantified using the curve of lutein; β-cryptoxanthin epoxides, α-cryptoxanthin, and zeinoxanthin using the curve of β-cryptoxanthin; β-carotene epoxides by the curve of β-carotene; and the (*Z*)-isomers using the curve of the corresponding (all-*E*)-isomer.
26. Results are usually expressed as μg/g, μg/100 g, mg/g, or mg/100 g of fresh weight (fw) or dry weight (dw).
27. NAS-IOM conversion factors: 12 μg of dietary (all-*E*)-β-carotene corresponding to 1 μg of retinol activity equivalent (RAE) and 24 μg of α-carotene or β-cryptoxanthin from food equivalent to 1 μg RAE [45].
28. High values of r^2 are required, always ≥0.99.
29. LOD and LOQ values are expressed in different units in different studies, but the most common one is μg/mL.
30. Applying Eqs. 3 and 4, typical values of LOD are 0.1–0.91 μg/mL and of LOQ are 0.5–2.77 μg/mL for diverse carotenes and free xanthophylls [25, 28, 49]. Different concentration ranges of the carotenoid calibration curves influence these calculated limits.
31. At least three to five different levels are necessary when these experiments are also used to evaluate selectivity (Subheading 3.7.1).
32. Levels of 5.16, 10.3, and 14.72 μg/g of lutein; 1.89, 3.76, and 7.54 μg/g of β-cryptoxanthin; and 0.61, 1.23, and 3.08 μg/g

of β-carotene were added to a freeze-dried pulp of orange in the green ripening stage [35]; concentrations of zeaxanthin (20 and 80 μg/mL), β-cryptoxanthin (10 and 40 μg/mL), and β-carotene (5 and 20 μg/mL) were added to 1 g *Lycium barbarum* Linnaeus powder sample [50].

33. Extract the original sample at least as three replicates when these experiments are also used to determine method repeatability (Subheading 3.7.4).
34. Acceptable recovery is a function of the concentration and the purpose of the analysis, e.g., 75 to 120% at μg/g and 80 to 115% at 10 μg/g [5].
35. Typical recovery values reported are 85–88% for (all-*E*)-lutein, 83–86% for (all-*E*)-β-cryptoxanthin, and 83–85% for (all-*E*)-β-carotene applying a method with saponification included [35] and 92% for zeaxanthin, 92% for β-cryptoxanthin, and 87% for β-carotene without saponification, while with saponification values dropped to 83, 74, and 74%, respectively [50]; value is around 82% for (all-*E*)-β-carotene when saponification step is included and from 92 to 107% without alkaline hydrolysis [47], 82% for (9*Z*)-neoxanthin, and 75% for (all-*E*)-violaxanthin with the saponification step [51].
36. For repeatability, acceptable RSD values depend on the compound concentration in the sample, 8% at μg/g and 6% at 10 μg/g [5].
37. Typical RSD values among replicates (complete analysis) are 0.6% to 11.8% for different carotenoids, free and esters, in the concentration range of 0.3–5.5 μg/g fw or 0.6–5.1 μg/g dw [16, 30, 35, 52].
38. A repeatability standard deviation can also be estimated by pooling several small data sets, e.g., $n = 2$, from different days [7].
39. Typical RSD values among injections are 1.3–2.3% intra-day and 1.8–3.7% inter-days [46, 50].

Acknowledgments

The author thanks CNPq (grant 309182/2018-2) and FAPESP (grants 2013/07914-8, 2018/23752-1).

References

1. Borsarelli CD, Mercadante AZ (2009) Thermal and photochemical degradation of carotenoids. In: Landrum J (ed) Carotenoids: physical, chemical, and biological functions and properties. CRC Press, Boca Raton, pp 229–253

2. Mercadante AZ (2008) Analysis of carotenoids. In: Socaciu C (ed) Food colorants: chemical and functional properties. CRC Press, Boca Raton, pp 447–478
3. Dias MG, Camões MF, Oliveira L (2012) Measurement uncertainty of food carotenoid determination. Accred Qual Assur 17:183–189
4. Kimura M, Rodriguez-Amaya DB, Godoy HT (1990) Assessment of the saponification step in the quantitative determination of carotenoids and provitamins A. Food Chem 35:187–195
5. AOAC International (2013) Appendix K: guidelines for single laboratory validation methods for dietary supplements and botanicals. AOAC Official Methods of Analysis, 32 p., Available at http://www.eoma.aoac.org/app_k.pdf. Accessed on 18 May 2021
6. International Conference on Harmonization of Technical Requirements for the Registration of Pharmaceuticals for Human Use (ICH) (2005) ICH quantification and method validation 371 harmonised tripartite guideline, validation of analytical procedures: text and methodology Q2(r1), p 13
7. Eurachem (2014) Eurachem Guide: the fitness for purpose of analytical methods—a laboratory guide to method validation and related topics. In: Magnusson B, Ornemark U (ed), 2nd edn, 70 p. Available at http://www.eurachem.org/images/stories/Guides/pdf/MV_guide_2nd_ed_en.pdf. Accessed on 18 May 2021
8. Thompson M, Ellison SLR, Wood R (2002) Harmonized guidelines for single-laboratory validation of methods of analysis. Pure Appl Chem 74:835–855
9. Bratinova S, Raffael B, Simoneau C (2009) Guidelines for performance criteria and validation procedures of analytical methods used in controls of food contact materials. EUR 24105 EN, 1st edn. p 74. Available at https://www.bfr.bund.de/cm/343/guidelines_for_performance_criteria_and_validation_procedures_of_analytical_methods.pdf. Accessed on 17 May 2021
10. Emenhiser C, Sander LC, Schwartz SJ (1995) Capability of a polymeric C_{30} stationary phase to resolve cis-trans carotenoid isomers in reversed-phase liquid chromatography. J Chromatogr A 707:205–216
11. Van Breemen RB, Dong LL, Pajkovic ND (2012) Atmospheric pressure chemical ionization tandem mass spectrometry of carotenoids. Int J Mass Spectrom 312:163–172
12. Mercadante AZ, Rodrigues DB, Petry FC, Mariutti LRB (2017) Carotenoid esters in foods – a review and practical directions on analysis and occurrence. Food Res Int 99:830–850
13. Mariutti LRB, Pereira DM, Mercadante AZ, Valentão P, Teixeira N, Andrade PB (2012) Further insights on the carotenoid profile of the echinoderm Marthasterias glacialis L. Mar Drugs 10:1498–1510
14. Liaaen-Jensen SG (1995) Combined approach: identification and structure elucidation of carotenoids. In: Britton G, Liaaen-Jensen S, Pfander H (eds) Carotenoids, vol. 1B: Spectroscopy. Birkhauser, Basel, pp 343–354
15. Petry FC, Mercadante AZ (2019) In: Mercadante AZ (ed) Quantification and method validation. Carotenoid esters in foods: physical, chemical and biological properties. The Royal Society of Chemistry, Cambridge, pp 351–371
16. Costa GA, Mercadante AZ (2018) *In vitro* bioaccessibility of free and esterified carotenoids in cajá frozen pulp-based beverages. J Food Compos Anal 68:53–59
17. Abdel-Aal SM, Rabalski I (2015) Composition of lutein ester regioisomers in Marigold flower, dietary supplement, and herbal tea. J Agric Food Chem 63:9740–9746
18. Chacón-Ordónéz T, Esquivel P, Jiménez VM, Carle R, Schweiggert RM (2016) Deposition form and bioaccessibility of keto-carotenoids from mamey sapote (Pouteria sapota), red bell pepper (Capsicum annuum), and sockeye salmon (Oncorhynchus nerka) filet. J Agric Food Chem 64:1989–1998
19. Lu Q, Huang X, Lv S, Pan S (2017) Carotenoid profiling of red navel orange “Cara Cara” harvested from five regions in China. Food Chem 232:788–798
20. Atienza SG, Ballesteros J, Martín A, Hornero-Méndez D (2007) Genetic variability of carotenoid concentration and degree of esterification among Tritordeum (Tritordeum Ascherson et Graebner) and durum wheat accessions. J Agric Food Chem 55:4244–4251
21. Breithaupt DE, Bamedi A (2001) Carotenoid esters in vegetables and fruits: a screening with emphasis on β-cryptoxanthin esters. J Agric Food Chem 49:2064–2070
22. Mariutti LRB, Rodrigues E, Mercadante AZ (2013) Carotenoids from Byrsonima crassifolia: identification, quantification and in vitro scavenging capacity against peroxyl radicals. J Food Compos Anal 31:155–160
23. de Rosso VV, Mercadante AZ (2007) Identification and quantification of carotenoids, by HPLC-PDA-MS/MS, from Amazonian fruits. J Agric Food Chem 55:5062–5072
24. Petry FC, Nadai FB, Cristofani-Yaly M, Latado RR, Mercadante AZ (2019) Carotenoid

biosynthesis and quality characteristics of new hybrids between tangor (*Citrus reticulata* x *C. sinensis*) cv. 'Murcott' and sweet orange (*C. sinensis*) cv. 'Pêra'. Food Res Int 122: 461–470

25. Petry FC, Mercadante AZ (2017) Impact of in vitro digestion phases on the stability and bioaccessibility of carotenoids and their esters in mandarin pulps. Food Funct 8:3951–3963
26. Petry FC, Mercadante AZ (2016) Composition by LC-MS/MS of new carotenoid esters in mango and citrus. J Agric Food Chem 64: 8207–8224
27. Rodrigues DB, Mariutti LRB, Mercadante AZ (2016) Two-step cleanup procedure for the identification of carotenoid esters by liquid chromatography-atmospheric pressure chemical ionization-tandem mass spectrometry. J Chromatogr A 1457:116–124
28. Chisté RC, Mercadante AZ (2012) Identification and quantification, by HPLC-DAD-MS/MS, of carotenoids and phenolic compounds from the Amazonian fruit Caryocar villosum. J Agric Food Chem 60:5884–5892
29. Facundo HVV, Gurak PD, Mercadante AZ, Lajolo FM, Cordenunsi BR (2015) Storage at low temperature differentially affects the colour and carotenoid composition of two cultivars of banana. Food Chem 170:102–109
30. Faria AF, Hasegawa PN, Chagas EA, Purgatto RPE, Mercadante AZ (2009) Cultivar influence on carotenoid composition of loquats from Brazil. J Food Compos Anal 22:196–203
31. Faria AF, de Rosso VV, Mercadante AZ (2009) Carotenoid composition of jackfruit (*Artocarpus heterophyllus*), determined by HPLC-PDA-MS/MS. Plant Foods Human Nutr 64:108–115
32. Assunção RB, Mercadante AZ (2003) Carotenoids and ascorbic acid from cashew apple (*Anacardium occidentale* L.) – variety and geographic effects. Food Chem 81:495–502
33. Martins GF, Fabi JP, Mercadante AZ, de Rosso VV (2016) The ripening influence of two papaya cultivars on carotenoid biosynthesis and radical scavenging capacity. Food Res Int 81:197–202
34. Nunes TP, Martins CG, Faria AF, Bíscola V, Souza KLO, Mercadante AZ, Cordenunsi BR, Landgraf M (2013) Changes in total ascorbic acid and carotenoids in minimally processed irradiated arugula (*Eruca sativa* Mill) stored under refrigeration. Radiat Phys Chem 30: 125–130
35. Petry FC, Mercadante AZ (2018) New method for carotenoid extraction and analysis by HPLC-DAD-MS/MS in freeze-dried citrus and mango pulps. J Braz Chem Soc 29:205–215
36. Xavier AAO, Domingos LD, Viotto WH, Mercadante AZ (2012) Desenvolvimento e validação de método espectrofotométrico para determinação de corante à base de luteína adicionado em iogurte desnatado. Quím Nova 35: 2057–2062
37. Dachtler M, Glaser T, Kohler K, Albert K (2001) Combined HPLC-MS and HPLC-NMR on-line coupling for the separation and determination of lutein and zeaxanthin stereoisomers in spinach and in retina. Anal Chem 73:667–674
38. Rodrigues DB, Mercadante AZ, Mariutti LRB (2019) Marigold carotenoids: much more than lutein esters. Food Res Int 119:653–664
39. Turcsia E, Nagya V, Deli J (2016) Study on the elution order of carotenoids on endcapped C18 and C30 reverse silica stationary phases. A review of the database. J Food Compos Anal 47:101–112
40. Britton G (1995) UV/visible spectroscopy. In: Spectroscopy BG, Liaaen-Jensen S, Pfander H (eds) Carotenoids, vol 1B. Birkhauser, Basel, pp 13–62
41. Mariutti LRB, Mercadante AZ (2019) Identification of carotenoids and carotenoid esters. In: Mercadante AZ (ed) Carotenoid esters in foods: physical, chemical and biological properties. The Royal Society of Chemistry, Cambridge, pp 322–350
42. Mercadante AZ (2017) Identificación de carotenoides. In: Carotenoides en agroalimentación y salud. Editora: Programa Iberoamericano de Ciencia y Tecnología para el Desarrollo, p 79–95
43. Britton G, Liaaen-Jensen S, Pfander H (2004) Handbook: carotenoids. Birkhauser Verlag, Basel, p 612
44. Rivera SM, Canela-Garayoa R (2012) Analytical tools for the analysis of carotenoids in diverse materials. J Chromatogr A 1224:1–10
45. NAS-IOM (2001) Dietary reference intakes for vitamin a, vitamin K, arsenic, boron, chromium, copper, iodine, iron, manganese, molybdenum, nickel, silicon, vanadium, and zinc. National Academy Press, Washington DC, p 92
46. Chisté RC, Yamashita F, Gozzo FC, Mercadante AZ (2011) Simultaneous extraction and analysis by high performance liquid chromatography coupled to diode array and mass spectrometric detectors of bixin and phenolic compounds from annatto seeds. J Chromatogr A 1218:57–63

47. Stinco CM, Benítez-González AM, Hernanz D, Vicario IM, Meléndez-Martínez AJ (2014) Development and validation of a rapid resolution liquid chromatography method for the screening of dietary plant isoprenoids: carotenoids, tocopherols and chlorophylls. J Chromatogr A 1370:162–170
48. Rodrigues E, Mariutti LRB, Mercadante AZ (2013) Carotenoids and phenolic compounds from *Solanum sessiliflorum*, an unexploited Amazonian fruit, and their scavenging capacity against reactive oxygen and nitrogen species. J Agric Food Chem 61:3022–3029
49. Silva NA, Rodrigues E, Mercadante AZ, de Rosso VV (2014) Phenolic compounds and carotenoids from four fruits native from the Brazilian Atlantic Forest. J Agric Food Chem 62:5072–5084
50. Inbaraj BS, Lu H, Hung CF, Wu WB, Lin CL, Chen BH (2008) Determination of carotenoids and their esters in fruits of Lycium barbarum Linnaeus by HPLC-DAD-APCI-MS. J Pharm Biomed Anal 47:812–818
51. Delpino-Rius A, Eras J, Marsol-Vall A, Vilaró F, Balcells M, Canela-Garayoa R (2014) Ultra performance liquid chromatography analysis to study the changes in the carotenoid profile of commercial monovarietal fruit juices. J Chromatogr A 1331:90–99
52. Breithaupt DE, Bamedi A (2002) Carotenoids and carotenoid esters in potatoes (Solanum tuberosum L.): new insights into an ancient vegetable. J Agric Food Chem 50:7175–7181

Chapter 25

Analysis of Anthocyanins: Extraction, Quantification, and Identification

Marcella Camargo Marques and Lilian Regina Barros Mariutti

Abstract

Anthocyanins are colored bioactive compounds found in superior plants' leaves, flowers, and fruits. The analysis of anthocyanins is important for both the characterization of plants and the commercial application of these compounds. The extraction methods are divided into traditional methods (static extraction and extraction over agitation) and non-traditional (e.g., microwave-assisted extraction (MAE), ultrasonic-assisted extraction, and accelerated-solvent extraction (ASE)). Depending on the complexity of the extract and the aim of the analysis, the extract may be purified through solid-phase extraction (SPE). The pH differential method for quantifying anthocyanin does not require identifying the anthocyanins' profile. Still, the most accurate method for anthocyanin quantification is the determination of the individual anthocyanin content by external calibration. Commonly, the chromatographic separation occurs in a C18 column before UV-vis (DAD) and mass spectrometry detection (MS), which allows the identification of the anthocyanins.

Key words Anthocyanins, Natural pigments, Color, Fruits, Vegetables

1 Introduction

Anthocyanins are hydrosoluble pigments widely distributed in plants, conferring color to leaves, flowers, and fruits [1]. In the cytoplasm, anthocyanins are synthesized and later transported into the vacuole or vesicles [2]. Less commonly, intensely colored intravacuolar bodies, called anthocyanin vacuolar inclusions (AVIs), can be observed in leaves and petals [3, 4]. These non-membrane-bound globular inclusions may be protein matrices associated with high levels of aromatically acylated anthocyanin [3].

Scientific interest in anthocyanins has grown in recent years due to their colorant capacity and bioactive potential [5]. Analyzing anthocyanins is essential to characterizing, understanding, and applying them.

Sample preparation is the first step in a representative determination of anthocyanin. Solid samples need extensive preparation,

Lilian Mariutti and Frank Vriesekoop (eds.), *Food Chemistry: Basic Protocols*, Methods and Protocols in Food Science, https://doi.org/10.1007/978-1-0716-4678-6_25,

including fractionation, reduction of particle size, and homogenization [6]. Seeds and grains are commonly crushed, pulverized [6], or ground [7], while lyophilization and subsequent grinding can be used for most plant tissues [5, 8–10]. Ground freeze-dried material can be easily handled and provides an enhanced superficial area, contributing to the extraction [6, 11].

The extraction of anthocyanins from the matrix relies on solubilizing these compounds from the vegetable tissue. Hence, the optimal extraction method maximizes the anthocyanin yield with minimal degradation and profile alterations [12]. Therefore, extraction is an essential step for quantifying and correctly characterizing anthocyanins from a matrix.

Extraction methods can be classified into traditional and modern or non-traditional [12]. The static method [13] is one of the most prevalent traditional methods. This method comprises the maceration of the plant material in mildly acidified solvent for 4 to 48 h under refrigeration [14, 15]. Important factors such as solvent type, temperature, time, and agitation have been studied to advance the extraction and minimize the decomposition of pigments [12, 16]. Therefore, extraction under agitation [17] became more popular [18–22].

The most common solvent for anthocyanin extraction is methanol due to its polar character, although ethanol and water are also employed, aiming at a more sustainable process [12]. Some procedures require the use of acetone, which may induce the formation of pyranoanthocyanins as extraction artifacts [16].

Acidified solvents improve the anthocyanins' recovery from the matrices and promote their stability since they are more stable in the flavylium cation form [12]. Although widely used, some mineral acids promote acid hydrolysis of acyl residues, particularly aliphatic acids [12, 15, 16]. For new samples, it is recommended to test the extraction procedures using acidified solvent and compare it to the use of neutral solvents to verify the hydrolyze of anthocyanins [11] or the use of trifluoroacetic acid (TFA) or acetic acid alternatively to hydrochloric acid (HCl) [16].

Non-traditional methods were introduced to meet environmental-friendly, low- or non-toxic solvent, less time-consuming, and cost-efficient demands [23, 24]. Microwave-assisted extraction (MAE) [25], ultrasonic-assisted extraction [26], and accelerated solvent extraction (ASE) [27] are technologies successfully applied to anthocyanin extraction and provide high yields and efficiency.

Various methods can be used to determine anthocyanins' content depending on the aim of the analysis, the sample's characteristics, and available techniques. UV-vis spectrophotometric methods are commonly used since the instrumentation is simple and allows for non-expensive and rapid determination [6].

The pH differential method [28, 29] relies on two concepts: (i) the absorbance of species is directly proportional to their concentration, and (ii) the structural conformation of anthocyanins changes according to pH conditions in pH 1 (colored flavylium cation form) and pH 4.5 (colorless hemiketal forms) [28]. The conformational conversion depends on anthocyanins' structure due to the intra-molecular co-pigmentation in anthocyanins acylated with cinnamic acids [16].

The quantification of anthocyanins is affected by numerous compounds, such as non-anthocyanin phenolic compounds through co-pigmentation, products of anthocyanins' degradation, and non-enzymatic browning reactions [13]. Thus, the individual anthocyanin content is the ideal method for quantifying anthocyanins [12]. Although highly accurate and sensitive, this method demands the chromatographic separation and identification of anthocyanins and requires the use of standards and an external calibration curve [12, 16].

Often, the extracts present a complex anthocyanin profile, and the chromatographic separation may be compromised [16]; thus, preparative techniques could provide proper resolution and prevent band-overlap [30]. Among the clean-up methods for anthocyanins, solid phase extraction (SPE) is widely used for removing interfering compounds and pre-fractionate anthocyanins before isolation depending on the sorbent [31, 32].

Since the cation flavilium form of anthocyanins is positively charged, one of the options is using a cation-exchange sorbent, such as Oasis MCX [32, 33]. The high purity level achieved using Oasis MCX [33] due to the ability to segregate anthocyanins from other phenolic compounds relies on both cation-exchange and reversed-phase adsorption mechanisms [31]. Therefore, this technique increases the purity of anthocyanin extracts but cannot be used to selectively isolate anthocyanins [31].

The Sephadex LH-20 is another sorbent commonly used to partially remove non-anthocyanin phenolic compounds and fractionate polymeric proanthocyanins [33]. This resin combines size exclusion gel with polar–polar interaction mechanisms, resulting in a semi-purified extract, which improves the HPLC separation [33].

Anthocyanins are commonly separated using a reversed-phase column (C18) at low pH operating conditions and detected by diode array (DAD) and mass spectrometry (MS) [17–19, 34–36]. Although UV-vis detection provides some information regarding the structure, such as the UV-vis spectrum, visible maximum wavelength, and A_{440}/A_{vis} value [37, 38], the identification is not definitive. The UV-vis spectrum, through the visible maximum wavelength and A_{440}/A_{vis} value, suggests the anthocyanidin identity, the sugar unit position, and the possibility of acylation of the sugar moiety [38]. The MS analysis provides the *m/z* information, and after the molecular ion (M^+) or protonated molecule

($[M + H]^+$) fragmentation, it is possible to determine the anthocyanins' structure, and positive identification is feasible through the confirmation with authentic standards [12].

Thus, we recommend methods for quantitative and qualitative analysis of anthocyanins from different sources and applications. This chapter comprises the most used techniques for sample preparation [6], anthocyanin extraction [13, 25–27, 36], quantification [7, 28], purification [33], and characterization [19].

2 Material

Prepare all solutions using distilled or ultra-pure water (prepared by purifying deionized water to attain a sensitivity of 18 MΩ.cm^{-1} at 25 °C), analytical grade reagents, and HPLC grade reagents as specified. Prepare and store all reagents at room temperature (unless indicated otherwise).

2.1 Extraction

Static Extraction

1. Powdered freeze-dried plant material (*see* **Note 1**).
2. Extracting solvent, HCl in methanol 0.1–1% (v/v) (*see* **Note 2**): transfer 27–270 mL of HCl 12 N (ACS grade) to a volumetric flask containing about 400 mL of methanol. Make up to 1 L with methanol. Store at 4 °C.
3. Whatman no. 1 filter paper.
4. Buchner funnel.
5. Rotary evaporator with vacuum pump, temperature ≤ 30 °C.
6. 0.5% (v/v) HCl in water: transfer 135 mL of HCl 12 N (ACS grade) to a volumetric flask containing about 400 mL of distilled water. Make up to 1 L with water. Store at 4 °C.

Extraction with Agitation

1. Powdered freeze-dried plant material (*see* **Note 1**).
2. HCl in methanol 0.1–1% (v/v) (*see* **Note 2**): transfer 27–270 mL of HCl 12 N (ACS grade) to a volumetric flask containing about 400 mL of methanol analytical grade. Make up to 1 L with methanol. Store at 4 °C.
3. Stirring homogenizer with stainless steel homogenizer.
4. Whatman no. 1 filter paper.
5. Buchner funnel.
6. Rotary evaporator with vacuum pump, temperature ≤ 30 °C.

7. 0.5% (v/v) HCl in water: transfer 135 mL of HCl 12 N (ACS grade) to a volumetric flask containing about 400 mL of distilled water. Make up to 1 L with water. Store at 4 °C.

Microwave-Assisted Extraction

1. Powdered freeze-dried plant material (*see* **Note 1**).
2. 40% methanol in water (v/v): transfer 400 mL of methanol analytical grade to a volumetric flask containing about 400 mL of distilled water. Make up to 1 L with water.
3. Teflon flask.
4. Microwave extractor.

Ultrasound-Assisted Extraction

1. Crush frozen fruit samples into pieces (3–4 mm).
2. HCl 1.5 M in ethanol/water (85:15, v/v): transfer 125 mL of HCl 12 N (ACS grade) to a volumetric flask containing about 400 mL of ethanol 85%. Make up to 1 L with ethanol 85%.
3. Double-neck flask with a cooling system.
4. Ultrasonic cell disruptor with a total nominal power of 650 W.

Pressurized Fluid-Assisted Extraction or Accelerated Solvent Extraction

1. Accelerated solvent extractor equipped with 22 mL stainless steel extraction cell.
2. Powdered freeze-dried plant material (*see* **Note 1**).
3. Cellulose extraction Cell Filters, Type D28.
4. Acid-washed sea sand.
5. Acetic acid:methanol:water (7:75:18) (v/v).

2.2 Total Monomeric Anthocyanins—pH Differential Method

1. Potassium chloride buffer 0.025 M (pH 1.0): weigh 1.86 g of KCl into a beaker, and add about 980 mL of distilled water. Adjust pH to 1.0 (± 0.05) with HCl (about 6.3 mL). Transfer to a volumetric flask, and make up to 1 L with distilled water.
2. Sodium acetate buffer, 0.4 M (pH 4.5): weigh 54.43 g of $CH_3CO_2Na{\cdot}3H_2O$ in a beaker, and add about 980 mL of distilled water. Adjust pH to 4.5 (±0.05) with HCl (about 20 mL). Transfer to a 1 L volumetric flask, and make up to volume with distilled water.
3. Crude extracts.

4. Spectrophotometer with a variable-wavelength detector for absorbance measurement at 520 and 700 nm.
5. Quartz cells (1 cm path length).

2.3 Purification

LH-20 SPE

1. Polypropylene cartridge (6 mL).
2. Sephadex® LH-20 (Sigma–Aldrich Corp., St. Louis, MO).
3. Crude extract.
4. Ultrapure water/methanol (80:20, v/v): transfer 20 mL of methanol HPLC grade to a volumetric flask containing about 40 mL of ultrapure water. Make up to 100 mL with water.
5. Methanol/ultrapure water (70:30, v/v): transfer 30 mL of methanol HPLC grade to a volumetric flask containing about 40 mL of ultrapure water. Make up to 100 mL with water.
6. Acetone/ultrapure water (70:30, v/v): transfer 400 mL of acetone HPLC grade to a volumetric flask containing about 200 mL of ultrapure water. Make up to 1 L with water.

Cation-Exchange SPE

1. Polypropylene cartridge (6 mL).
2. Cation-exchange SPE Oasis® MCX (Waters Corp., Milford, MA).
3. Ultrapure water with 0.1% TFA: transfer 101 μL of methanol HPLC grade to a volumetric flask containing about 40 mL of ultrapure water. Make up to 100 mL with ultrapure water.
4. Crude extracts in aqueous solution (12 mL).
5. Methanol with 0.1% TFA: transfer 101 μL of TFA (⩾99%) to a volumetric flask containing about 40 mL of methanol HPLC grade. Make up to 100 mL with methanol HPLC grade.
6. Methanol containing 1% NH_4OH: transfer 3.5 mL of NH_4OH (ACS grade) to a volumetric flask containing about 40 mL of methanol HPLC grade. Make up to 100 mL with methanol HPLC grade.
7. Water/methanol (40:60, v/v) containing 1% NH_4OH: transfer 3.5 mL of NH_4OH (ACS grade) to a volumetric flask containing about 40 mL of methanol/water (40:60, v/v). Make up to 100 mL with methanol/water.
8. Formic acid (⩾99%).

2.4 Characterization

Standard Preparation

1. Cyanidin-3-glucoside chloride analytical standard (≥96.0%, HPLC) (*see* **Notes 3** and **4**).
2. Pelargonidin-3-O-glucoside chloride, analytical standard (≥95.0%, HPLC).
3. Peonidin-3-O-glucoside chloride, analytical standard (≥95.0%, HPLC).
4. Malvidin-3-O-glucoside chloride, analytical standard (≥95.0%, HPLC).
5. Petunidin-3-O-glucoside, chloride analytical standard (≥95.0%, HPLC).
6. Delphinidin-3-O-glucoside chloride (≥95.0%, HPLC).
7. Other anthocyanin standards available for HPLC.
8. Potassium chloride buffer, 0.025 M (see Subheading 2.3, item 1).
9. Water with 1% HCl: transfer 2.7 mL of HCl 12 N (ACS grade) to a volumetric flask containing about 4 mL of ultrapure water. Make up to 10 mL with water.
10. Methanol with 0.1% HCl: transfer 0.27 mL of HCl 12 N (ACS grade) to a volumetric flask containing about 4 mL of methanol HPLC grade. Make up to 10 mL with methanol.
11. Spectrophotometer.
12. Quartz cells (1 cm path length).

HPLC-DAD-MS Analysis

1. HPLC system with a binary pump for solvent gradients and a DAD detector.
2. Reversed-phase C18 column (Shim-pack CLC-ODS, 4.6 × 250 mm, 5 μm) (*see* **Note 5**).
3. 0.45 μm polypropylene syringe filter.
4. Membrane filter, ≤ 0.45 μm pore size.
5. Mobile phase A: ultrapure water with 5% formic acid (v/v)—transfer 101 mL of formic acid (⩾99%) to a volumetric flask containing about 400 mL of ultrapure water. Make up to 1 L with water.
6. Mobile phase B: methanol with 5% formic acid (v/v)—transfer 101 mL of formic acid (⩾99%) to a volumetric flask containing about 400 mL of methanol HPLC grade. Make up to 1 L with methanol.
7. Filter all the mobile phases in a membrane filter, ≤ 0.45 μm pore size.
8. Mass spectrometer (MS) equipped with electrospray ionization (ESI) and ion trap analyzer (*see* **Note 6**).

3 Methods

3.1 Extraction

Static Extraction

1. Weigh the powdered freeze-dried sample with 0.01 g precision in a caped flask.
2. Transfer the extracting solvent 1:50 (dw/v, g/mL) to the flask.
3. Homogenize at room temperature for 3 min.
4. Macerate overnight at 4 °C.
5. Filter the extract through a Buchner funnel with a Whatman No. 1 filter paper.
6. Wash the flask quantitatively with the extracting solvent.
7. Wash the residue/cake repeatedly with the extracting solvent until the filtrate has no color.
8. Concentrate the extract in a rotary evaporator at 40 °C under vacuum.
9. Transfer the concentrated extract to a volumetric flask, and make up to a known volume with 0.01% HCl-acidified water.
10. Store the extract at −80 °C until further analysis.

Extraction with Agitation

1. Weight the powdered freeze-dried sample with 0.01 g precision in a screw-cap tube.
2. Add the extracting solvent 1:15 (dw/v, g/mL).
3. Homogenize in a vortex for 30 s at room temperature.
4. Sonicate for 5 min.
5. Keep at room temperature for 5 s and agitate for 30 s in a vortex.
6. Centrifuge at 4550 *g* for 10 min and remove the supernatant.
7. Repeat items 2–7 and combine the supernatants.
8. Concentrate the extract in a rotary evaporator at 40 °C under vacuum.
9. Transfer the concentrated extract to a volumetric flask, and make up to a known volume with 0.01% HCl-acidified water.
10. Store the extract at −80 °C until further analysis.

Microwave-Assisted Extraction

1. Weigh the powdered freeze-dried sample with 0.01 g precision in a microwave Teflon vessel.
2. Add the extracting solvent 2:25 (fw/v, g/mL) and the magnetic stir bar.

3. Close the Teflon vessel.
4. Proceed the MAE at 100 °C for 5 min under stirring, using 500 W as the maximum system power.
5. Filter the extract through a Buchner funnel with a Whatman No. 1 filter paper.
6. Wash the flask quantitatively with the extracting solvent.
7. Wash the residue/cake repeatedly with the extracting solvent until the filtrate has no color.
8. Concentrate the extract in a rotary evaporator at 40 °C under vacuum.
9. Transfer the concentrated extract to a volumetric flask, and make up to a known volume with 0.01% HCl-acidified water.
10. Store the extract at −80 °C until further analysis.

Ultrasonic-Assisted Extraction

1. Weight the sample with 0.01 g precision in a double-neck flask with a cooling system.
2. Add the extracting solvent 1:4 (fw/v, g/mL).
3. Introduce the ultrasonic cell disruptor, and proceed with the extraction for 3.3 min (total nominal power of 650 W, operating at 22 kHz frequency), maintaining the temperature below 40 °C.
4. Filter the extract through a Buchner funnel with a Whatman No. 1 filter paper.
5. Collect the cake and repeat the procedures described in items 3–4.
6. Combine the filtrates and transfer them to a volumetric flask.
7. Store the extract at −80 °C until further analysis.

Pressurized Fluid-Assisted Extraction or Accelerated Solvent Extraction

1. Place three cellulose filters in the bottom of a 22 mL stainless steel extraction cell.
2. Cover the surface with acid-washed sea sand.
3. Weigh 0.5 g of powdered freeze-dried sample with 0.01 g precision.
4. Mix the freeze-dried sample with the acid-washed sea sand (1: 1, w/w).
5. Load the mixture of sample and acid-washed sea sand into the cell.
6. Close the extraction cell tightly.

7. Preheat the solvent for 5 min at 80–100 °C.
8. Proceed the extraction at a constant pressure of 1500 psi for three cycles of 50 min.
9. Collect the extract in amber glass vials, and adjust the volume to 50 mL.
10. Store the extract at −80 °C until further analysis.

3.2 Total Monomeric Anthocyanins—pH Differential Method

Dilution Factor Determination

1. In a 50 mL volumetric flask, 10 mL of the anthocyanin extract is added, and the volume is completed with the pH 1.0 buffer.
2. In a spectrophotometer, read the absorbance at 520 nm, which must be within the linear range (between 0.2 and 1.4 AU for most spectrophotometers).
3. Adjust the dilution if necessary.
4. Record the dilution factor.

Test Assay

1. Using the dilution factor, prepare two dilutions of the test sample with pH 1.0 buffer and pH 4.5 buffer in a 50 mL volumetric flask.
2. Determine the absorbance at 520 and 700 nm.
3. Calculate the anthocyanin concentration:

$$TMC = \frac{\left((Abs_{520nm} - Abs_{700nm})_{pH1.0} - (Abs_{520nm} - Abs_{700nm})_{pH4.5}\right) * MW * DF * 10^3}{\varepsilon * 1}$$

in mg/L of cyanidin-3-glucoside equivalent.

where

TMC = total monomeric anthocyanin content.

MW = molecular weight, 449.2 g/mol for cyanidin-3-glucoside.

DF = dilution factor.

ε = molar extinction coefficient, in $L.mol^{-1}.cm^{-1}$, 26,900 for cyanidin-3-glucoside.

1 = pathlength in cm.

10^3 = conversion factor from g to mg.

3.3 Purification

LH-20 SPE

1. Apply the crude extract to the cartridge containing LH-20 SPE.

2. Wash the hydrophilic compounds with 20% methanol.
3. Using 70% methanol, elute and collect the anthocyanin fraction.
4. Remove the solvent of the fraction in a rotary evaporator at ≤35 °C.
5. Transfer the concentrated extract to a volumetric flask, and make up to a known volume with 0.01% HCl-acidified water.

Cation-Exchange SPE Oasis MCX

1. Apply 12 mL of crude extract in aqueous solution.
2. Wash the cartridge with 12 mL of 0.1% TFA.
3. Collect the other-phenol fraction with 12 mL of methanol (0.1% TFA).
4. Collect the anthocyanin fraction with 6 mL of methanol followed by 6 mL of water/methanol (40:60, v/v) containing 1% NH_4OH.
5. Remove the solvent of the fraction in a rotary evaporator at 35 ° C.
6. Transfer the concentrated extract to a volumetric flask, and make up to a known volume with 0.01% HCl-acidified water.

3.4 Characterization

Standards and External Calibration

1. Solubilize the cyanidin-3-glucoside standard in potassium chloride buffer, 0.025 M (pH 1.0) (*see* **Note 3**).
2. In a spectrophotometer, read the absorbance at 510 nm.
3. Use the Lambert–Beer's law and the molar absorptivity (ε = 26,900) for calculations.
4. Dilute the standard solutions to seven concentrations using the mobile phase (*see* **Note 4**).
5. Filter the standard solutions through 0.45 μm, and inject them into the HPLC under the same conditions as the samples (see item *LC separation)*.
6. Repeat the procedure described in items 1–4 for all the available anthocyanin standards.
7. Calculate the linear regression equation, Pearson correlation index (linearity), detection, and quantification limits.
8. Calculate each anthocyanin content using the linear regression equation, considering their respective calibration curve and peak area. If the standard is not available, calculate the anthocyanin content as cyanidin-3-glucoside equivalents.

HPLC-PDA-MS Analysis: LC Separation and Anthocyanin Identification

1. Set the C18 column to the initial condition: 0.9 mL/min of mobile phases A and B (90:10).
2. Filter the crude or purified extract through a syringe filter ≤0.45 μm, and inject a known volume into the HPLC.
3. Separate the anthocyanins using a linear gradient of A/B mobile phases from 90:10 to 60:40 in 20 min, passing to 20: 80 in 15 min and keeping this proportion for 5 min.
4. Acquire the UV-vis spectra between 200 and 600 nm, and process the chromatogram at 520 nm.
5. Split the flow from the column, allowing only 0.15 mL/min into the ESI source.
6. Proceed the ionization with capillary voltage at 95.2 V, an endplate offset at 500 V, drying gas (N_2) temperature of 325 °C, and flow of 8 L/min, nebulizer at 30 psi.
7. Acquire MS and MS/MS using total ion scan (from *m/z* 200 to 1000) (*see* **Note 5**).
8. Extract the selective ion chromatogram for the MS/MS, selectively for *m/z* 271 (pelargonidin), *m/z* 287 (cyanidin), *m/z* 301 (peonidin), *m/z* 303 (delphinidin), *m/z* 317 (petunidin), and *m/z* 331 (malvidin).
9. Use the retention time of the authentic standards, elution order on the C18 column (*see* **Note 7**), comparison with the literature, and the mass spectra interpretation (Fig. 1) for the correct identification.

4 Conclusions

The analysis of anthocyanins is essential for the characterization of biodiversity and a better understanding of the bioactivity of these compounds and their commercial applications. Multiple extraction techniques are available, while the traditional methods are simple, reliable, and cost-effective, non-traditional methods are fast and less solvent-consuming. The quantification of anthocyanins using the pH differential method is simple but does not reveal information regarding the anthocyanin profile. Alternatively, the individual anthocyanin content demands the identification of the compounds; therefore, complex instrumentation is required.

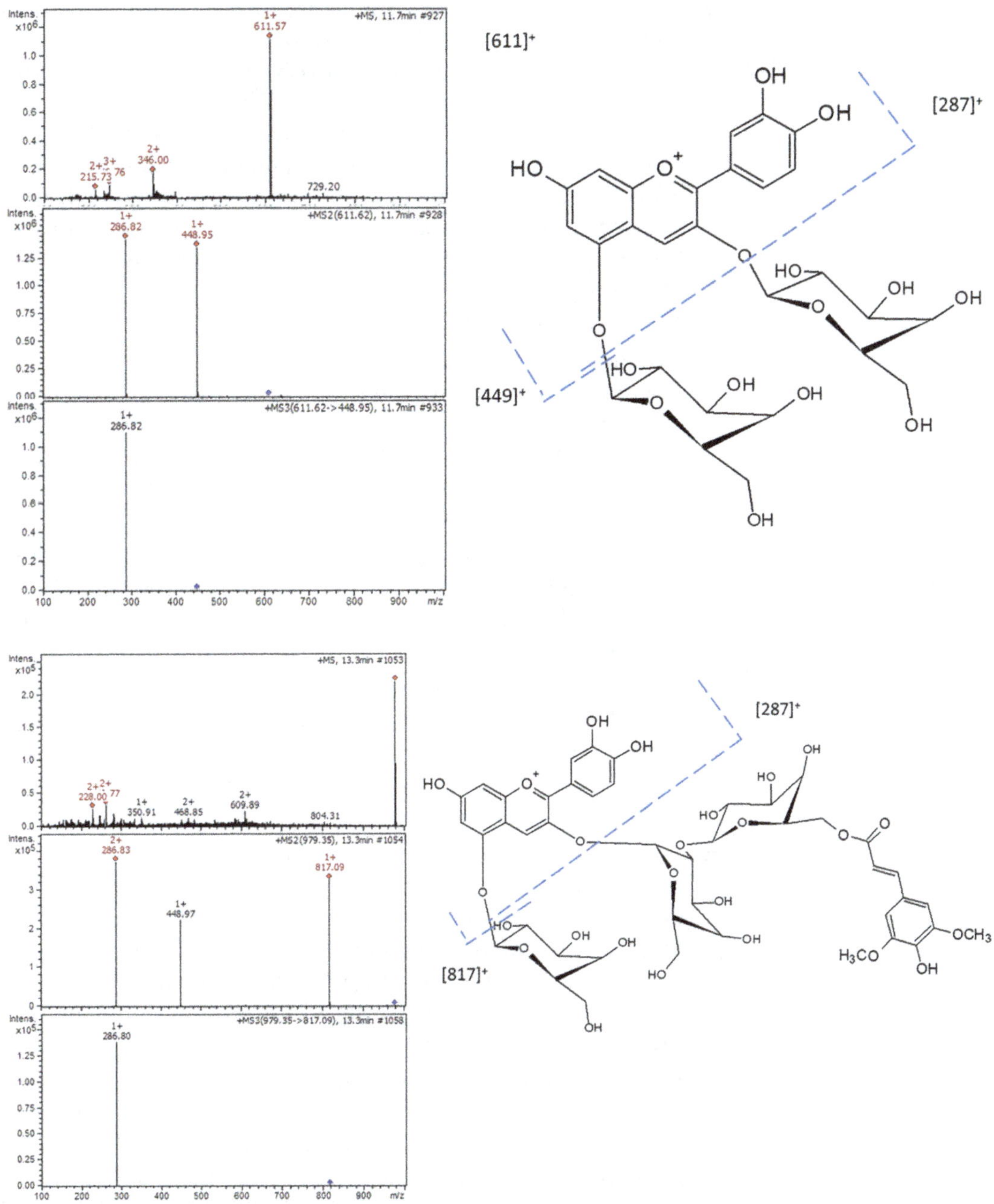

Fig. 1 Mass spectra and fragmentation pattern of anthocyanins: A, cyanidin 3,5-diglucoside; B, cyanidin-3-(sinapoyl-diglucoside)-5-(glucoside) from red cabbage

5 Notes

1. Lyophilized plant material, after dried, ground, and homogenized, can be easily distributed and stored. The low moisture also allows for a larger weight for extraction, which results in a more concentrated extract [6].

2. In the plant tissue, anthocyanins are found in anthocyanoplasts and anthocyanin vacuolar inclusion [3, 4]. When in solution, anthocyanins are shown to be involved in reversible structural transformations as a function of pH. The flavylium cation form is predominant and stable at a highly acidic medium pH around 1.0 [16]. Therefore, solvents containing mineral or organic acids for anthocyanin extraction are generally applied. Once the extraction of anthocyanins under high acid conditions can lead to partial or total hydrolysis of acyl groups, lower acid concentrations may be advised for matrices with acylated anthocyanins [12, 15, 16, 39].
3. Commercial anthocyanin standards are costly and scarce. Therefore, the anthocyanin content can be determined through the external calibration of the major compound standard. The most used is cyanidin 3-glucoside [7, 15]. The injection of multiple standards informs the retention time of compounds, which can help further identification.
4. Due to the high cost and low availability of commercial anthocyanin standards, external calibration is often reduced to injecting three dilution points to establish a linear relation between detector response and concentration [7].
5. Various chromatographic columns can be used for the separation of anthocyanins. The most common is the reversed-phase column (C18) on a silica base, although the polymer-based columns provide better resolution for acylated anthocyanins [12].
6. Numerous mass analyzers are used for *m/z* determination, but tandem mass spectrometry (MS/MS) has been proven as an option for anthocyanin structural analysis due to the sequential *m/z* analysis and ion fragment acquisition. Ion trap [19], triple quadrupole [40], quadrupole-time-of-flight [41], and quadrupole-orbitrap [42] mass spectrometers can be used for the sequential mass analysis.
7. Different anthocyanins present similar UV-vis characteristics, the same molecular weight, and oftentimes indistinct MS fragmentation spectra. Therefore, the elution order in the reversed-phase column is an important factor for anthocyanin identification. This is the case of cyanidin-3-galctoside and cyanidin-3-glucoside; both present M^+ and ion fragments at *m/z* 449 and 287, respectively [17, 35]. On the other hand, sometimes the identity assignment without a standard is possible due to MS/MS spectrum interpretation, in which compounds with the same molecular weight have different fragmentation patterns and distinct retention times. This is the case of cyanidin-3,5-diglucoside and cyanidin 3-diglucoside, which present M^+ at *m/z* 611 and fragment ions at *m/z* 449/287 and only 287, respectively [17, 35].

Acknowledgments

This study was financed in part by the Coordenação de Aperfeiçoamento de Pessoal de Nível Superior—Brasil (CAPES)—Finance Code 001. LRBM thanks CNPq for the fellowship (305778/2022-6).

References

1. Mercadente AZ, Bobbio FO (2008) Anthocyanins in foods: occurrence and physicochemical properties. In: Socaciu C (ed) Food colorants: chemical and functional properties. CRC Press, Boca Raton, pp 241–268
2. Nozzolillo C, Ishikura N (1988) An investigation of the intracellular site of anthocyanoplasts using isolated protoplasts and vacuoles. Plant Cell Rep 7:389–392
3. Kallam K, Appelhagen I, Luo J et al (2017) Aromatic decoration determines the formation of Anthocyanic vacuolar inclusions. Curr Biol 27:945–957
4. Markham KR, Gould KS, Winefield CS et al (2000) Anthocyanic vacuolar inclusions — their nature and significance in flower colouration. Phytochemistry 55:327–336
5. He J, Wallace TC, Keatley KE et al (2009) Stability of black raspberry anthocyanins in the digestive tract lumen and transport efficiency into gastric and small intestinal tissues in the rat. J Agric Food Chem 57:3141–3148
6. Mazza G, Cacace JE, Kay CD (2004) Methods of analysis for Anthocyanins in plants and biological fluids. J AOAC Int 87:129–145
7. Lao F, Giusti MM (2016) Quantification of purple corn (Zea mays L.) Anthocyanins using spectrophotometric and HPLC approaches: method comparison and correlation. Food Anal Methods 9:1367–1380
8. Albuquerque BR, Pinela J, Barros L et al (2020) Anthocyanin-rich extract of jabuticaba epicarp as a natural colorant: optimization of heat- and ultrasound-assisted extractions and application in a bakery product. Food Chem 316:126364
9. Jordheim M, Calcott K, Gould KS et al (2016) High concentrations of aromatic acylated anthocyanins found in cauline hairs in Plectranthus ciliatus. Phytochemistry 128:27–34
10. Wiczkowski W, Szawara-Nowak D, Topolska J (2013) Red cabbage anthocyanins: profile, isolation, identification, and antioxidant activity. Food Res Int 51:303–309
11. Takeoka G, Dao L (2002) Anthocyanins. In: Hurst WJ (ed) Methods of analysis for functional foods and nutraceuticals. CRC Press
12. Giusti M, Jing P (2007) Analysis of Anthocyanins. In: Socaciu C (ed) Food colorants: chemical and functional properties. CRC Press, Boca Raton, pp 479–506
13. Fuleki T, Francis FJ (1968) Quantitative methods for Anthocyanins. J Food Sci 33:78–83
14. Degenhardt A, Knapp H, Winterhalter P (2000) Separation and purification of anthocyanins by high-speed countercurrent chromatography and screening for antioxidant activity. J Agric Food Chem 48:338–343
15. Revilla E, Ryan JM, Martin-Ortega G (1998) Comparison of several products used for the extraction of Anthocyanins from red grapes. J Agric Food Chem 46:4592–4597
16. Andersen ØM, Jordheim M (2013) Basic anthocyanin chemistry and dietary sources. In: Wallac TC, Giusti MM (eds) Anthocyanins in health and disease. CRC Press, pp 13–89
17. Wu X, Prior RL (2005) Identification and characterization of anthocyanins by high-performance liquid chromatography-electrospray ionization-tandem mass spectrometry in common foods in the United States: vegetables, nuts, and grains. J Agric Food Chem 53:3101–3113
18. Tarone AG, Goupy P, Ginies C et al (2021) Advanced characterization of polyphenols from Myrciaria jaboticaba peel and lipid protection in in vitro gastrointestinal digestion. Food Chem 359:129959
19. Faria AF, Marques MC, Mercadante AZ (2011) Identification of bioactive compounds from jambolão (Syzygium cumini) and antioxidant capacity evaluation in different pH conditions. Food Chem 126:1571–1578
20. Morales P, Barros L, Dias MI et al (2016) Non-fermented and fermented jabuticaba (Myrciaria cauliflora Mart.) pomaces as valuable sources of functional ingredients. Food Chem 208:220–227
21. Alezandro MR, Dubé P, Desjardins Y et al (2013) Comparative study of chemical and phenolic compositions of two species of jaboticaba: Myrciaria jaboticaba (Vell.) Berg and

Myrciaria cauliflora (Mart.) O. Berg. Food Res Int 54:468–477

22. Sousa De Brito E, De AMCP, Alves RE et al (2007) Anthocyanins present in selected tropical fruits: Acerola, jambolão, jussara, and guajiru. J Agric Food Chem 55:9389–9394
23. Turner C (2006) Overview of modern extraction techniques for food and agricultural samples. In: Turner C (ed) Modern extraction techniques. ACS publication, pp 3–19
24. Kaufmann B, Christen P (2002) Recent extraction techniques for natural products: microwave-assisted extraction and pressurised solvent extraction. Phytochem Anal 13:105–113
25. Liazid A, Guerrero RF, Cantos E et al (2011) Microwave assisted extraction of anthocyanins from grape skins. Food Chem 124:1238–1243
26. Chen F, Sun Y, Zhao G et al (2007) Optimization of ultrasound-assisted extraction of anthocyanins in red raspberries and identification of anthocyanins in extract using high-performance liquid chromatography-mass spectrometry. Ultrason Sonochem 14:767–778
27. Truong VD, Hu Z, Thompson RL et al (2012) Pressurized liquid extraction and quantification of anthocyanins in purple-fleshed sweet potato genotypes. J Food Compos Anal 26:96–103
28. Lee J, Durst RW, Wrolstad RE (2005) Determination of total monomeric anthocyanin pigment content of fruit juices, beverages, natural colorants, and wines by the pH differential method: collaborative study. J AOAC Int 88:1269–1278
29. Giusti MM, Wrolstad RE (2001) Characterization and measurement of Anthocyanins by UV-visible spectroscopy. In: Current protocols in food analytical chemistry. Wiley, pp 1–13
30. Frøytlog C, Slimestad R, Andersen ØM (1998) Combination of chromatographic techniques for the preparative isolation of anthocyanins – applied on blackcurrant (Ribes nigrum) fruits. J Chromatogr A 825:89–95
31. Ahmadiani N, Sigurdson GT, Robbins RJ et al (2019) Solid phase fractionation techniques for segregation of red cabbage anthocyanins with different colorimetric and stability properties. Food Res Int 120:688–696
32. Diaconeasa Z, Rugină D, Socaciu C (2016) High-purity Anthocyanins isolation using solid phase extraction techniques. Bull Univ Agric Sci Vet Med Cluj-Napoca Food Sci Technol 73:1–6
33. He J, Giusti MM (2011) High-purity isolation of anthocyanins mixtures from fruits and vegetables – a novel solid-phase extraction method using mixed mode cation-exchange chromatography. J Chromatogr A 1218:7914–7922
34. Inada KOP, Leite IB, Martins ABN et al (2021) Jaboticaba berry: a comprehensive review on its polyphenol composition, health effects, metabolism, and the development of food products. Food Res Int 147:110518
35. Wu X, Prior RL (2005) Systematic identification and characterization of Anthocyanins by HPLC-ESI-MS/MS in common foods in the United States: fruits and berries. J Agric Food Chem 53:2589–2599
36. Wu X, Gu L, Prior RL et al (2004) Characterization of Anthocyanins and Proanthocyanidins in some cultivars of Ribes, Aronia, and Sambucus and their antioxidant capacity. J Agric Food Chem 52:7846–7856
37. Giusti MM, Rodríguez-Saona LE, Wrolstad RE (1999) Molar absorptivity and color characteristics of acylated and non-acylated pelargonidin-based anthocyanins. J Agric Food Chem 47:4631–4637
38. Hong V, Wrolstad RE (1990) Use of HPLC separation/photodiode Array detection for characterization of Anthocyanins. J Agric Food Chem 38:708–715
39. Bakker J, Timberlake CF (1985) The distribution of Anthocyanins in grape skin extracts of port wine cultivars as determined by high performance liquid chromatography. J Sci Food Agric 36(12):1315–1324
40. Zou T, Wang D, Guo H et al (2012) Optimization of microwave-assisted extraction of Anthocyanins from mulberry and identification of Anthocyanins in extract using HPLC-ESI-MS. J Food Sci 77:46–50
41. Zhang J (2018) Original article Isolation and identification of two major acylated anthocyanins from purple sweet potato (Ipomoea batatas L. cultivar Eshu No . 8) by UPLC-QTOF-MS / MS and NMR. 1932–1941
42. Hong HT, Netzel ME, Hare TJO (2020) Optimisation of extraction procedure and development of LC – DAD – MS methodology for anthocyanin analysis in anthocyanin-pigmented corn kernels. Food Chem 319:126515

Chapter 26

Betalains: Chemistry and Analysis

Patricia Esquivel, Andrea Irías-Mata, and Víctor M. Jiménez

Abstract

Betalains are a group of plant pigments mainly found in the Caryophyllales order. According to their structure and color, they have been divided into two groups: the betacyanins with red-violet hues and the yellow-orange betaxanthins. Because of their chemical properties and stability, there is particular interest in the food, cosmetic, and pharmaceutical industries in exploring using betalains as ingredients or pigments to be included in their formulations. The profile of individual betalains can largely vary according to the (plant) source, and, therefore, an increasing amount of literature describing the characteristics of extracts from different plants is now available. This has been possible due to the use of modern analytical methods. In this chapter, besides presenting an overview of the importance of this pigment group, the chemical characteristics, and relevant sources, emphasis is given to the methods employed to analyze these compounds in different plant structures. Most relevant results from reports on betalain analysis are summarized below, and the betalain concentration and profile in the different sources are emphasized, as well as the list of species in which each of the individual compounds has been identified. Finally, information on the bioaccessibility and bioavailability of this interesting group of pigments is given.

Key words Betalains, Betacyanins, Betaxanthins, Betalamic acid

1 Introduction

Betalains are water-soluble and nitrogen-containing pigments that are classified into betacyanins and betaxanthins. These pigments are commonly present in plants of the order Caryophyllales, where they accumulate in edible roots and fruits and bracts, stems, flowers, and other nonedible structures [1]. Betalains are one of the most important plant-derived pigments used industrially, providing red-purple and yellowish colors. Besides, evidence about the positive effects of these pigments on human health has been recently published, bringing more attention to the importance of betalains at the industrial and nutritional levels [2].

Currently, several protocols are available for extraction, isolation, and analysis aiming at the identification and quantitation of betalains in several plant structures, some currently being used as

Lilian Mariutti and Frank Vriesekoop (eds.), *Food Chemistry: Basic Protocols*, Methods and Protocols in Food Science, https://doi.org/10.1007/978-1-0716-4678-6_26,

foodstuff or with the potential to become one. Ultraviolet-visible (UV-vis) spectroscopy is a widely used technique to quantitate the two main betalain groups, while high-performance liquid chromatography (HPLC) coupled to diode-array detector (DAD) or mass detector is the most-used technique for betalain profiling. Nuclear magnetic resonance (NMR) spectroscopy is also a complementary analytical tool for unambiguous betalain identification [3].

This chapter aims to describe the chemistry of betalains and summarize the analytical approaches used for these compounds with examples of their main plant sources. Likewise, the biological activities of betalains and their role in human health, as well as the assessment of their bioavailability and bioaccessibility, are mentioned.

2 Chemistry of Betalains

Betalamic acid (I) is the precursor of all betalains. Its conjugation with cyclo-DOPA (cyclo-3,4-dihydroxyphenylalanine) or its glucosyl derivates forms the red-violet betacyanins (II), while its condensation with amino compounds, such as amino acids, amines, or derivatives, leads to the formation of yellow-orange betaxanthins (III) (Fig. 1) [4, 5].

Betanidin is the simplest betacyanin, with two hydroxyl groups located at the *o*-position (C5 and C6) of the cyclo-DOPA moiety. The substitution of different glucosyl groups to the oxygen atoms of one or both hydroxyl groups results in forming four structural betacyanin types: betanin, gomphrenin, amaranthin, and bougainvillein (Fig. 2). Betanin-type group (I) is characterized by a glucosyl group substitution in C5, whereas gomphrenin-type group (II) by the same substitution in C6. Amaranthin-type group (III) has two contiguous glucosyl groups attached at C5. Bougainvillein (IV), also known as 2-descarboxybetanin-type group, comprises decarboxylated as well as carboxylated betanins. Different moieties

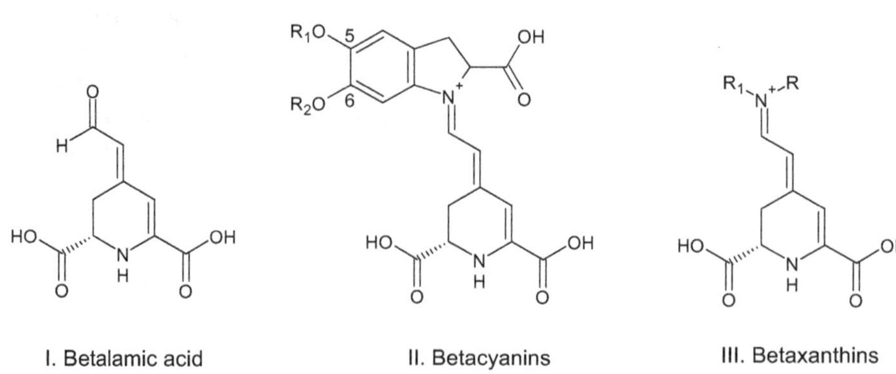

Fig. 1 Basic structures of betalamic acid (I), betacyanins (II), and betaxanthins (III)

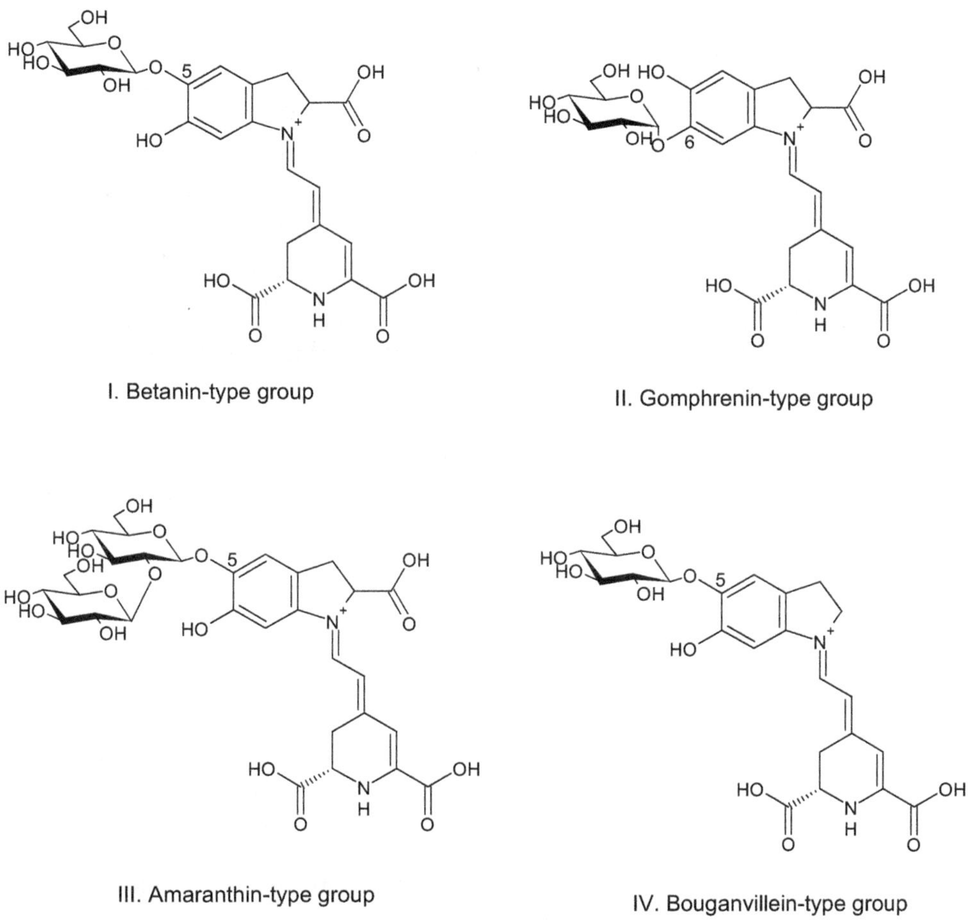

Fig. 2 Structural betacyanin types

define the type of betaxanthins, with proline, glutamine, methionine sulfoxide, tyramine, DOPA, and 5-hydroxynorvaline as common substitutes [4].

It is well known that the conjugated double bonds of the betalamic acid are the chromophore of all betalains. Betaxanthins exhibit a maximum absorbance at around 470 nm, whereas the aromatic ring's double bond of cyclo-DOPA in betacyanins raises their maximum absorbance at 50–70 nm, around 540 nm [4, 5].

3 Sources of Betalains

As previously mentioned, betalains are mainly found in several plants of the Caryophyllales order, some with edible structures. Their occurrence has been reported in *Amaranthus* spp. [6, 7], *Alternanthera* spp. [6, 8, 9], *Chenopodium* spp. (quinoa) [6, 10], *Celosia* spp. [6, 11], *Bougainvillea* sp. [12], *Portulaca grandiflora*

[13], *Phytolacca americana* [14], *Gomphrena globosa* [12], *Boerhavia erecta* L. [7], *Basella* spp. (Malabar spinach) [15, 16], *Beta vulgaris* (red beet, yellow beet, Swiss chard) [17–20], *Opuntia* spp. (prickly cactus pear) [18, 20–23], *Stenocereus* spp. (pitaya) [24], and *Hylocereus* spp. (red-purple dragon fruit) [25, 26]. These pigments have also been reported in some genera of higher fungi, such as *Amanita*, *Hygrocybe*, and *Hygrophorus* [27].

4 Biological Activities of Betalains and Their Role in Human Health

Betalains have been reported to be metabolites with beneficial effects by preventing oxidative stress and lipid peroxidation, and they are also characterized by their anti-genotoxic and cytotoxic actions and their anti-inflammatory and chemopreventive properties [1, 28]. Evidence about their positive influence on cardiovascular, gastrointestinal, and metabolic events has also been published [1].

Recent investigations on the application of betalain-rich plant extracts from *Beta vulgaris*, *Opuntia ficus-indica*, and *Hylocereus polyrhizus* on different cell lines, from lymphoma, ovarian, cervical epithelial, and bladder cancer cells, melanoma cancer cells, and leukemia cells, have shown a chemopreventive potential of these extracts, since they effectively inhibit cell growth and induce apoptosis [2].

However, a recent investigation failed to find a specific correlation between betalain content, extracted from *Beta vulgaris* and *Opuntia ficus-indica*, and their chemopreventive potential. Since the plant extracts were shown to stimulate cytotoxic effects and activation of detoxification enzymes in colon adenocarcinoma cells, it is therefore suggested that this effect must be due to an interaction between different phytochemicals present in the extracts rather than a unique betalain effect [29].

Besides, a direct correlation between betalain content and antioxidant activity has been reported in several publications (e.g., in *Hylocereus* sp. [30], *Chenopodium quinoa* [31], and *Opuntia ficus-indica* [32]). In addition, extracts of *Beta vulgaris* and *Opuntia ficus-indica* showed a strong correlation with the antioxidant capacity, measured by spectrophotometric methods as Folin–Ciocalteu and DPPH radicals and by cellular antioxidant activity in a colon adenocarcinoma cell line [29]. In another publication, antioxidant capacity was measured by 4 in vitro assays in 13 varieties of *Beta vulgaris*, enriched in betanin, isobetanin, and glutamine-betaxanthin. Extracts demonstrated a radical scavenging ability; this antiradical potential was positively and significantly correlated with their betalain composition [17].

Betanin is the most common of all betalains in the plant kingdom and, therefore, the most widely studied. A review of its properties described betanin as a potential scavenger of reactive oxygen species. Betanin may also induce phase II enzymes and endogenous antioxidant defense mechanisms via gene regulatory systems. Therefore, betalains may be related to preventing low-density lipoproteins (LDL) oxidation and DNA damage [33].

It is important to highlight that several studies related to the benefits of betalains on health have been conducted in vitro in cultured cells, where the betalain concentrations applied did not reflect those commonly acquired through the diet [33]. Therefore, betalain bioavailability needs to be considered for extrapolations to the organism level.

5 Betalain Analysis

5.1 Overview

For betalain analysis, the first step involves the extraction of the pigments, trying to avoid the presence of impurities that might interfere with the following determination. The extraction method depends on the plant matrix to be analyzed. Fruit juices may be used directly without any extraction processes. Other plant materials, such as inflorescences, stems, or petioles, require extraction processes using polar solvents, such as aqueous methanol or ethanol solutions. The plant material can be frozen in liquid nitrogen and pulverized. After adding the solvent, the extraction may take around 30 min. Ascorbic acetate may be added to reduce oxidation processes during the analysis. After extraction time, the samples are filtered or centrifuged. The extraction may be repeated several times, depending on the material under study. The plant material to solvent ratio depends on the betalain concentration.

This initial step is usually followed by the quantification of total betalains. After extraction, samples should be diluted with a citrate–phosphate buffer (pH 6.0). If the concentration of betalains is low, extraction solvent may be eliminated using rotary evaporation under vacuum followed by reconstitution with the buffer. The betalainic extracts or juices should be diluted so that the absorbance measurement can be between 0.9 and 1.1 AU at the respective absorption maxima. After estimating the total absorbance using UV-vis spectroscopy, total betalains are quantified using theoretical molar extinction coefficients, as discussed below.

The separation of individual betalains in each sample should be achieved using a C18 reverse phase column in an HPLC system coupled to MS. For compound identification, ion mass fragment patterns are used, which are enlisted afterward. Moreover, retention times and absorbance maxima are required. After peak identification, individual betalain concentrations may be calculated considering molar equivalences and percentual peak areas in relation to the total content of betalains.

5.2 Extraction and Isolation

Several protocols for betalain extraction have been published, with variations depending on the plant source. All of them aim to eliminate interfering agents and concentrate pigments prior to qualitative and quantitative analyses. This section describes extraction and isolation procedures with recent examples.

Extraction of betalains from *Alternanthera sessilis* (red) leaves and stems has been performed using freeze-dried and ground samples sieved through a 1 mm mesh. Samples were subsequently extracted with ethyl acetate (ratio 1:20 sample to solvent) for 24 h at 30 °C in an incubator shaker and then centrifuged. The resulting supernatant was filtered, and the solvent was removed by rotary evaporation at 40 °C. For further analysis, the crude extract was redissolved in dimethyl sulfoxide and stored at −20 °C [8].

Betalains from *Alternanthera brasiliana* and *Alternanthera tenella* were obtained from leaves and stems dried in an air oven for 5 days at 35 °C. Samples were then extracted with acetonitrile: water acidified with 0.1% formic acid (1:1 v/v) by vortexing for 60 s, centrifuged, and filtered through 0.45 μm membranes to conduct the analysis [9] directly.

Extraction of betalains from orange, red, and purple inflorescences of *Gomphrena globosa* and yellow, orange, red, and purple-colored *Bougainvillea* sp. was done using cryogenic milled samples to preclude enzymatic degradation during conventional grinding. Samples of *Gomphrena globosa* were extracted with an aqueous solution of 60% methanol with 50 mM sodium ascorbate under 30 min of continuous stirring. In contrast, for *Bougainvillea* the betalains were extracted for 15 min, stirring in an aqueous solution with 50 mM sodium ascorbate. Plant residues were removed by filtering, and the resulting extracts were concentrated under reduced pressure at 30 °C. Extracts were redissolved in purified water, filtered through a membrane filter (0.45 μm), and stored at −80 °C [12].

Analysis of betalains from stem bark of *Amaranthus spinosus* and *Boerhavia erecta* was performed using oven-dried (30 °C, 36 h) and ground samples. They were extracted with purified water, filtered, and vacuum-concentrated for direct betalain quantitation by UV-vis. For HPLC analysis, the extracts were adjusted to pH 1.5 with trifluoroacetic acid, and betalains were separated from lipophilic compounds after three times fractionation with ethyl acetate. Colored aqueous fractions containing betalains were adjusted to pH 6 with aqueous ammonia to ensure betalain stability, vacuum concentrated, and analyzed [7].

Betalains were extracted from grains of black, yellow-orange, and red-violet *Chenopodium quinoa* varieties with an aqueous solution of sodium acetate and sodium ascorbate (ratio 1/10 sample to solvent). Extracts were directly used to quantify betalains by UV-vis and for identification by HPLC-DAD-MS/MS [10].

For the extraction of betalain from differently pigmented inflorescences of *Celosia argentea* var. *cristata* and var. *plumosa*, samples were frozen in liquid nitrogen, ground with mortar, and subsequently extracted with 80% aqueous methanol containing ascorbate. The samples were then centrifuged, and the supernatant was directly used for betalain quantitation by UV-vis and identification by HPLC-DAD-MS/MS [11].

For flowers of four *Portulaca grandiflora* genotypes, violet, red, pale-yellow, and deep-yellow, samples were frozen in liquid nitrogen and ground to a fine powder, and betalains extracted with a mixture of methanol and aqueous solution acidified with HCl. Extraction was done twice, and supernatants were recovered by centrifugation, pooled together, and vacuum dried. Before UV-vis quantitation and HPLC-DAD quantitation, extracts were resuspended in water [13].

Novel techniques are used nowadays to fractionate and isolate higher quantities of betalains, such as ion-pair high-speed countercurrent chromatography (HSCCC). This technique was used to extract betalains from dark violet *Phytolacca americana* berries. For that purpose, the fruits were blended in a mixture of acidified water with trifluoroacetic acid containing acetonitrile and ascorbic acid, and the extract was directly frozen and then freeze-dried. For further clean-up, the extract was dissolved in acidified water with trifluoroacetic acid, passed through a C18 solid-phase extraction column, and subsequently rinsed with an aqueous solution of trifluoroacetic acid. Betalains were then eluted using a solvent mixture of acetonitrile:acidified water (with trifluoroacetic acid) 4:96 (v/v). Betalain extract was deep-frozen, lyophilized, and ground to a fine powder. Subsequently, betalains were fractionated by HSCCC, applying a biphasic solvent system of 1-butanol-acetonitrile-water with trifluoroacetic acid in a proportion of 5:1:6 (v/v/v). The resulting fractions were dried with nitrogen to remove the solvent and were analyzed by HPLC-DAD-MS/MS [14].

For betalain extraction, purple-ripened fruits of *Basella alba* and *Basella rubra* were manually squeezed to remove seeds, ground with a mortar and pestle in distilled water, and centrifuged. The resulting supernatant was further spectrophotometrically analyzed for betalain quantitation and HPLC-UV-MS/MS identification [15, 16].

Several options have been tested for the extraction of betalains from *Beta vulgaris* according to the plant material (including sub-species). In yellow beetroot (*Beta vulgaris* L. ssp. *vulgaris*), these pigments were extracted from the flesh of peeled beets. Samples were macerated in a blender with an aqueous solution of 50% methanol with 50 mM sodium ascorbate. Plant residues were removed by filtering; the filter was rinsed twice with the extraction solution and once with methanol, and the resulting extracts were concentrated under reduced pressure at 30 °C. Extracts were redissolved in purified water, flushed with nitrogen, and frozen stored at −80 °C until further HPLC-DAD-MS/MS analysis [20].

Betalains from differently colored purple, red-purple, yellow-orange, and yellow Swiss chard (*Beta vulgaris* L. ssp. *cicla*) were extracted from the petioles. Samples were ground to powder using liquid nitrogen in a Waring blender and further extracted by adding an aqueous solution of 60% methanol with 50 mM sodium ascorbate followed by 10 min of continuous shaking under ice cooling. Plant residues were removed by filtering. The filter was rinsed with methanol, and the resulting extracts were concentrated under reduced pressure at 30 °C. Extracts were resuspended in purified water, filtered through a membrane filter (0.45 μm), and stored frozen at −26 °C until UV-vis quantitation and HPLC-DAD-MS/MS identification [19].

In the case of red beetroot (*Beta vulgaris* L. ssp. *vulgaris*), the same protocol described above was used for the extraction of their betalains, with the inclusion of an additional extraction step with 60% methanol with 50 mM sodium ascorbate for 20 min. This protocol was also optionally used to extract betalains from yellow beetroot [18]. In another protocol for the extraction of betalains from red beetroot, samples were lyophilized, pulverized, and extracted five times using a mixture of water/methanol/formic acid (84.95/15/0.05v/v/v) by consecutive agitation with vortex and sonication. Supernatants were collected and centrifuged, and the upper phase was directly used for the analysis with HPLC-DAD-MS/MS [17].

For extraction of betalains from *Opuntia* sp. pulp, fruits were manually peeled, seeds removed, the pulp squeezed, and the obtained juice filtered, flushed with nitrogen, and directly stored at −80 °C [20, 23]; or the filtered juice was mixed with an aqueous solution of 60% methanol with 50 mM sodium ascorbate and concentrated under reduced pressure at 30 °C, and the extract was redissolved in purified water, flushed with nitrogen, filtered through membrane filter (0.45 μm), and stored frozen at −80 °C until further UV-vis and HPLC-DAD-MS/MS analysis [18]. Another published protocol used freeze-dried powdered samples of *Opuntia* sp. The pulp was subsequently stirred with water for 10 min and centrifuged, and the supernatant was filtered through 0.45 μm before spectrophotometric and HPLC-DAD-MS analyses [22].

Red and orange flesh of *Stenocereus pruinosus* and red flesh of *Stenocereus stellatus* pitaya fruits endemic to Mexico were homogenized and freeze-dried before betalain extraction. Samples were mixed with methanol and 1% trifluoroacetic acid in water in a mixture of 80:20 (v/v), agitated in an ultrasonic bath for 30 min, incubated at −20 °C for 3 h, centrifuged, and the supernatant recovered. Extractions were repeated two consecutive times, and the pooled supernatants were concentrated under reduced pressure. Extracts were redissolved in purified water and frozen until further HPLC-DAD-MS/MS analysis [24].

The betalain profile in *Hylocereus*, another genus of pitaya, known as dragon fruit endemic to Mesoamerica, has been addressed in commercial fruits and genotypes cultivated in Costa Rica [25, 26]. For that purpose, ripe fruits were manually peeled, and pulp was homogenized in a food processor. Homogenized pulp was filtered to obtain a clarified juice and directly used for betalain quantitation by UV-vis or membrane filtered (0.45 μm) for betalain identification by HPLC-UV-MS/MS [25].

5.3 Identification and Structural Elucidation

5.3.1 UV-Vis Spectroscopy for Betalain Analysis

Total betalain content can be analyzed using UV-vis spectroscopy by measuring the absorbance at specific wavelengths according to the betalain classification (viz., betacyanins and betaxanthins). Quantitation using this method is particularly useful when complex betalain profiles are present in the samples and reference standards are not available for individual compounds.

With this method, general reference compounds, such as betanin to report betacyanins and betaxanthin for betaxanthins, are commonly used. In consequence, even if the maximum absorbance of each betalain in a sample is unknown, this approach allows one to obtain quantitative equivalents of the other betalains by applying the mean of the molar extinction coefficient and the molecular weight of the target betalains [12].

The calculation of betalain contents by UV-vis can be performed using the equation:

$$\mathrm{BC[mg/L]} = \frac{A \times \mathrm{MW} \times \mathrm{DF} \times 1000}{\varepsilon \times L}$$

where A is the absorbance at the specific wavelength; MW is the molecular weight of the betalains; DF is the dilution factor applied to reach an absorbance between 0.9 and 1.1; ε is the molar extinction coefficient of the betalains; and L is the path length (usually 1 cm) of the cuvette [8, 12].

For example, for the analysis of betalains in *Alternanthera sessilis*, as well as in *Portulaca grandiflora*, sample extracts were measured at 536 nm for betacyanin content (betanin and amaranthin) and at 480 nm for betaxanthin content [8, 13]. In the case of *Amaranthus spinosus*, the absorbance of amaranthin at 538 nm was used to report total betalain content, whereas for *Boerhavia erecta*, the absorbance of betanin and neobetanin at 538 nm and 476 nm, respectively, was employed. The decision about which compounds should be used as standards was based on the betalain pattern measured by HPLC, which defined the most abundant betalains [7].

Table 1
Molecular weight and molar extinction coefficient in water of some betalains [7, 8, 10–13, 15, 16, 18]

Betalains		Molecular weight (g/mol)	Molar extinction coefficient in water (L/mol × cm)
Betacyanins	Betanin	550	60,000
	Neobetanin	548	18,200
	Amaranthin	727	56,600
	Betanidin	388	54,000
Betaxanthins	Betaxanthin	308	48,000
	Histidine-bx	348	48,000
	Dopa-bx	390	48,000
	Proline-bx	308	50,000
	Glycine-bx	268	38,900
	Glutamine-bx	339	48,000

For varieties of *Chenopodium quinoa*, total betalain quantitation was done based on the absorbance of betaxanthin at 480 nm and on that of betanidin, betanin, and amaranthin at 536 nm [10]. For inflorescences of *Celosia argentea* var. *cristata* and var. *plumosa*, 540 nm was used for betacyanins (with amaranthin as a reference compound) and 475 nm for betaxanthins quantitation [11]. In the case of *Basella rubra*, fruit extracts were measured at 535 nm and 477 nm, for betacyanins (betanin) and betaxanthins, respectively [15].

Beta vulgaris, *Opuntia* spp., and *Hylocereus* spp. have their total betacyanins quantitated with this method based on the absorbance of betanin at 538 nm and of betaxanthins at 480 nm using glutamine-bx as reference for *Beta vulgaris* and proline-bx for *Opuntia* spp. and *Hylocereus* spp. [19, 23, 34].

Table 1 summarizes the molecular weight and molar extinction coefficient of the most common betalains used for quantitation, while the total content of betaxanthins and betacyanins reported in some plant structures by means of UV/vis spectroscopy is summarized in Table 2.

5.3.2 High-Performance Liquid Chromatography (HPLC) Coupled to Mass Spectrometry (MS) for Betalain Analysis

Reversed-phase HPLC coupled to MS fitted with electrospray ionization (ESI) source operating in positive mode is one of the most used techniques to identify betalains. This detection technique pairs with UV-vis detection and retention times of reference compounds to confirm the identification of the individual betalains. Their further quantitation is done using calibration curves of authentic standards, and for compounds without authentic standards available, their content is calculated using the calibration curve of related structural compounds.

Table 2
Content of betaxanthins and betacyanins in some plant structures as reported by UV-vis spectroscopy

Species	Plant structure	Structure color	Extraction solvent	Reference compound(s)	Content (mg/g) dry weight		Reference
					Betaxanthins	Betacyanins	
Alternanthera sessilis	Leaf extract	Red	Ethyl acetate	Betanin	67.08	53.92	[8]
	Stem extract				2.57	3.66	
Gomphrena globosa L.[a]	Inflorescences	Orange	60% aqueous ethanol	Betaxanthin/betanin	0.0454	0.0306	[12]
		Red			0.0750	0.0937	
		Purple			tr	0.5568	
Bougainvillea sp.	Inflorescences	Yellow	Distilled water	Betaxanthin/betanin	3.9218	0.4080	[12]
		Orange			3.9605	1.3277	
		Red			2.7305	2.7871	
		Purple			tr	2.5941	
Amaranthus spinosus L.	Stem	Red	Distilled water	Amaranthine	n.r	0.239	[7]
Boerhavia erecta L.	Stem	Red	Distilled water	Betanin	n.r	1.855	[7]
				Neobetanin	n.r	0.5621	
Chenopodium quinoa	Grains	Red-violet	Sodium acetate buffer	Betaxanthin/betanin	0.2749	0.0931	[10]
		Yellow-orange			0.0033	0.0083	
		Black			0.0128	0.0009	
Celosia argentea var. *cristata* L.[b]	Inflorescences	Yellow	80% aqueous methanol	Betaxanthin/amaranthine	76.5	3.5	[11]
		Orange			96.4	96.8	
		Red			12.2	373.6	
Celosia argentea var. *plumosa* L.[b]	Inflorescences	Yellow	80% aqueous methanol	Betaxanthin/amaranthine	436.9	tr	[11]
		Orange-red			402.2	208.3	
Portulaca grandiflora	Flower buds	Violet	80% aqueous methanol	Indicaxanthin/betanin	0.59	17.8	[13]
		Red			5.00	12.5	
		Pale yellow			1.98	n.r	
		Deep yellow			9.36	n.r	

(continued)

Table 2 (continued)

Species	Plant structure	Structure color	Extraction solvent	Reference compound(s)	Content (mg/g) dry weight		Reference
					Betaxanthins	Betacyanins	
Basella rubra L.[a]	Fruits	Purple	60% aqueous methanol	Betaxanthin/betanin	0.1916	1.2418	[15]
Beta vulgaris L. ssp. *cicla*[a]	Petioles	Purple Red-purple Yellow-orange Yellow	60% aqueous methanol	Betanin/vulgaxanthin I	0.0243 0.0249 0.0220 0.0497	0.0511 0.0257 0.0116 n.r	[19]
Opuntia sp.[c]	Fruits	Green Orange Red Purple	Distilled water	Betanin/indicaxanthin	0.4 76.3 67.9 195.8	0.1 6.6 120.0 431.0	[22]
Stenocereus pruinosus	Fruits	Red Orange	Methanol/TFA/water (79:1:20)	Betanin/indicaxanthin	17.707 22.053	5.423 0.162	[24]
Stenocereus stellatus	Fruits	Red	Methanol/TFA/water (79:1:20)	Betanin/indicaxanthin	21.681	2.526	[24]
Hylocereus sp. from Costa Rica[c]	Fruits	Red-purple	Distilled water	Betanin	n.r	474.1–717.1[c]	[25]

[a]Content expressed in fresh weight
[b]Content expressed in nmol/g of fresh weight
[c]Content expressed in mg/L juice
n.r not reported, *tr* traces

For chromatographic separation, the stationary phase commonly used is a silica-packed column, whose hydroxyl groups are derivatized with octyl (C8) or octadecyl (C18) silyl groups. Mobile phase is usually a combination of acidic aqueous solutions, which ensures a low pH value, mixed with an organic solvent such as acetonitrile or methanol. Due to polarity, betaxanthins tend to elute prior to betacyanins. Stereochemistry and polarity are the main factors behind the separation of betalains. Their complete separation is usually accomplished via gradient elution [3].

For example, the chromatographic separation of betalains from *Alternanthera brasiliana* and *Alternanthera tenella* was carried out in an Agilent Zorbax Eclipse XDB-C18 column, eluted with a mobile phase composed of acidified water and acidified acetonitrile both with 0.1% formic acid using gradient flow. Mass spectra were acquired with a hybrid mass spectrometer quadrupole-time of flight using an ESI source in positive mode over the 10–1000 *m/z* [9] range.

Betaxanthin and betacyanin compounds of different colored extracts from *Gomphrena globosa* and *Bougainvillea* sp. were analyzed by HPLC-DAD-MS. Betalains were separated with an Atlantic dC18 column (5 μm) and aqueous formic acid and acetonitrile/water (80/20) mixture as eluents using gradient flow. With the DAD detector, betaxanthins were monitored at 470 nm and betacyanins at 535 nm, and their mass spectra were obtained with an ion trap mass spectrometer fitted with an ESI source operating in positive mode [12].

Betalain patterns from *Amaranthus spinosus* and *Boerhavia erecta* were also determined by HPLC-DAD-MS. Separation of all betalains was achieved using a Luna C18 column with a particle size of 5 μm (Phenomenex) eluted with an aqueous formic acid and acetonitrile/water (80/20) mixture using gradient flow. Betaxanthins and betacyanins were monitored at 476 nm and 538 nm, respectively. Mass spectra of betalain acquisition were performed with an ion trap mass spectrometer and ESI source in positive mode with a scanning range of 50–1000 *m/z* [7].

For *Chenopodium quinoa*, betaxanthins and betacyanins were separated using a Kinetex C-18 column (Phenomenex), operated at 25 °C, and a mixture of acidic water (A) and acidic acetonitrile (B) with 0.05% trifluoroacetic acid eluted on a linear gradient from 0% B to 50% B in 35 min. DAD followed betacyanins at 536 nm and betaxanthins at 480 nm. An ion trap spectrometer and ESI source operated in positive mode were used to acquire the mass spectra using a scanning range of 50–1200 *m/z* [10].

Betalains from *Celosia argentea* var. *cristata* and var. *plumosa* were separated with HPLC-DAD analysis by using a 5 μM Nucleosil C18 column and a mixture of aqueous phosphoric acid with acetonitrile eluted in gradient flow, with compound monitoring between 400 and 650 nm. For HPLC-MS, betalains were separated in minor particle size C18 column (3 μM), using a mixture of water and acetonitrile acidified with acetic acid as eluent, and mass data for the compounds were measured using a time-of-flight mass spectrometer with ESI source in positive mode [11].

In the case of flowers of *Portulaca grandiflora,* chromatographic separation was done with a 3 μm Hypersil ODS C18 column and a mixture of aqueous acetic acid with acetonitrile eluted in gradient flow. Betalains were identified with a DAD detector between 280 and 536 nm [13, 35]. Betalains isolated from *Phytolacca americana* were separated in an Onix C18 column and eluted in a gradient flow with a mixture of aqueous 4% acetic acid and acetonitrile. Monitoring of betaxanthins and betacyanins was done between 310 nm and 538 nm. Mass spectra of betalains acquisition were performed with a mass spectrometer operating with an ESI source in positive mode [14].

To analyze betalains from fruits of *Basella* sp., sample extracts were injected in a 5 μm Hypersil ODS C18 column and run using a solvent system containing aqueous trifluoroacetic acid and methanol. Separation of betalains was achieved with a gradient flow, and pigments were detected by absorbance at 280 nm and 530 nm [15, 16]. Mass spectra for betalains were obtained with a mass spectrometer with ESI source and positive ion polarity [16].

Betalains from *Beta vulgaris* (red beetroot, yellow beetroot, Swiss chard) were separated using a 5 μm C18-ODS column operated at 30 °C, and a mixture of acidic water with 1% formic acid and aqueous acetonitrile (80/20, v/v) eluted with two different gradient flows to separate betacyanins and betaxanthins [18, 19]. Another HPLC-DAD-MS/MS protocol for the analysis of betalains from red beetroot used a 3.5 μm XBridge C18 column (Waters) operated at 45 °C and a solvent gradient system consisting of 0.012% formic acid aqueous solution with 5 mM ammonia and 5% acidic water (0.012% formic acid) in acetonitrile with 5 mM ammonia [17]. DAD followed betacyanins at 538 nm and betaxanthins at 470 nm. An ion trap spectrometer [18, 19] or a MS Triple TOF (triple quadrupole coupled with time of flight) [17] and ESI source operated in positive mode were used for the acquisition of the mass spectra [17–19].

Separation of betalains from *Opuntia* sp. was achieved on an Atlantis 5 μm C18 column at 25 °C eluted with aqueous 0.2% formic acid and acetonitrile mixture using gradient flow [23]. Another chromatographic separation was performed using a Symmetry 3.5 μm C18 column and a mixture of methanol and water as elution solvents for a gradient flow [22]. Simultaneous monitoring of betaxanthins and betacyanins was done between 476 nm and 538 nm [22, 23]. Mass spectra for betalain acquisition were performed with an ion trap mass spectrometer [23] or a time of flight (TOF) mass spectrometer [22], operating with an ESI source in positive mode [22, 23].

In the case of fruits from *Stenocereus* sp., chromatographic separation was done with a 5 μm Phenomenex C18 column operated at 35 °C and a mixture of 0.1% trifluoroacetic acid in water with acetonitrile eluted in gradient flow. DAD followed betacyanins at 530 nm and betaxanthins at 480 nm. An ion trap-triple quadrupole spectrometer and ESI source operated in positive mode were used for the acquisition of the mass spectra using a scanning range of 100–1000 *m/z* [24].

For *Hylocereus* sp., betalain separation was achieved using a 5 μm Sunfire C18 column operated at 30 °C and eluted with 0.2% formic acid in water mixed with methanol:acetonitrile in a proportion of 80:20 (v/v) using gradient flow. Pigments were detected by absorbance at 470 nm and 536 nm. An ion trap spectrometer and ESI source operated in positive mode were used for the acquisition of the mass spectra [25].

The maximum wavelength absorbance and mass spectrometric data of the betaxanthins and betacyanins reported in several plant structures are summarized in Tables 3 and 4, respectively.

For all betalains, the mass-to-charge value of the parent ions was as expected for the protonated molecular ions of each compound. For betaxanthins, a reduction of 44 *m/z* between the parent ions and the partition ions (MS^2) was observed in all betaxanthin-type compounds, corresponding to the loss of a carboxylic acid group [10, 11]. Another typical change is that of 210 *m/z* due to the loss of an amine group. In addition, a fragment of 211 *m/z*, which belongs to the aminated betalamic acid, is usually displayed in betaxanthin fragmentation [11].

In betacyanins, the partition ions of this family of betalains were commonly formed by the loss of a glucuronic acid moiety and a glucose moiety with *m/z* of 176 and 162, respectively [10, 12]. In some cases, the loss of a diglucose moiety was detected with a *m/z* of 324 [16]. For example, amaranthin, with a molecular ion of 727 *m/z*, is fragmented into 551 *m/z* by the loss of the glucuronic acid and into 389 *m/z* due to an additional loss of glucose. The first fragment of 551 *m/z* matches the structure of betanin, and the further fragment of 389 *m/z* belongs to the molecular ion mass of betanidin [10]. For gomphrenin I, a molecular ion of 551 *m/z* with a fragment of 389 *m/z* reflects the loss of a glucose moiety, whereas for bougainvillein a molecular ion of 713 *m/z* and two product ion fragments, at 551 *m/z* and 389 *m/z*, indicates the loss of one glucose and two glucose moiety, respectively [16].

Table 3
Maximum wavelength absorbance and mass spectrometric data of betaxanthins reported in several plant species

Betaxanthin(s)	λ_{max} (nm)	m/z $[M + H]^+$	MS^2 m/z	Plant species [Reference]
Histidine-bx	478, 472	349	305	*Gomphrena globosa* L. [12], *Bougainvillea* sp. [12], *Beta vulgaris* L. ssp. *cicla* [18, 19], *Beta vulgaris* L. ssp. *vulgaris* [18], *Opuntia* sp. [18, 22, 23]
Arginine-bx	469	368	324	*Gomphrena globosa* L. [12]
Glutamine-bx	470	340	296	*Gomphrena globosa* L. [12], *Bougainvillea* sp. [12], *Portulaca grandiflora* [13], *Beta vulgaris* L. ssp. *cicla* [18, 19], *Beta vulgaris* L. ssp. *vulgaris* [17, 18, 20], *Opuntia* sp. [18, 20, 22, 23]
Glutamic acid-bx	469	341	297	*Beta vulgaris* L. ssp. *cicla* [18, 19], *Beta vulgaris* L. ssp. *vulgaris* [18], *Opuntia* sp. [18, 22]
Lysine-bx	458	340	296	*Gomphrena globosa* L. [12], *Bougainvillea* sp. [12], *Beta vulgaris* L. ssp. *cicla* [18], *Beta vulgaris* L. ssp. *vulgaris* [18]
Isoleucine-bx, isoleucine-isobx	470	325	281	*Gomphrena globosa* L. [12], *Phytolacca americana* [14], *Beta vulgaris* L. ssp. *cicla* [18, 19], *Beta vulgaris* L. ssp. *vulgaris* [18, 20], *Opuntia* sp. [18, 20, 22]
Leucine-bx, leucine-isobx	469	325	281	*Bougainvillea* sp. [12], *Phytolacca americana* [14], *Beta vulgaris* L. ssp. *cicla* [18, 19], *Beta vulgaris* L. ssp. *vulgaris* [17, 18, 20], *Opuntia* sp. [18, 20, 22]
Tryptophan-bx	473	398	354	*Gomphrena globosa* L. [12], *Bougainvillea* sp. [12], *Celosia argentea* [11], *Beta vulgaris* L. ssp. *cicla* [18, 19], *Beta vulgaris* L. ssp. *vulgaris* [17, 18, 20], *Opuntia* sp. [18, 22]
Putrescine-bx	461	282	238	*Bougainvillea* sp. [12]
Proline-bx, proline-isobx	479	309	265, 221, 188	*Bougainvillea* sp. [12], *Phytolacca americana* [14], *Chenopodium quinoa* [10], *Beta vulgaris* L. ssp. *cicla* [18, 19], *Beta vulgaris* L. ssp. *vulgaris* [18, 20], *Opuntia* sp. [18, 20–23], *Stenocereus* sp. [24], *Hylocereus* sp. [25]
Dopa-bx	472	391	347	*Bougainvillea* sp. [12], *Chenopodium quinoa* [10], *Portulaca grandiflora* [13], *Beta vulgaris* L. ssp. *cicla* [18], *Beta vulgaris* L. ssp. *vulgaris* [18]
	471	375	331	

(continued)

Table 3 (continued)

Betaxanthin(s)	λ_{max} (nm)	m/z $[M + H]^+$	MS^2 m/z	Plant species [Reference]
Tyrosine-bx, tyrosine-isobx				*Bougainvillea* sp. [12], *Portulaca grandiflora* [13], *Phytolacca americana* [14], *Beta vulgaris* L. ssp. *cicla* [18, 19], *Beta vulgaris* L. ssp. *vulgaris* [18], *Opuntia* sp. [18]
3-Methoxytyramine-bx	457, 462	361	317, 211, 151	*Bougainvillea* sp. [12], *Alternanthera brasiliana* [9], *Alternanthera tenella* [9], *Celosia argentea* [11], *Beta vulgaris* L. ssp. *cicla* [18, 19], *Beta vulgaris* L. ssp. *vulgaris* [17, 18]
Phenylalanine-bx, phenylalanine-isobx	472, 467	359	315	*Bougainvillea* sp. [12], *Phytolacca americana* [14], *Beta vulgaris* L. ssp. *cicla* [18, 19], *Beta vulgaris* L. ssp. *vulgaris* [17, 18, 20], *Opuntia* sp. [18, 20, 22]
Dopamine-bx	457, 462	347	303, 211, 137.	*Alternanthera brasiliana* [9], *Alternanthera tenella* [9], *Chenopodium quinoa* [10], *Celosia argentea* [11], *Portulaca grandiflora* [13], *Beta vulgaris* L. ssp. *cicla* [18, 19], *Beta vulgaris* L. ssp. *vulgaris* [18, 20]
Glycine-bx	468	269	225	*Portulaca grandiflora* [13], *Beta vulgaris* L. ssp. *cicla* [18, 19], *Beta vulgaris* L. ssp. *vulgaris* [18], *Opuntia* sp. [22]
Asparagine-bx	470	326	282	*Beta vulgaris* L. ssp. *cicla* [18, 19], *Beta vulgaris* L. ssp. *vulgaris* [17, 18], *Opuntia* sp. [22]
Serine-bx	468	299	255	*Beta vulgaris* L. ssp. *cicla* [18, 19], *Beta vulgaris* L. ssp. *vulgaris* [18, 20], *Opuntia* sp. [18, 20, 22]
Histamine-bx	468	305	261	*Beta vulgaris* L. ssp. *cicla* [18, 19]
Alanine-bx	468	283	239	*Beta vulgaris* L. ssp. *cicla* [18, 19], *Beta vulgaris* L. ssp. *vulgaris* [18]
γ-aminobutyric acid-bx	459	297	253	*Beta vulgaris* L. ssp. *cicla* [18, 19], *Beta vulgaris* L. ssp. *vulgaris* [17, 18, 20], *Opuntia* sp. [18, 20, 22, 23]
Valine-bx, Valine-isobx	470	311	267	*Beta vulgaris* L. ssp. *cicla* [18, 19], *Beta vulgaris* L. ssp. *vulgaris* [17, 18, 20], *Opuntia* sp. [18, 20, 22]
Tyramine-bx	461	331	287	*Beta vulgaris* L. ssp. *cicla* [18, 19], *Beta vulgaris* L. ssp. *vulgaris* [17, 18]

(continued)

Table 3 (continued)

Betaxanthin(s)	λ_{max} (nm)	m/z $[M + H]^+$	MS^2 m/z	Plant species [Reference]
Aspartic acid-bx	469	327	283	*Beta vulgaris* L. ssp. *cicla* [18], *Beta vulgaris* L. ssp. *vulgaris* [17, 18]
Ethanolamine-bx	460	255	211	*Beta vulgaris* L. ssp. *cicla* [18], *Beta vulgaris* L. ssp. *vulgaris* [18]
Threonine-bx	469	313	269	*Beta vulgaris* L. ssp. *cicla* [18], *Beta vulgaris* L. ssp. *vulgaris* [17, 18]
Methionine-bx	470	343	299	*Beta vulgaris* L. ssp. *cicla* [18], *Beta vulgaris* L. ssp. *vulgaris* [18], *Opuntia* sp. [18, 22, 23]
Hydroxyproline-bx	483	325	281	*Opuntia* sp. [22]
Phenethylamine	475	315	271	*Opuntia* sp. [22]

5.3.3 NMR Spectroscopy for Betalain Analysis

NMR spectroscopy of proton and carbon 13 (^{1}H-NMR and ^{13}C NMR, respectively) are additional techniques for identification of individual betaxanthins and betacyanins. NMR is required to elucidate unambiguous betalain structures, such as isomers with the same molecular weight and possibly similar mass spectra measured by HPLC-MS/MS because NMR is capable of differentiating H-spin and C13-spin systems [3, 36].

^{1}H-NMR data on betalain pigments has been more widely documented than ^{13}C NMR data. Measurements have been done in deuterated methanol, dimethyl sulfoxide, and acetic acid. It is recommended to slightly acidify the betalain solutions to get stable zwitterions of betalains and, therefore, obtain a clearer spectrum; however, highly acidic solutions have also shown to be detrimental for betalain stability, due to hydrolysis of betalain bonds and possible exchange among deuterium from acidic solvent and proton from betalain [3], as well as rapid and undesirable C-15 and C-12/C-13 isomerizations [36].

A solvent system consisting of deuterated water alone or mixed with purified water and slightly acidified with 0.01–0.05% trifluoroacetic acid at 25 °C to assure a pH range of 5–7 has been reported as a valuable option that provides better stability to betalains and allows improved NMR data acquisition [36, 37].

To assure reliable results, measurements by NMR require isolation and purification of the compounds of interest by experimental protocols that include removal of undesirable impurities of the plant extract by solid-phase extraction, followed by preparative HPLC coupled to DAD to separate and isolate individual pigments using a C18 column and eluents such as aqueous 0.5% acetic acid

Table 4
Maximum wavelength absorbance and mass spectrometric data of betacyanins reported in several plant species

Betacyanin(s)	λ_{max} (nm)	$[M + H]^+$ m/z	MS^2 m/z	Plant species [Reference]
Amaranthin, isoamaranthin	535, 538, 541	727	551, 389	*Gomphrena globosa* L. [12], *Alternanthera brasiliana* [9], *Alternanthera tenella* [9], *Amaranthus spinosus* L. [7], *Chenopodium quinoa* [10], *Celosia argentea* [11]
17-Descarboxy-amaranthin	502	683	507, 345	*Gomphrena globosa* L. [12]
Betanin, isobetanin	529, 538	551	389	*Gomphrena globosa* L. [12], *Bougainvillea* sp. [12], *Alternanthera brasiliana* [9], *Alternanthera tenella* [9], *Amaranthus spinosus* L. [7], *Boerhavia erecta* L. [7], *Chenopodium quinoa* [10], *Portulaca grandiflora* [13], *Phytolacca americana* [14], *Beta vulgaris* L. ssp. *cicla* [19], *Beta vulgaris* L. ssp. *vulgaris* [17], *Opuntia* sp. [20–23], *Hylocereus* sp. [25, 26]
17-Descarboxy-betanin, 17-descarboxy-isobetanin	505	507	345	*Phytolacca americana* [14], *Beta vulgaris* L. ssp. *vulgaris* [17]
2-Descarboxy-betanin, 2-descarboxy-isobetanin	533	507	345	*Phytolacca americana* [14], *Beta vulgaris* L. ssp. *vulgaris* [17], *Stenocereus* sp. [24]
2,17-Bidescarboxy-betanin, 2,17-bidescarboxy-isobetanin	507	463	301	*Beta vulgaris* L. ssp. *vulgaris* [17]
2′-Apiosylbetanin, 2′-apiosylisobetanin	539	683	551, 389	*Phytolacca americana* [14], *Hylocereus* sp. [25]
6′-O-feruloyl-betanin, 6′-O-feruloyl-isobetanin	532	727	389	*Beta vulgaris* L. ssp. *vulgaris* [17]
6′-O-malonyl-2-descarboxy-betanin, 6′-O-malonyl-2-descarboxy-isobetanin	512, 516	593	549, 507, 345	*Stenocereus* sp. [24]
Neobetanin	473	549	387, 343	*Boerhavia erecta* L. [7], *Phytolacca americana* [14], *Beta vulgaris* L. ssp. *vulgaris* [17], *Opuntia* sp. [22, 23], *Hylocereus* sp. [25]

(continued)

Table 4 (continued)

Betacyanin(s)	λ_{max} (nm)	$[M + H]^+$ *m/z*	MS^2 *m/z*	Plant species [Reference]
17-Descarboxy-neobetanin	485	505	343, 297	*Beta vulgaris* L. ssp. *vulgaris* [17]
2-Descarboxy-neobetanin	485	505	343, 297	*Beta vulgaris* L. ssp. *vulgaris* [17]
2,17-Bidescarboxy-neobetanin	456	461	299	*Beta vulgaris* L. ssp. *vulgaris* [17]
Prebetanin, isoprebetanin	538	631	551, 389	*Beta vulgaris* L. ssp. *vulgaris* [17]
Gomphrenin I, isogomphrenin I	537	551	389, 321	*Gomphrena globosa* L. [12], *Bougainvillea* sp. [12], *Phytolacca americana* [14], *Basella* sp. [16], *Opuntia* sp. [22, 23], *Stenocereus* sp. [24], *Hylocereus* sp. [25]
Betanidin, isobetanidin	541	389	345, 150	*Gomphrena globosa* L.[12], *Bougainvillea* sp. [12], *Phytolacca americana* [14], *Beta vulgaris* L. ssp. *cicla* [19], *Beta vulgaris* L. ssp. *vulgaris* [17], *Opuntia* sp. [22, 23], *Hylocereus* sp. [25]
17-Descarboxy-betanidin, 17-descarboxy-isobetanidin	433	345	301	*Beta vulgaris* L. ssp. *vulgaris* [17]
15-Descarboxy-betanidin, 15-descarboxy-isobetanidin	433	345	301	*Beta vulgaris* L. ssp. *vulgaris* [17]
Celosianin II, isocelosianin II	540	903	551, 389	*Gomphrena globosa* L. [12]
Gomphrenin II, isogomphrenin II	544	697	551, 389	*Gomphrena globosa* L. [12]
Gomphrenin III, isogomphrenin III	544	727	551, 389	*Gomphrena globosa* L. [12]
Bougainvillein, isoboungainvillein	537	713	551, 389	*Bougainvillea* sp. [12], *Opuntia* sp. [22], *Hylocereus* sp. [25, 26]
Lampranthin II, isolampranthin II	533	727	551, 389	*Bougainvillea* sp. [12], *Beta vulgaris* L. ssp. *cicla* [19]
Phyllocactin, isophyllocactin	538	637	593, 551, 389	*Opuntia* sp. [23], *Stenocereus* sp. [24], *Hylocereus* sp. [25, 26]
Butyrylbetanin, isobutyrylbetanin	534	637	551, 389	*Hylocereus* sp. [25, 26]
Hylocerenin, isohylocerenin	532	695	551, 389	*Hylocereus* sp. [25, 26]

mixed with acetonitrile:water 50:50 (v/v). Collected fractions of each betalain need to be vacuum-concentrated at room temperature and directly lyophilized to remove residues of water and acid and, therefore, prevent artifact generation and degradation of the pigments before NMR [36, 37].

Complete ^{1}H-NMR spectra, as well as partial ^{13}C NMR data, have been produced for betanin, isobetanin, phyllocactin, and hylocerenin isolated from red-purple pitaya (*Hylocereus polyrhizus*) [36], and complete ^{13}C NMR spectra of proline-bx isolated from yellow-orange cactus pear fruits (*Opuntia ficus-indica*) and dopamine-bx extracted from yellow Swiss chard petioles (*Beta vulgaris* L. ssp. *cicla*) [37]. Measurements were achieved using 0.5–2 mg of freeze-dried betalains at 25 °C with 500 MHz for ^{1}H and 125 MHz for ^{13}C, with solvent suppression in the pulse sequences to prevent undesired isomerizations [36, 37].

6 Assessment of Bioavailability and Bioaccessibility of Betalains

In vivo studies with model animals, ex vivo bioaccessibility experiments for simulation of human absorption, and human studies to evaluate bioavailability are useful tools to determine the possible role of betalains on human health [2]. Betanin from *Beta vulgaris* was extracted and purified by semipreparative HPLC-MS/MS to evaluate its absorption and metabolism using a human digestion simulation and ex vivo colon fermentation. Results showed that 50% of the betanin was recovered in the small intestine fluid, whereas no traces were found after colon fermentation, and interestingly, the antioxidant ability of the betanin remained constant after simulated small intestine digestion [38].

A more accurate assessment of the mechanisms of absorption and metabolism of betalains has been conducted through animal and human studies. Since they have been previously reviewed [28], only the most representative examples will be mentioned here. In an animal study, Wistar rats received a dose of 50 mg betalains/kg body weight by means of a fermented *Beta vulgaris* juice directly infused into the stomach. One hour after administration, rats were killed, and blood samples were withdrawn and used to analyze betalains. Native betalains, betanin, isobetanin, betanidin, isobetanidin, neobetanin, 17-descarboxy-betanin, 17-descarboxy-isobetanin, and 2-descarboxy-neobetanin, as well as their metabolites, 15-descarboxy-betanin and 2,17-bidescarboxy-betanin, were detected in the rat plasma samples, where betanin and its isomer were found at the highest concentrations [39].

In the case of human studies, betalain bioavailability has been reported to be rather low. In a study where betalains from *Opuntia* sp. were supplemented, it was reported that betanin reached the maximum concentration within the range of 0.2 μmol/L in plasma 3 h after administration, whereas the proline-betaxanthin reached the highest concentration of around 7 μmol/L. There is evidence that betanin is mainly excreted via urine; however, at a concentration lower than 4% of the supplied dose. Therefore, it is postulated that betanin is being isomerized, and isobetanin represents the main metabolite in urine. Other betalains, such as betanidin or glucuronides, sulfates, or methylated betalains conjugates, have not been detected in plasma or urine [40].

7 Conclusions

Betalains are considered of importance for the food industry due to their chemical properties, stability, and associated health benefits. They are classified in betacyanins, mostly related to red and purple colors and betaxanthins, representing yellow hues. In nature, the combination of both types of betalains gives a broad spectrum of colors going from purple to yellow.

Due to their potential for coloring foodstuff and the variety of plant sources from which they can be obtained, several betalain-containing plants have been studied to assess profiles and total contents. Therefore, several extraction methods have been described. Compared to the requirements of other compounds, betalain extraction methods are relatively simple, differing basically on the solvents used, where methanol and ethanol solutions are the most commonly used, with the addition of ascorbate to reduce oxidation processes.

Total contents of betalain are mostly determined by detecting absorbance of the samples through UV-vis spectroscopy and quantitation using theoretical molar extinction coefficients, due to the limitations in the number of standards available. These absorbance values are mostly determined after dilution of the extracts or juices. Due to their differences in absorbance maxima, betacyanins and betaxanthins must be assessed separately, using the same quantitative calculation but with the corresponding molar extinction coefficient, depending on the reference compound of interest.

Betalain profiles and contents vary largely in different plant species and structures. Reversed-phase HPLC coupled to MS fitted with ESI source operating in positive mode is one of the most used techniques to identify and assess betalain profiles. Individual contents are commonly estimated considering the percentual peak area related to the total contents of betalains, estimated using quantitative equivalents and theoretical molar extinction coefficients, again generally due to the lack of available standards. The mobile phase corresponds usually to combinations of acidic aqueous solutions, which ensure a low pH value, mixed with organic solvents, such as acetonitrile or methanol. Stereochemistry and polarity are the main factors behind the separation of betalains, and their complete separation is usually accomplished via gradient elution. NMR spectroscopy of proton and carbon 13 (^{1}H-NMR and ^{13}C NMR, respectively) is an additional technique for unambiguous identification of individual betaxanthins and betacyanins.

References

1. Coy-Barrera E (2020) Chapter 17 – Analysis of betalains (betacyanins and betaxanthins). In: Sanches Silva A, Nabavi SF, Saeedi M, Nabavi SM (eds) Recent advances in natural products analysis. Elsevier, pp 593–619
2. Gandía-Herrero F, Escribano J, García-Carmona F (2016) Biological activities of plant pigments Betalains. Crit Rev Food Sci Nutr 56:937–945. https://doi.org/10.1080/10408398.2012.740103
3. Akbar Hussain E, Sadiq Z, Zia-Ul-Haq M (2018) Analysis of betalains. In: Akbar Hussain E, Sadiq Z, Zia-Ul-Haq M (eds) Betalains: biomolecular aspects. Springer, Cham, pp 139–164
4. Slimen IB, Najar T, Abderrabba M (2017) Chemical and antioxidant properties of betalains. J Agric Food Chem 65:675–689. https://doi.org/10.1021/acs.jafc.6b04208
5. Rodriguez-Amaya DB (2019) Update on natural food pigments – a mini-review on carotenoids, anthocyanins, and betalains. Food Res Int 124:200–205. https://doi.org/10.1016/j.foodres.2018.05.028
6. Miguel MG (2018) Betalains in some species of the Amaranthaceae family: a review. Antioxidants 7:53. https://doi.org/10.3390/antiox7040053
7. Stintzing FC, Kammerer D, Schieber A, Adama H, Nacoulma OG, Carle R (2004) Betacyanins and phenolic compounds from *Amaranthus spinosus* L. and *Boerhavia erecta* L. Z Für Naturforschung C 59:1–8. https://doi.org/10.1515/znc-2004-1-201
8. Hazli UHAM, Abdul-Aziz A, Mat-Junit S, Chee CF, Kong KW (2019) Solid-liquid extraction of bioactive compounds with antioxidant potential from *Alternanthera sesillis* (red) and identification of the polyphenols using UHPLC-QqQ-MS/MS. Food Res Int 115:241–250. https://doi.org/10.1016/j.foodres.2018.08.094
9. Deladino L, Alvarez I, De Ancos B, Sánchez-Moreno C, Molina-García AD, Schneider Teixeira A (2017) Betalains and phenolic compounds of leaves and stems of *Alternanthera brasiliana* and *Alternanthera tenella*. Food Res Int 97:240–249. https://doi.org/10.1016/j.foodres.2017.04.017
10. Escribano J, Cabanes J, Jiménez-Atiénzar M, Ibañez-Tremolada M, Gómez-Pando LR, García-Carmona F, Gandía-Herrero F (2017) Characterization of betalains, saponins and antioxidant power in differently colored quinoa (*Chenopodium quinoa*) varieties. Food Chem 234:285–294. https://doi.org/10.1016/j.foodchem.2017.04.187
11. Schliemann W, Cai Y, Degenkolb T, Schmidt J, Corke H (2001) Betalains of *Celosia argentea*. Phytochemistry 58:159–165. https://doi.org/10.1016/S0031-9422(01)00141-8
12. Kugler F, Stintzing FC, Carle R (2007) Characterisation of betalain patterns of differently coloured inflorescences from *Gomphrena globosa* L. and *Bougainvillea* sp. by HPLC–DAD–ESI–MSn. Anal Bioanal Chem 387: 637–648. https://doi.org/10.1007/s00216-006-0897-0
13. Trezzini GF, Zrÿd J-P (1991) Two betalains from *Portulaca grandiflora*. Phytochemistry 30:1897–1899. https://doi.org/10.1016/0031-9422(91)85035-X
14. Jerz G, Skotzki T, Fiege K, Winterhalter P, Wybraniec S (2008) Separation of betalains from berries of *Phytolacca americana* by ion-pair high-speed counter-current chromatography. J Chromatogr A 1190:63–73. https://doi.org/10.1016/j.chroma.2008.02.093
15. Kumar SS, Manoj P, Shetty NP, Prakash M, Giridhar P (2015) Characterization of major betalain pigments -gomphrenin, betanin and isobetanin from *Basella rubra* L. fruit and evaluation of efficacy as a natural colourant in product (ice cream) development. J Food Sci Technol 52:4994–5002. https://doi.org/10.1007/s13197-014-1527-z
16. Lin S-M, Lin B-H, Hsieh W-M, Ko H-J, Liu C-D, Chen L-G, Chiou RY-Y (2010) Structural identification and bioactivities of red-violet pigments present in *Basella alba* fruits. J Agric Food Chem 58:10364–10372. https://doi.org/10.1021/jf1017719
17. Sawicki T, Bączek N, Wiczkowski W (2016) Betalain profile, content and antioxidant

capacity of red beetroot dependent on the genotype and root part. J Funct Foods 27: 249–261. https://doi.org/10.1016/j.jff.2016.09.004

18. Kugler F, Graneis S, Stintzing FC, Carle R (2007) Studies on betaxanthin profiles of vegetables and fruits from the Chenopodiaceae and Cactaceae. Z Für Naturforschung C 62:311–318. https://doi.org/10.1515/znc-2007-5-601
19. Kugler F, Stintzing FC, Carle R (2004) Identification of betalains from petioles of differently colored Swiss chard (*Beta vulgaris* L. ssp. *cicla* [L.] Alef. Cv. Bright lights) by high-performance liquid chromatography–electrospray ionization mass spectrometry. J Agric Food Chem 52:2975–2981. https://doi.org/10.1021/jf035491w
20. Stintzing FC, Schieber A, Carle R (2002) Identification of betalains from yellow beet (*Beta vulgaris* L.) and cactus pear [*Opuntia ficus-indica* (L.) Mill.] by high-performance liquid chromatography–electrospray ionization mass spectrometry. J Agric Food Chem 50:2302–2307. https://doi.org/10.1021/jf011305f
21. Mena P, Tassotti M, Andreu L, Nuncio-Jáuregui N, Legua P, Del Rio D, Hernández F (2018) Phytochemical characterization of different prickly pear (*Opuntia ficus-indica* (L.) Mill.) cultivars and botanical parts: UHPLC-ESI-MSn metabolomics profiles and their chemometric analysis. Food Res Int 108: 301–308. https://doi.org/10.1016/j.foodres.2018.03.062
22. Castellanos-Santiago E, Yahia EM (2008) Identification and quantification of betalains from the fruits of 10 Mexican prickly pear cultivars by high-performance liquid chromatography and electrospray ionization mass spectrometry. J Agric Food Chem 56:5758–5764. https://doi.org/10.1021/jf800362t
23. Stintzing FC, Herbach KM, Mosshammer MR, Carle R, Yi W, Sellappan S, Akoh CC, Bunch R, Felker P (2005) Color, betalain pattern, and antioxidant properties of cactus pear (*Opuntia* spp.) clones. J Agric Food Chem 53:442–451. https://doi.org/10.1021/jf048751y
24. García-Cruz L, Dueñas M, Santos-Buelgas C, Valle-Guadarrama S, Salinas-Moreno Y (2017) Betalains and phenolic compounds profiling and antioxidant capacity of pitaya (*Stenocereus* spp.) fruit from two species (*S. pruinosus* and *S. stellatus*). Food Chem 234:111–118. https://doi.org/10.1016/j.foodchem.2017.04.174
25. Esquivel P, Stintzing FC, Carle R (2007) Pigment pattern and expression of colour in fruits from different *Hylocereus* sp. genotypes. Innov Food Sci Emerg Technol 8:451–457. https://doi.org/10.1016/j.ifset.2007.03.022
26. Stintzing FC, Schieber A, Carle R (2002) Betacyanins in fruits from red-purple pitaya, *Hylocereus polyrhizus* (Weber) Britton & Rose. Food Chem 77:101–106. https://doi.org/10.1016/S0308-8146(01)00374-0
27. Velíšek J, Cejpek K (2011) Pigments of higher fungi – a review. Czech J Food Sci 29:87–102. https://doi.org/10.17221/524/2010-CJFS
28. Rahimi P, Abedimanesh S, Mesbah-Namin SA, Ostadrahimi A (2019) Betalains, the nature-inspired pigments, in health and diseases. Crit Rev Food Sci Nutr 59:2949–2978. https://doi.org/10.1080/10408398.2018.1479830
29. Koss-Mikołajczyk I, Kusznierewicz B, Wiczkowski W, Sawicki T, Bartoszek A (2019) The comparison of betalain composition and chosen biological activities for differently pigmented prickly pear (*Opuntia ficus-indica*) and beetroot (*Beta vulgaris*) varieties. Int J Food Sci Nutr 70:442–452. https://doi.org/10.1080/09637486.2018.1529148
30. Esquivel P, Stintzing FC, Carle R (2007) Phenolic compound profiles and their corresponding antioxidant capacity of purple pitaya (*Hylocereus* sp.) genotypes. Z Für Naturforschung C 62:636–644. https://doi.org/10.1515/znc-2007-9-1003
31. Abderrahim F, Huanatico E, Segura R, Arribas S, Gonzalez MC, Condezo-Hoyos L (2015) Physical features, phenolic compounds, betalains and total antioxidant capacity of coloured quinoa seeds (*Chenopodium quinoa* Willd.) from Peruvian Altiplano. Food Chem 183:83–90. https://doi.org/10.1016/j.foodchem.2015.03.029
32. Albano C, Negro C, Tommasi N, Gerardi C, Mita G, Miceli A, De Bellis L, Blando F (2015) Betalains, phenols and antioxidant capacity in cactus pear [*Opuntia ficus-indica* (L.) Mill.] fruits from Apulia (South Italy) genotypes. Antioxidants 4:269–280. https://doi.org/10.3390/antiox4020269
33. Esatbeyoglu T, Wagner AE, Schini-Kerth VB, Rimbach G (2015) Betanin—a food colorant with biological activity. Mol Nutr Food Res 59: 36–47. https://doi.org/10.1002/mnfr.201400484
34. Stintzing FC, Schieber A, Carle R (2003) Evaluation of colour properties and chemical quality parameters of cactus juices. Eur Food Res Technol 216:303–311. https://doi.org/10.1007/s00217-002-0657-0
35. Trezzini GF, Zrÿb J-P (1991) Characterization of some natural and semi-synthetic betaxanthins. Phytochemistry 30:1901–1903.

https://doi.org/10.1016/0031-9422(91)85036-Y

36. Stintzing FC, Conrad J, Klaiber I, Beifuss U, Carle R (2004) Structural investigations on betacyanin pigments by LC NMR and 2D NMR spectroscopy. Phytochemistry 65:415–422. https://doi.org/10.1016/j.phytochem.2003.10.029
37. Stintzing FC, Kugler F, Carle R, Conrad J (2006) First ^{13}C-NMR assignments of betaxanthins. Helv Chim Acta 89:1008–1016. https://doi.org/10.1002/hlca.200690077
38. da Silva DVT, Baião DDS, de Oliveira Silva F, Alves G, Perrone D, Del Aguila EM, VMF P (2019) Betanin, a natural food additive: stability, bioavailability, antioxidant and preservative ability assessments. Molecules 24:458. https://doi.org/10.3390/molecules24030458
39. Sawicki T, Juśkiewicz J, Wiczkowski W (2017) Using the SPE and micro-HPLC-MS/MS method for the analysis of betalains in rat plasma after red beet administration. Molecules 22:2137. https://doi.org/10.3390/molecules22122137
40. Frank T, Stintzing FC, Carle R, Bitsch I, Quaas D, Straß G, Bitsch R, Netzel M (2005) Urinary pharmacokinetics of betalains following consumption of red beet juice in healthy humans. Pharmacol Res 52:290–297. https://doi.org/10.1016/j.phrs.2005.04.005

Chapter 27

Analysis of Chlorophyll Pigments in Food Sources

María Roca and Antonio Pérez-Gálvez

Abstract

Chlorophylls are essential for life because of their functional role in light harvesting for photosynthesis and photochemistry. It is the most abundant pigment on Earth, widely distributed in the tissues of all photosynthetic organisms, from algae to higher plants. Chlorophylls undergo chemical and/or enzymatic reactions during natural metabolism, food processing, and storage, being crucial to the accurate determination of the chlorophyll profile. This chapter describes a detailed experimental procedure and approach to extract chlorophylls from foods and their analysis with HPLC-UV/vis or HPLC hyphenated with MS to characterize the chlorophyll profile. The protocol includes information to facilitate the chlorophyll identification and the isolation of standards.

Key words Chlorophylls, Chlorophyllide, Pheophorbide, Pheophytin, Extraction, Mass spectrometry, UV-vis, HPLC, APCI, ESI

1 Introduction

Chlorophylls are essential natural pigments for photosynthesis, a key process for life on Earth, harvesting efficiently light quanta and participating in the forward and backward electron transfer. They are ubiquitous in photosynthetic algae, bacteria, and higher plants, being therefore present in marine and terrestrial, traditional, and novel food sources in almost all geographical areas. Indeed, considering the current dietary recommendations from health agencies and policymakers [1], chlorophylls are significantly present in our daily diet, with an estimated intake of 200 mg/day [2]. The interest in the possible benefits of chlorophylls for human health has improved in the last decades through in vitro and in vivo studies [2–4] and considering several modes of action, including antioxidant [5], antimutagenic and antigenotoxic activities [6, 7], and anticarcinogenic [8]. These investigations run in parallel with the recent advances explaining the in vivo selective absorption and metabolism of chlorophylls [9, 10] since the first evidence [11]. From a chemical point of view, chlorophylls are cyclic

Lilian Mariutti and Frank Vriesekoop (eds.), *Food Chemistry: Basic Protocols*, Methods and Protocols in Food Science,
https://doi.org/10.1007/978-1-0716-4678-6_27,

Chlorophyll compound	Structure	R_1	R_2	R_3	R_4	13^2	15^1
Chlorophyllide *a*	A	Mg	CH_3	H	$COOCH_3$	-	-
13^2-hydroxy-pheophorbide *a*	A	2H	CH_3	H	$COOCH_3$	OH	-
Pheophorbide *a*	A	2H	CH_3	H	$COOCH_3$	-	-
15^1-hydroxy-lactone-chlorophyll *a*	B	Mg	CH_3	phytyl	$COOCH_3$	-	OH
Chlorophyll *b*	A	Mg	CHO	phytyl	$COOCH_3$	-	-
13^2-hydroxy-chlorophyll *a*	A	Mg	CH_3	phytyl	$COOCH_3$	OH	-
Chlorophyll *a*	A	Mg	CH_3	phytyl	$COOCH_3$	-	-
15^1-hydroxy-lactone-pheophytin *a*	B	2H	CH_3	phytyl	$COOCH_3$	-	OH
Pheophytin *b*	A	2H	CHO	phytyl	$COOCH_3$	-	-
13^2-hydroxy-pheophytin *a*	A	2H	CH_3	phytyl	$COOCH_3$	OH	-
Pheophytin *a*	A	2H	CH_3	phytyl	$COOCH_3$	-	-
Pyro-pheophytin *a*	A	2H	CH_3	phytyl	H	-	-

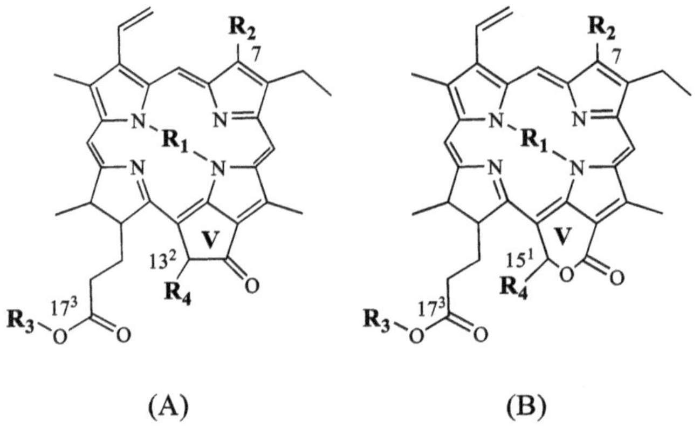

Fig. 1 Structures of common chlorophyll derivatives present in foods

tetrapyrrole derivatives with a system of coplanar conjugated double bonds that arrange an aromatic structure with delocalization of electron density in π-orbitals, carrying a characteristic isocyclic ring (ring V in Fig. 1), while the extended π-system is able to coordinate naturally with magnesium. Chlorophylls are grouped in families or series featured by a functional group or the arrangement of the unsaturation in the macrocycle. Thus, the methyl group at C7 is characteristic of the *a* series, while the *b* series arranges a formyl group in that position. The introduction of additional unsaturation (s) in the macrocycle produces the occurrence of chlorophyll c_1 and c_2 series, exclusively present in algae and phytoplankton [12]. Another feature of the structural arrangement of natural chlorophylls is the phytyl chain esterifying the propionic acid tail at $C17^3$ (*c* series is not esterified at this position), while chemical and enzymatic reactions in their natural physiological environment yield the corresponding chlorophyll derivatives of each series (pheophytins, chlorophyllides, pheophorbides, pyro-pheophytins...). Although there are naturally occurring more than 100 different chlorophyll structures, this protocol is focused on the most common chlorophylls present in food. Because of natural metabolism, or during food processing or storing, chlorophylls are prone to chemical changes. For example, the replacement of the central Mg atom by two hydrogen atoms originates pheophytins (from chlorophylls) and pheophorbides (from chlorophyllides). The de-phytylation at $C17^3$ forms chlorophyllides (from chlorophylls) and pheophorbides (from pheophytins). Other common reactions are the oxidations that take place at the isocyclic ring, as the introduction of one hydroxyl group at $C13^2$, producing 13^2-hydroxy-chlorophylls. Further oxidations generate a lactone group at $C15^1$, as in the case of 15^1-hydroxy-lactone chlorophylls. Heat or prolonged storage induces the loss of the carboxymethyl group at $C13^2$, yielding, for example, the pyro-pheophytins. Finally, epimerization at $C13^2$ is a common reaction of chlorophylls, being frequent their presence among the main pigments of the chlorophyll profile. Consequently, the application of analytical procedures to monitor changes in the chlorophyll profile enhances not only our knowledge of the distinctive chlorophylls featuring different food sources, the chemical and/or enzymatic processes that modulate the pigment content, and the appearance of chlorophyll derivates but also our perspective about the actual chlorophyll derivatives that are available for digestion/absorption, which might definitively develop biological actions and a beneficial impact in our health status.

2 Material

2.1 Equipment

1. Analytical balance.
2. Homogenizer type Ultra-Turrax® T 25.
3. Vortex mixer.
4. Refrigerated centrifuge with rotor for 15 mL and 1.5 mL centrifuge screw-capped tubes.
5. HPLC system interfaced with a photodiode array detector (PDA) with an absorption range between 300 and 700 nm. As an example, we use an Agilent 1260 Infinite chromatograph fitted with a 1290 Thermostat, 1260 DAD, 1260 quaternary pump, and controlled by HP ChemStation (Rev.A.05.04) software.
6. HPLC system hyphenated with a PDA detector in the 300–700 nm range, with auto-sample tray and a mass spectrometry instrument with high-resolution acquisition capability and ESI interface required for the analysis of polar

chlorophyll derivatives and APCI source for the analysis of phytylated chlorophyll derivatives. We use a Dionex Ultimate 3000RS U-HPLC (Thermo Fisher Scientific) with a micrO-TOF-QII™ high-resolution time-of-flight mass spectrometer (Bruker Daltonics) controlled by Bruker Compass HyStar software (Bruker Daltonics version 3.2).

7. Computer and software for MS data analysis; we use the Bruker Compass DataAnalysis (Bruker Daltonics version 4.2).
8. Rotary evaporator.
9. UV-vis spectrophotometer; we use a Shimadzu UV 1900i.

2.2 Solvents and Reagents

1. For chlorophyll extractions, analytical grade could be used, but the quality of the solvents should reach HPLC grade for HPLC determinations, while for MS analysis, the solvents must be LC-MS grade.
2. Acetone, methanol.
3. 1 M ammonium acetate, the solution should be filtered through a 0.22 μm nylon filter.

2.3 Chlorophyll Extraction

1. Centrifuge screw-capped tubes 15 mL.
2. 100 mL glass rotary evaporator flasks.
3. 1.5 mL microcentrifuge tubes.
4. 0.22 μm PTFE syringe filter.
5. HPLC vials and caps.

2.4 HPLC Analysis

1. Solvent A: mix ultrapure water (such as Milli-Q®), 1 M ammonium acetate, and methanol with a ratio of 1:1:8 (v/v/v).
2. Solvent B: mix methanol and acetone with a ratio of 1:1 (v/v).
3. Analytical reversed-phase HPLC column. This protocol used a 3 μm particle size C18 column (200 mm × 4.6 mm i.d., Teknokroma, Barcelona, Spain), protected with a guard column of the same packing material.
4. 0.22 μm nylon filter.
5. Volumetric flasks.
6. Pipettes and pipette tips.
7. 0.5/1.0 mL quartz cuvette.
8. Chlorophyll standards: each standard should be dissolved in the solvent where the specific absorption coefficient ($A^{1\%}_{1cm}$) has been determined. Table 1 shows some references and commercial suppliers where they can be purchased (*see* **Note 1**).

Table 1
Commercial suppliers and specific absorption coefficients for some chlorophyll standards

Chlorophyll	Commercial supplier	$A^{1\%}_{1cm}$	λ	Solvent	Reference
Chlorophyll *a*	Sigma, DHI, Wako, Parchem	127	428	Diethyl ether	[13]
Chlorophyll *b*	Sigma, DHI	173	453	Diethyl ether	[13]
Pheophorbide *a*	DHI, Sigma, Frontier Scientific	177	411	Tetrahydrofuran	[14]
Pheophytin *a*	DHI, Wako	139	411	Tetrahydrofuran	[15]

2.5 LC-MS Analysis

1. Calibration mix of synthetic polymer poly(ethylene glycol) (PEG 400, 10 μg/mL) and sodium formate (NaCOOH) (10 mM NaOH in 300 μL of formic acid), for APCI source and ESI interface, respectively, were used for calibration of the MS instrument.
2. Formic acid.

3 Methods

3.1 Chlorophyll Extraction

Among the different chlorophyll extraction protocols, the use of organic solvents is the most common method. However, the selection of solvent (or a mixture of solvents) depends mainly on the characteristics of the food matrix (*see* **Note 2**). Chlorophylls are relatively unstable, so special cautions should be taken when isolating them: green light illumination is suggested (or dim light as an alternative), working temperatures should be below 25 °C, and avoid prolonged contact with air or solvents (*see* **Notes 3** and **4**).

1. Weigh 1 g of fresh tissue (*see* **Note 5**) into a 15 mL screw-capped tube, add 10 mL of acetone (*see* **Note 6**), and break up the tissue with a homogenizer (*see* **Note 7**).
2. Spin the solution (16,000 ×*g* for 10 min), and transfer the supernatant into a 100 mL evaporator flask.
3. Re-dissolve the pellet with an additional 10 mL of acetone, vortex vigorously until the pellet has completely re-suspended, and repeat **step 2**.
4. Repeat the operations (*see* **steps 2** and **3**) until no more color is extracted from the remaining tissue (*see* **Note 8**).
5. Once all the acetone fractions are collected, evaporate them to dryness in a rotary evaporator at a temperature not higher than 30 °C (*see* **Note 9**).
6. Dissolve the residue in 100 μL acetone (*see* **Note 10**), and transfer to a 1.5 mL microcentrifuge tube.

7. Filter the extract through a 0.22 μm PTFE syringe filter (if the sample is analyzed through MS), or spin at 16,000 ×*g* for 10 min (for HPLC analysis), and transfer to an HPLC vial (*see* **Note 11**).

3.2 HPLC Chlorophyll Analysis

The selection of the analytical method depends obviously on the availability of the instrumental and the aim of the pigment determination. Sometimes, it is only necessary to know the total amount of chlorophyll *a* and *b*, and then rapid spectrophotometric methods should be applied (*see* **Note 12**). If it is required to determine the individual chlorophyll compounds of an extract, then a separation technique is required (as HPLC). If a more accurate determination is needed (or identification of unknown chlorophyll derivatives), considering that several chlorophylls exhibit the same absorption spectra, then it should be performed a mass spectrometry study (HPLC-MS).

1. Preparation of gradient solvents (*see* **Note 13**).
2. Turn on the HPLC and set the column temperature to 25 °C.
3. Collect the PAD signal from 300 to 700 nm, although you can display 430 and 666 nm to an online control.
4. Equilibrate the column, first with A solvent, then with B solvent, and finally with the initial solvent conditions (75% A and 25% B).
5. Set the flow rate at 1 mL/min.
6. Calculate the calibration curves: prepare stock solutions of the standards, and calculate the concentration (μg/mL = (Absorbance × 10^4)/$A^{1\%}{}_{1cm}$)) measuring absorbance with the UV-vis spectrophotometer at the wavelength where the specific absorption coefficient ($A^{1\%}{}_{1cm}$) has been calculated for each standard (Table 1 and *see* **Note 1**). Then, they are diluted (dissolving appropriated volumes of the stock solution with acetone) to cover the chlorophyll concentration range of the samples being analyzed. Inject three times each dilution, and represent the area's media (at the corresponding λ) versus the amount injected. For each chlorophyll standard, calculate the fitting curve (Fig. 2 shows an example of the calibration curve we use for pheophytin *a* standard).
7. Inject 50 μL of the sample into the HPLC.
8. Run the HPLC gradient shown in Table 2 (*see* **Note 14**).
9. Identify the chlorophyll pigments comparing the absorption spectra and retention times of peaks with the standard solutions (Table 3). As an example, Fig. 3 shows the HPLC chromatogram of a typical green food obtained with the described chromatographic conditions. Figure 4 shows the UV-vis spectra of the main chlorophyll compounds.

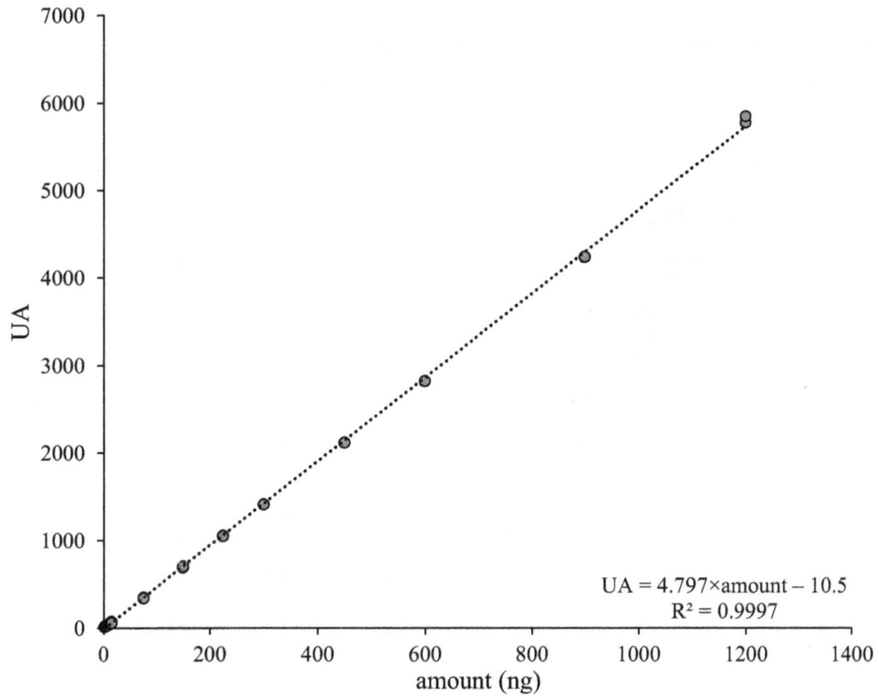

Fig. 2 Calibration curve for pheophytin *a* standard in the ng-μg range

Table 2
HPLC gradient elution for separating chlorophylls on a Mediterranean Sea C18 3 μm column

Time (min)	% A	% B
0	75	25
8	25	75
10	25	75
18	10	90
23	0	100
38	0	100
43	75	25

10. Record the area of each chlorophyll compound at the wavelength where you have developed the calibration curve. Using the calculated equations (*see* **step 6**), transform the area into amount/concentration (*see* **Note 15**).

Table 3
Chromatographic properties of the main chlorophylls present in foods

Peak number[a]	Chlorophyll	t_R[a]	UV-vis max I	UV-vis max II	UV/-vis spectrum[b]
1	Chlorophyllide *a*	4.68	430	666	a
2	13^2-OH-pheophorbide *a*	6.21	410	666	c
3	Pheophorbide *a*	7.36	410	666	c
4	Pheophorbide *a′*	8.23	410	666	c
5	15^1-OH-lactone chlorophyll *a*	14.9	420	656	e
6	Chlorophyll *b*	16.09	466	650	b
7	Chlorophyll *b′*	16.65	467	650	b
8	13^2-OH-chlorophyll *a*	16.91	432	666	a
9	Chlorophyll *a*	18.26	432	666	a
10	Chlorophyll *a′*	19.06	432	666	a
11	15^1-OH-lactone pheophytin *a*	20.36	400	670	f
12	Pheophytin *b*	22.59	435	654	d
13	13^2-OH-pheophytin *a*	23.40	410	666	c
14	Pheophytin *a*	25.55	410	666	c
15	Pheophytin *a′*	26.26	411	666	c
16	Pyro-pheophytin *a*	29.06	411	666	c

[a]See HPLC chromatogram in Fig. 3
[b]See Fig. 4 for a complete UV-vis spectrum of each chlorophyll derivative

3.3 LC-MS Data Acquisition

The following description is the practice routinely used in our laboratory [16, 17], considering the differences noted for the ionization technique (APCI source or ESI interface). The chromatographic conditions are described in Subheading 3.2.

1. Set up the LC-MS system with the chromatographic conditions selected for the separation of chlorophylls, and wait until the system is equilibrated to analyze the sequence of samples in the auto-sample tray.
2. A split post-column of 0.4 mL × $\min^{-1}$ is directly introduced on the ESI interface or in the APCI source, both operated in positive mode.
3. A calibration routine of the MS instrument is made on each analytical run (start-and-end mode) with the calibration mix that yields products in the entire mass range of the instrument (50–1500 Da) (*see* **Note 16**).

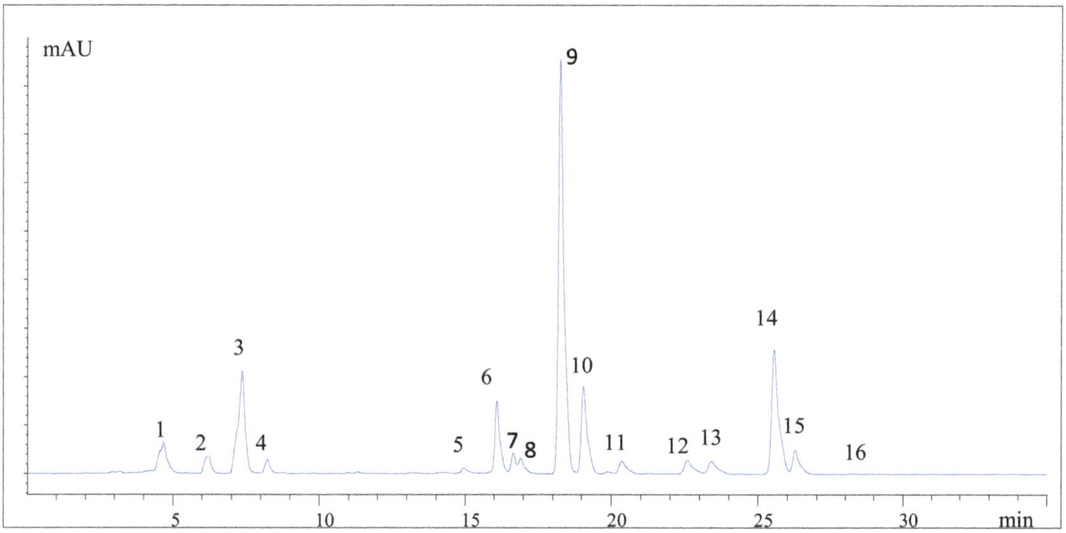

Fig. 3 HPLC chromatogram trace of a chlorophyll extract from a processed green food source, analyzed with the conditions described in this protocol. For peak number identification, see Table 3

4. The MS scan modes are tailored depending on the mass analyzer available at the user's lab (Q, IT, Q-IT, TOF, Q-TOF). Scanning of the protonated molecular ions is obtained through a conventional full-scan mode, while the convenient strategy is to obtain both protonated molecular ions and products through broadband collision-induced dissociation (bbCID) mode. When the user aims to obtain the products of a specific protonated molecular ion yields, then the precursor ion scan mode should be the corresponding analytical approach. In our laboratory, MS data are acquired in bbCID mode. The conditions of the ESI interface or the APCI source are optimized for each LC-MS system, while the setup of the mass analyzer function is correlated with the precondition of the ESI/APCI (*see* **Note 17**).

3.4 Peak Prediction and Annotation

The acquisition of the chlorophyll profile is the final aim of the protocol analysis, which, when no new pigment structure(s) are predicted, could be satisfactorily built with reasonable expertise and practice. At this protocol step, the user should consider the UV-vis features that characterize chlorophyll derivatives, the chromatographic properties, and the essential MS properties of common chlorophyll structures [18–20].

1. When a target profile is the data analysis approach, the user builds an *in-house* database containing the calculated *m/z* values and the elemental composition of the predicted chlorophylls in the analyzed extract. If the untargeted strategy is selected, then detailed information regarding the complete

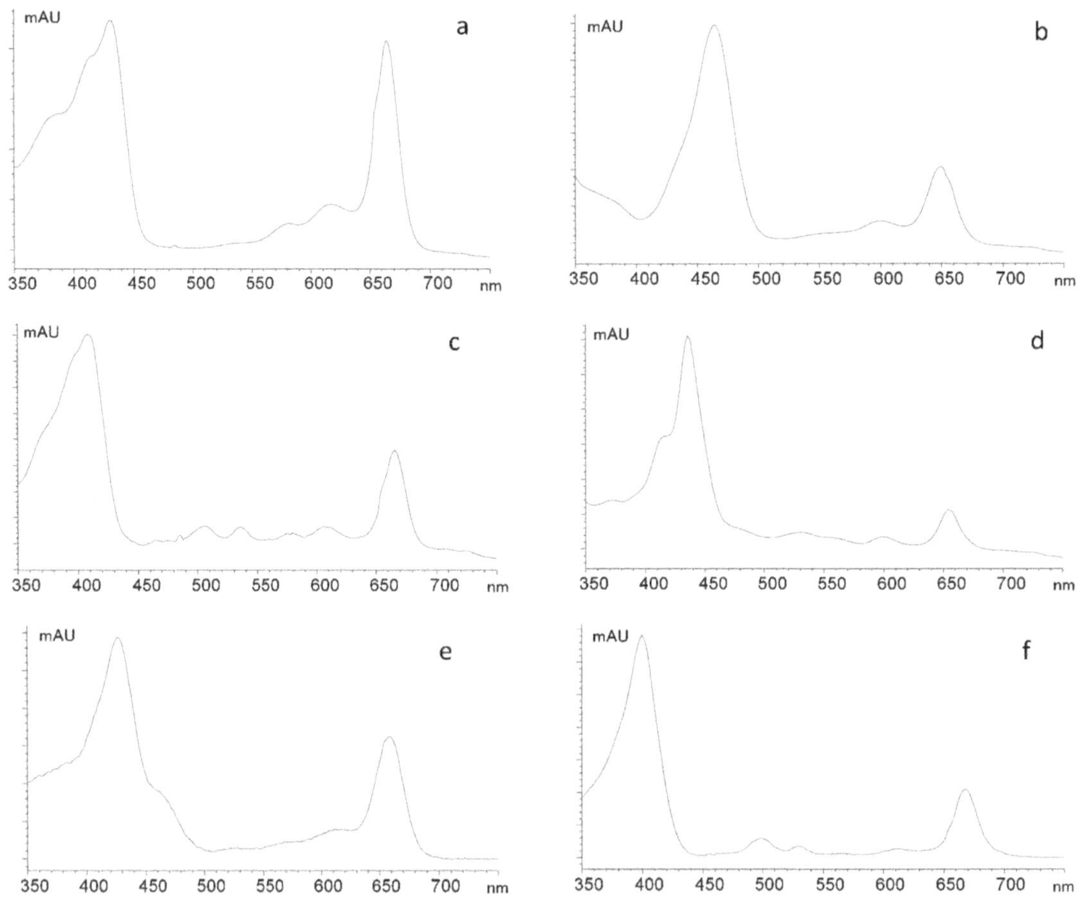

Fig. 4 UV-vis spectrum of chlorophyll compounds: (**a**) chlorophyll *a*; (**b**) chlorophyll *b*; (**c**) pheophytin *a*; (**d**) pheophytin *b*; (**e**) 15^1-OH-lactone chlorophyll *a*; (**f**) 15^1-OH-lactone pheophytin *a*

knowledge of chromatographic, spectroscopic, and mass spectrometry features of chlorophylls is compiled, and, if possible, the characteristics of the structures the user may predict. Table 4 includes the mass characteristics of the protonated molecular ion and the featured productions of common chlorophyll derivatives detected in food sources. Figure 5 depicts the tandem MS of pheophytin *a* standard with the protonated molecular ion and the main characteristic productions arising from its fragmentation.

2. Perform the mass screening with the software to all the raw data containing the protonated ions, applying the selected filtering rules to reach the required accuracy of the screening (*see* **Note 18**). The first rule is the mass filtering applied for the protonated molecular ions. The lower the mass error value, the higher the accuracy of the predicted match (below 5 ppm is reasonable, and below 1 ppm is optimal). The second rule is

Table 4
Elemental composition, exact mass, and characteristic product ions of common chlorophyll derivatives observed in food sources

Chlorophyll	Elemental composition	Exact mass $[M + H]^+$	Product ions (*m/z*)
Chlorophyll *a*	$C_{55}H_{72}MgN_4O_5$	893.5426	615.2452; 555.2241
13^2-OH-pheophytin *a*	$C_{55}H_{74}N_4O_6$	887.5681	593.2760; 533.2555
Chlorophyllide *a*	$C_{35}H_{34}MgN_4O_5$	615.2452	555.2241; 467.1716
13^2-OH-pheophorbide *a*	$C_{35}H_{36}N_4O_6$	609.2708	531.2390; 503.2420
Pheophorbide *a*	$C_{35}H_{36}N_4O_5$	593.2759	533.2547; 505.2234
Chlorophyll *b*	$C_{55}H_{70}MgN_4O_6$	907.5219	629.2245; 569.2033
13^2-OH-chlorophyll *a*	$C_{55}H_{72}MgN_4O_6$	909.5375	555.2241; 481.1870
Chlorophyll *a*	$C_{55}H_{72}MgN_4O_5$	893.5426	615.2452; 555.2241
15^1-OH-lactone-pheophytin *a*	$C_{55}H_{74}N_4O_7$	903.5630	609.2708; 549.2496
Pheophytin *b*	$C_{55}H_{72}N_4O_6$	885.5525	607.2552; 547.2340
13^2-OH-pheophytin *a*	$C_{55}H_{74}N_4O_6$	887.5681	593.2759; 533.2547
Pheophytin *a*	$C_{55}H_{74}N_4O_5$	871.5732	593.2759; 533.2547
Pyro-pheophytin *a*	$C_{53}H_{72}N_4O_3$	813.5677	535.2704; 461.2336

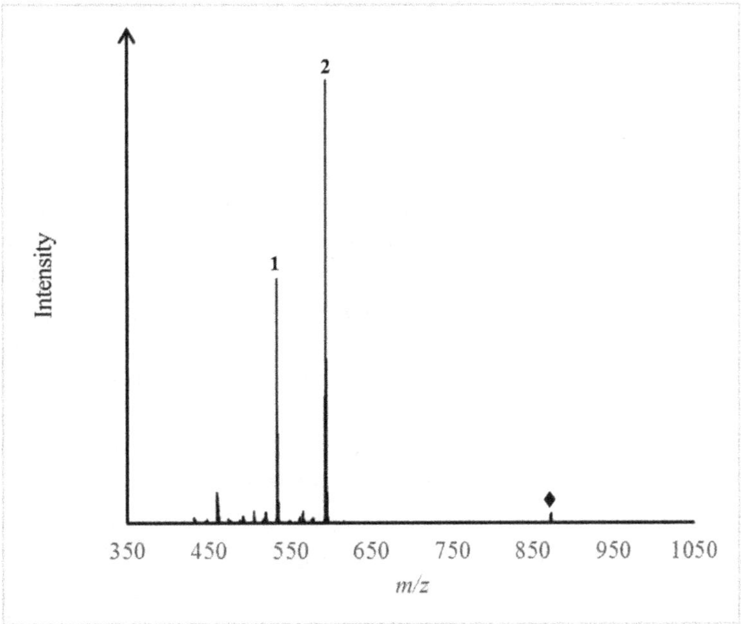

Fig. 5 Tandem MS of pheophytin *a* standard with the protonated molecular ion at $m/z = 871.5736$ (♦) and the main characteristic product ions denoted at $m/z = 533.2549$ (**1**) and at $m/z = 593.2762$ (**2**)

built through the isotopic pattern value of the calculated and experimental MS data. The lower the deviation between the calculated and experimental isotopic patterns, the higher the accuracy of the predicted match. With this experimental data, a first list is obtained including the tentatively identified chlorophylls in the food extract. The report is fused within a Microsoft Excel-based sheet where additional information (chromatographic properties, UV-vis maxima, peak area) is compiled.

3. A second screening is performed with the mass data for the product ions (bbCID mode) yielded from the protonated molecular ions of interest. The same filtering rules for mass error and isotopic pattern are applied to the product ions (*see* **Note 18**).

3.5 Peak Identification

The tentative list generated with the tools outlined in Subheading 3.4 is subsequently refined and checked with the final assessment of chromatographic data, UV-vis spectra, and MS information. This is a key point as the user must ensure that the tentative product ions (*see* Table 4) arise from the specific protonated molecular ion through (1) a reasonable structural arrangement of the protonated molecular ion and (2) statistically feasible fragmentation pathway (s) observed for chlorophylls [19]. With these issues in hand, the user might both theoretically and experimentally substantiate the identification of new chlorophyll structures.

1. Draw the structure of the protonated molecular ion of the pointed chlorophyll (*see* **Note 19**).
2. Obtain the list of product ions (*see* **Note 20**).
3. Compare the list of the in silico product ions with the list of experimental product ions for the selected chlorophyll structure. Then, review the structures proposed for each product ion, and keep noted the fragmentation pathway(s).
4. Review the proposed mechanisms for each product ion. The software might suggest several mechanisms to generate the same fragment ion structure, according to the introduced protonated molecular ion structure. Revise and select the mechanism, considering that it should be consistent with the chemical behavior of chlorophylls.
5. Examine the consistency of the isotopic pattern between product ions and the corresponding protonated molecular ion. This action is usually performed with tools implemented in the software for data analysis of the MS equipment. The rule to be applied is that the isotopic pattern should be consistent with both fragment ions and the precursor one.

6. Compare the in silico and experimental data of the tentatively identified chlorophylls with experimental data of authentic standards (if available) to increase the accuracy of the identification.

4 Notes

1. Only the main chlorophyll standards are commercially available; therefore the rest of chlorophyll compounds should be lab-produced. Recently, the protocols for obtaining/forming a wide range of chlorophyll derivatives have been optimized [16, 17].
2. *N*,*N*-dimethylformamide (DMF) is one of the solvents with the greatest extraction capacities, but its use requires the subsequent transfer to a solvent with a lower boiling point (commonly diethyl ether). Specifically, for lipophilic food matrices, as saponification degrades chlorophyll pigments, it is necessary to develop a solvent partition system [21], for example, between DMF (containing most of the chlorophyll compounds and xanthophylls) and the hexane organic layers (containing triacylglycerides, β-carotene, and pyropheophytins). Anyhow, several solvent mixtures have been assayed, allowing the user to select the one that suits better depending on the characteristic of the food matrix.
3. The extracting solvent should be saturated with $MgCO_3$ to neutralize the acid conditions after cell disruption, and avoid, consequently, the artificial formation of pheophytins.
4. Chlorophyllase is an ubiquitous enzyme responsible for chlorophyllide formation. Therefore, it is important to minimize its activity during the chlorophyll extraction, working as fast as possible and in cold conditions.
5. In some cases sample preparation is required. For example, fresh food samples should be chopped before the homogenization with solvent or freezing with liquid nitrogen. Working with phytoplankton material requires conventionally an initial biomass filtering step (usually a GF/F glass filter).
6. The proportion of tissue/solvent depends mainly on the chlorophyll content, food matrix, solvent extractability capacity, etc. A 10:1 solvent/tissue ratio would be enough, but some raw materials or dark green tissue could need a higher ratio.
7. An alternative useful method is to freeze the tissue with liquid nitrogen and disrupt the cell material with a mortar pestle (optionally adding sand) taking advantage of the ice crystals formed. But other methods have been also effectively applied as sonication, freeze-drying, etc.

8. Typically, three extractions should be enough to assure a complete chlorophyll extraction. If more extractions are needed, then the analyst could reduce the amount of sample or, alternatively, increase the solvent to tissue ratio.
9. Optionally, the solvent extract can be evaporated under an N_2 stream. If this option is selected, then the acetone extracts can be joined in a 50 mL screw-capped tube instead of a rotary evaporator flask.
10. The final volume should be as concentrated as possible but avoiding saturation of the signal intensity during UV-vis detection.
11. Samples can be kept at −20 °C for several days until analysis.
12. Among the diverse equations developed for the spectrophotometric determination of chlorophylls, the most cited is the one published by Porra et al. [22], displayed for four different solvents and verified by atomic absorption spectroscopy. However, numerous equations, algorithms, and more recently chemometric methods and multivariate analysis have been developed. The selection should be made considering the pro and cons of each approach.
13. The different chlorophylls display diverse polarities, from polar chlorophylls (those de-phytylated chlorophyll derivatives) to non-polar chlorophylls (phytylated ones). To increase the separation between polar chlorophyll compounds, an ion pair reagent is needed.
14. As the HPLC mobile phase contains salts, it is advisable to clean properly the HPLC equipment and column, running 100% methanol for at least 30 min before switching off the system.
15. To eliminate the interference of carotenoids, the analyst could focus on the extracted UV-vis chromatogram at 666 nm, where the carotenoids do not absorb.
16. With this practice, it is not necessary to obtain external calibration files, while any modification of both the ion source/interface and the optics of the instrument is individually collected for each sample. Acquisition of the calibration curves for all the samples should be performed with the identical algorithm for linear/nonlinear regression modelling, and the standard deviation should be below 0.001 ppm for high-resolution analysis.
17. In our laboratory, the APCI conditions commonly applied are the following: dry gas at 4 L/min, dry heater 220 °C, capillary voltage 4000 V, nebulizer 2.5 bar, APCI heater 300 °C, and dry heater 220 °C. ESI conditions are dry gas 8 L/min, dry heater 200 °C, capillary voltage 4500 V, and nebulizer 1.2 bar. Collision energy is estimated dynamically based on appropriate

values for the mass and stepped across a ± 10% magnitude range to ensure good quality fragmentation spectra.

18. The filtering rules are independent, and complementary physicochemical properties available at the experimental and theoretical level, including chromatographic behavior, UV-vis spectrum, m/z values (for both $[M + H]^+$ and the product ions), and isotopic pattern (for both $[M + H]^+$ and the product ions) that are once organized and compared (experimental vs. theoretical values), allow the characterization within a database of targeted compounds. Producers of MS equipment offer up-to-date software for instrument control and tools for data analysis. These tools may be embedded as a module or option in the main software distributed with the MS instrument. We apply the DataAnalysis™ software with the TargetAnalysis™ and SmartFormula3D™ utilities from Bruker, but other alternatives exist (Xcalibur, Analyst, MassHunter, MSFact, XCMS2).
19. Several chemical drawing software tools are freely available on the Internet (ChemDraw Prime, ChemSketch, BKChem, Avogadro).
20. Predictive software tools commonly applied in this step are MassFrontier™, HighChem, Mass Fragmenter, Advanced Chemistry Development, and EPIC. Alternatively, open versions can be found on the Internet, as OpenMS-Simulator, MetFragCL, CFM-ID, MAGMa+, or MS-FINDER.

Acknowledgments

This work was supported by the project PID2021-127990B-I00 funded by MCIN/AEI/10.13039/501100011033 and by "ERDF A way of making Europe".

References

1. Joint FAO/WHO workshop on fruit and vegetables for health (2005). Available at: http://apps.who.int/iris/bitstream/10665/43143/1/9241592818_eng.pdf?ua=1&ua=1. Last accessed 11/01/2021
2. Roca M, Pérez-Gálvez A (2024) Application of EFSA EU menu database and R computing language to calculate the green chlorophyll intake in the European population. Food Chem 461:140912
3. Vogel J, Jonker-Termont DS, Van Lieshout EM, Katan MB, Van der Meer R (2005) Green vegetables, red meat and colon cancer: chlorophyll prevents the cytotoxic and hyperproliferative effects of haem in rat colon. Carcinogenesis 26:387–393
4. Ferruzzi MG, Blakeslee J (2007) Digestion, absorption, and cancer preventative activity of dietary chlorophyll derivatives. Nutr Res 27:1–12
5. Solymosi K, Mysliwa-Kurdziel B (2017) Chlorophylls and their derivatives used in food industry and medicine. Mini Rev Med Chem 17:1194–1222
6. Negishi T, Rai H, Hayatsu H (1997) Antigenotoxicity of natural chlorophylls. Mutat Res 376:97–100

7. Kimm S, Park S, Kang S (1982) Antimutagenic activity of chlorophyll to direct- and indirect-acting mutagens and its contents in the vegetables. Korean J Biochem 14:1–8
8. McQuistan TJ, Simonich MT, Pratt MM, Pereira CB, Hendricks JD, Dashwood RH, Williams DE, Bailey GS (2012) Cancer chemoprevention by dietary chlorophylls: a 12,000-animal dose-dose matrix biomarker and tumor study. Food Chem Toxicol 50: 341–352
9. Viera I, Chen K, Ríos JJ, Benito I, Pérez-Gálvez A, Roca M (2018) First-pass metabolism of chlorophylls in mice. Mol Nutr Food Res 62:1800562
10. Viera I, Benito I, Pérez-Gálvez A, Roca M (2025) Authentic composition of in vivo absorbed copper chlorophyllins using a mice model. Food Res Int 211:116438
11. Egner PA, Stansbury KH, Snyder EP, Rogers ME, Hintz PA, Kensler TW (2000) Identification and characterization of chlorin e4 ethyl ester in sera of individuals participating in the chlorophyllin chemoprevention trial. Chem Res Toxicol 13:900–906
12. Garrido JL, Zapata M, Muniz S (1995) Spectral characterization of new chlorophyll *c* pigments isolated from *Emiliania huxbi* (Prymnesiophyceae) by high performance liquid chromatography. J Phycol 31:761–768
13. Hynninen PH, Lötjönen S (1983) Large-scale preparation of crystalline. (10S)-chlorophylls a and b. Synthesis 9:705–708
14. Hynninen PH, Lötjönen S (1980) Preparation of phorbin derivatives from chlorophyll mixture utilizing the principle of selective hydrolysis. Synthesis 7:539–541
15. Lötjönen S, Hynninen PH (1983) An improved method for the preparation of (10R)- and (10S)-pheophytins a and b. Synthesis 9:708–710
16. Chen K, Ríos JJ, Pérez-Gálvez A, Roca M (2015) Development of an accurate and high-throughput methodology for structural comprehension of chlorophylls derivatives. (I) Phytylated derivatives. J Chromatogr A 1406:99–108
17. Chen K, Ríos JJ, Roca M, Pérez-Gálvez A (2015) Development of an accurate and high-throughput methodology for structural comprehension of chlorophylls derivatives. (II) Dephytylated derivatives. J Chromatogr A 1412:90–99
18. Pérez-Gálvez A, Ríos JJ, Roca M (2015) A new probe for tracking the presence of E141i food colorant. Food Control 51:240–243
19. Viera I, Roca M, Pérez-Gálvez A (2017) Mass spectrometry of non-allomerized chlorophylls *a* and *b* derivatives from plants. Curr Org Chem 21:1–35
20. Pérez-Gálvez A, Viera I, Roca M (2020) Development of an accurate and direct method for the green food colorants detection. Food Res Int 136:109484
21. Mínguez-Mosquera MI, Garrido-Fernández J (1989) Chlorophyll and carotenoid presence in olive fruit (Olea europaea L.). J Agric Food Chem 37:1–7
22. Porra RJ, Thompson WA, Kriedemann PE (1989) Determination of accurate extinction coefficients and simultaneous equations for assaying chlorophylls a and b extracted with four different solvents: verification of the concentration of chlorophyll standards by atomic absorption spectroscopy. Biophys Biochim Acta 975:384–394

Part VIII

Others

Chapter 28

Analysis of Volatile Compounds: Quantification with Gas Chromatography Flame Ionization Detector and Profiling with Gas Chromatography Mass Spectroscopy

Sebnem Kurhan and Frank Vriesekoop

Abstract

Volatile compounds are low-molecular-weight components that can naturally occur in raw food material as well as develop during food processing, i.e., roasting, smoking, baking, fermentation, and aging. They significantly contribute to food sensorial properties in terms of aroma and flavor characteristics and play an important role in consumer acceptance and appreciation of foods. On the other hand, unwanted chemical and microbiological changes also yield off-odor compounds that can be used as a spoilage indicator in foods.

The following chapter describes quantitation and aroma profile screening methods for the volatile compounds with gas chromatography flame ionization detector (GC-FID) and gas chromatography mass spectroscopy (GC-MS).

Key words Volatiles, Headspace, Solid phase microextraction (SPME), Gas chromatography, Flame ionization detector (FID), Mass spectroscopy (MS)

1 Introduction

Volatiles are relatively low-molecular-weight compounds that exist in different forms and molecular structures stemming from activator/inducer reactions (i.e., thermal treatments, microbial activity, biological metabolism), reaction pathway (i.e., decarboxylation, oxidation), and transformation origin (carbohydrates, proteins, phenolics, fatty acids). Volatiles such as alcohols, aldehydes, organic acids, esters, organic sulfides, and carbonyl compounds are more related to fermentation and microbiological activity, while terpenes are usually plant-derived natural secondary metabolism products, and pyrazines and furans are prone to form after extensive thermal treatments; however, some biological pathways also lead their production in plants [1, 2]. These compounds easily translocate into vapor phase by physical intervention or at ambient temperatures

Lilian Mariutti and Frank Vriesekoop (eds.), *Food Chemistry: Basic Protocols*, Methods and Protocols in Food Science, https://doi.org/10.1007/978-1-0716-4678-6_28,

and contribute to the aroma characteristics of many foods. In some cases, they can also be used as a fingerprint for specific products and play a role in consumer acceptance on foods [1].

Fermented products gain many of their authentic aroma and flavor characteristics throughout the fermentation process. Aroma compounds differ based on the substrate(s) utilized, the fermentation conditions, and the starter culture (yeast, lactic acid bacteria, fungi) used. Alcohol is the major fermentation product in yeast-driven fermentations but is also formed by intrinsic biological pathways of living cells in different chemical forms (cholesterol, retinol). Aldehydes and carboxylic acids are often the oxidation products of primary alcohols, while oxidation of secondary alcohols typically yields ketones [3]. Alcohols having more than two carbon atoms in the molecule structure are known as higher- or fusel alcohols that are found in alcoholic beverages at relatively low amounts that are often produced by yeast amino acid catabolism and are significant for taste and flavor development. Fusel alcohols at high concentrations may lead to undesirable off-odors, while at low concentrations they contribute to the fermented food and beverages aroma. Similarly, esters can be derived from fusel alcohol and simple organic acids such as acetic acid or other short-chained fatty acids and ethanol as the alcohol.

Thermal food processes (i.e., cooking, roasting, baking) facilitate reactions like lipid–protein interactions, decomposition of the carbohydrates, and decarboxylation of the phenolics [4] and result in a wide variety of flavor and aroma compounds. Pyrazines and furans are the most common thermal sourced reaction products that have very low sensory threshold for food flavor. These compounds affect overall food flavor characteristics [5, 6]. Furans also naturally exist in fresh fruits (like arctic bramble, cranberries). Together with pyrazines, furan is also detected throughout the food/beverage aging and formed by microbial activity as well as biological pathways. Pyrroles are another nitrogen containing heterocyclic flavor compounds like pyrazines; their formation is strictly dependent on thermal treatments and/or microbial activity. Pyrroles are not found in raw/unprocessed food on the contrary to pyrazines and furans [7].

The detection and quantification of these volatile compounds are most commonly facilitated by GC-FID and/or GC-MS [8–11].

1.1 Methodologies of Analysis

1.1.1 Sampling and Storage

The concentrations of volatile compounds in the samples are proportionally lower than other main constituents found in foods. Furthermore, these volatile compounds are sensitive to temperature changes even at ambient levels; thus, sampling is a key factor to maintain the original sample content and reveal the composition by the analytical methods employed. Relevant samples should preferably be collected just before the analysis into air-tight containers, and if storage is needed, the containers should be stored at relatively low temperatures, −80 °C is preferred, but −20 °C will often suffice.

1.1.2 Extraction Techniques

Several extraction methods can be applied to extract volatiles. Steam distillation, solvent extraction, supercritical fluid, and simultaneous distillation-solvent extraction are well-known and commonly used approaches. Although liquid-based extractions provide better recovery for less volatile and higher molecular weight analytes. Solid phase extraction (SPE) can also be used for sample extraction, and this technique works well with semi-volatile analytes. Disadvantages of these extraction techniques are that they require high volume/weight of sample and may lead to losses on highly volatile compounds during any concentration step [12]. Headspace techniques including static and/or dynamic headspace sampling and solid-phase microextraction (SPME) provide a solvent-free analysis that can overcome the losses of highly volatile compound losses sometimes observed during solvent extraction [12, 13]. Based on specifically selected SPME fibers, analyte selectivity can be reduced or enhanced. Furthermore, headspace sampling is also open to environmental contamination if the appropriate sampling environment is not provided.

If the experiment is designed to profile the array of volatile compounds, multiple extraction methods are often recommended.

Solvent Extractions

Steam Distillation Steam distillation is a generic technique to extract flavor and aroma constituents such as essential oils from plant materials. This extraction method relies on steam being forced into the raw material, disrupting the cell structure and releasing the oils including the various volatile compounds which require capture through the inclusion of a condenser.

Liquid-Liquid or Solid-Liquid Extraction The principle of this extraction method is to trap volatile compounds in such a solvent (like dichloromethane, chloroform) having a low boiling point and higher density than water or solvents like hexane, which have a lower than water density and are not miscible with water. A set volume of liquid sample is combined with the organic extraction solvent and mixed with a mechanical shaker/rotator/vortex for a fixed time and speed. After the extraction is completed, the organic solvent phase is separated. This extraction can be repeated several times, until no compounds of interest remain in the original sample. After several solvent extractions, the extracts are combined and concentrated under gentle N_2 steam or vacuum evaporator [14].

An identical extraction procedure can be applied to solid samples. Samples are weighed and diluted with a degassed buffer at a particular pH or degassed/deodorized water. The sample is then homogenized in the selected buffer/water or directly subjected to extraction with an organic solvent as described above.

Simultaneous Distillation-Solvent Extraction This procedure is a common technique to isolate volatile organic compounds from complex liquid samples using a Likens-Nickerson apparatus, even when present at very low concentrations. The Likens-Nickerson method combines distillation with a simultaneous co-extraction in the vapor phase with an immiscible solvent [15, 16].

Aroma Extract Dilution This method is used for gas chromatography-olfactometry analysis in order to define the odor characteristics of the relevant analytes. The individual sample is diluted in degassed water, [17] while an aromatic extract is diluted with a compatible solvent (generally same solvent used for extraction, i.e., dichloromethane) in linearly increasing order. Followingly, the solvent in each dilution is distilled off, and hence the sample is concentrated [18, 19]. The dilution factor of the compound of interest is calculated from the number of serial dilutions, until the compound is no longer detectable in olfactometry.

Solvent-Free Extractions: Headspace Sampling

Headspace analysis relies on the evaporation (extraction) of the volatiles into the headspace of a hermetically sealed GC vial with the aid of temperature and collection of the vapor using a specific tool. The addition of NaCl (salting out) of the sample increases the sample's ionic strength and aids to increase extraction efficiency by displacing more volatiles into the headspace of the vials.

Static Headspace A fixed quantity of solid or liquid sample is transferred into a glass GC vial (10 or 20 mL) and closed with a crimp or screw cap containing PTFE septa. Subsequently, the vial is placed in a dry block heater or in an oven at a set temperature for a fixed period. Volatile analytes present in the samples evaporate and are accumulated in the headspace of the vial. These analytes can be collected either with a solid phase microextraction sorbent fiber (see below) or an air-tight syringe* at a particular volume (1 mL) and injected into the GC injection port.

*Air-tight syringe should also be heated to the sample extraction temperature or higher.

Stirring Plate Headspace Extraction Although sampling is identical to the static headspace technique mentioned above, the only addition into the sample vial is a magnetic stirrer, to help to release less volatile analytes from the sample matrix into the headspace.

Dynamic Headspace (Purge and Trap) As mentioned above, the number and the content of analytes in the vapor phase increase during heating and agitation. The dynamic headspace technique involves constantly sweeping/purging the headspace of the sample by diffusion onto an absorbent trap at specific time intervals. Following repeated extractions (purge and trap), a glass trap is later thermally desorbed in the GC injection port. Dynamic headspace traps do not consist of stationary phase; therefore, the analyte selectivity is low.

Solid Phase Microextraction (SPME) The technique relies on analyte adsorption on particles embedded in a polymeric film or absorption onto the polymeric film coated fiber. The absorbent type fibers include polydimethylsiloxane (PDMS), polyacrylate (PA), and polyethylene glycol (PEG), while the adsorbent (particle) fibers typically contain divinylbenzene (DVB), Carboxen® (CAR), or a combination of both on a PDMS binder.

Static and dynamic headspace extraction techniques are easily adapted to SPME method of which having varied types of stationary phase/coating material also provides analyte selectivity. Before sample collection, fiber should be thermally conditioned either in the injection or conditioning port of an autosampler. Following fiber conditioning and sample heating, fiber is inserted into the vial headspace through PTFE septa and let adsorb/absorb volatile analytes for a set time. It also provides an analyte enrichment. Subsequently, fiber material is withdrawn into the needle, removed from the vial, and inserted into the injection port either manually or automatically based on the existing infrastructure. Any liner in the injection port should be compatible with SPME injection and not contain glass wool. If the fiber comes in contact with glass wool, the fiber and fiber coating may be damaged or broken.

The fiber material to be employed should be selected according to the target analyte's molecular weight, polarity, concentration level, and the complexity of the sample matrix. The chosen SPME fiber significantly affects the number of analytes analyzed with the abovementioned methods.

Headspace extraction methods require thermal preconditioning of the sampling tool (air-tight syringes or fibers) to avoid volatile condensation and remove any air-borne contaminations. Extraction and collection of the targeted or untargeted volatile compounds by means of temperature must be standardized to achieve high accuracy and precision (validation).

Other Extractions

Solid-phase extraction (SPE) is an extractive technique which relies on the affinity of the analytes to the stationary phase, which is typically presented on a cartridge. The stationary phase quantity varies with the cartridge volume. Furthermore, stationary phase material is a significant parameter to consider for targeted analyte selectivity. The most used SPE cartridges contain ethylvinyl benzene-divinylbenzene or styrene-divinylbenzene. It has been demonstrated that ethylvinyl benzene-divinylbenzene has a stronger retention with regard to most volatiles [11]. Volatile extraction requires approximately 40–50 mL of sample. Analytes are retained on the stationary phase which concentrates the analytes that are subsequently eluted from the stationary phase material with an appropriate eluent.

1.1.3 Instrumentation

Volatile compounds can be easily analyzed with several gas chromatography systems (Table 1). In this section, GC compartments and their functions and mechanisms are briefly described.

Gas chromatography consists of three essential parts: injection port, column, and detector.

Injection Port Autosamplers have recently become the default compartment in GC as the injection port. Multi-function autosamplers are also available for liquid, headspace, and SPME injections. Some autosamplers can also be equipped with thermal incubation ports with stirring function for headspace analysis. The sample is introduced to the GC system via an injection port. Upon injection liquid samples are vaporized, or gas constituents maintain their gaseous status in the injection port. Typically, the temperature of the injection port is adjusted 10 to 20 °C higher than the targeted compound(s) boiling point(s). Beyond the introduction of the sample in a vapor phase at the injection port, a controlled volume introduction of the vapors onto the subsequent column can be facilitated via a split/splitless injection mode, which is a significant feature that is controlled at the injection port. In the splitless injection mode, the injected sample volume is completely transferred to the GC column, while in the split mode, the injected sample is split, and only a fraction of the vapors is introduced to the column. A commonly used split ratio is 1:50, which means that only 1/50th of the sample vapors are transferred onto the column with the help of the chosen carrier gas.

Table 1
Gas chromatography systems with various detectors: *acronyms and full names*

Acronym	Instrument full name
GC-FID	Gas chromatography-flame ionization detector
GC-MS	Gas chromatography mass spectrometry
GC-MS/MS	Gas chromatography tandem mass spectrometry
GC-QTOF-MS	Gas chromatography-quadrupole-time of flight-mass spectrometry
GC-HRMS	Gas chromatography-high-resolution mass spectrometry

Helium is the most common carrier gas used in GC systems. Alternatively, hydrogen and nitrogen can also be used as carrier gases. The carrier gas acts as an inert mobile phase and provides sample transportation from the injection port to the detector through the column without interacting with the analytes.

Column The analytes, which are transferred from injection port, are separated on the column. There are two types of columns that are used: packed columns and capillary columns. Packed columns are filled with an inert solid support, typically consisting of diatomaceous earth or small glass beads, and have a 2–5 mm internal diameter and are 2–10 meter long. Packed columns are typically used for larger-scale, preparative separation of compounds. Fused silica-coated capillary columns are the most common columns used in quantitative GC systems. The injected sample is carried by a carrier gas through the column, and each molecular analyte is separated through the interaction with the column stationary phase which is the film coated on the inner wall of the capillary column. Column properties are defined as "Brand – % functional group" (length (mm) × internal diameter (ID) (mm), film thickness (μm)).

Analyte separation relies on "likes dissolves likes" principle; therefore, the stationary phase should be compatible with the analyte chemical properties. Using non-polar columns (i.e., 100% dimethyl polysiloxane coated capillary columns), only non-polar analytes (alkanes) can be efficiently analyzed, while more polar compounds such as alcohols, amines, carboxylic acids, diols, esters, ethers, ketones, and thiols are typically analyzed using polar columns. The stationary phase is polarized with the addition of functional groups (phenyl or cyclopropyl phenyl) to a methyl siloxane skeleton. As the percentage of the functional group ascends, stationary phase polarity increases. The length of capillary columns varies between 15 meters and 100 meters, and length should be selected depending on the sample complexity. Generally, 30 meter columns are sufficient. If isomeric (cis/trans) separation is required, 100 meter columns are recommended. The internal diameter of capillary columns varies from 0.1 mm to 0.53 mm. The internal diameter affects the detected analyte concentration that can be effectively separated. The more commonly used internal diameter for capillary columns is 0.25 mm and is generally considered as an optimal value; however, it must be selected based on the application to be performed. Film thickness should be selected based on targeted analyte's boiling point. Thicker film columns are more suitable for highly volatile compound analysis that also requires lower column oven temperature. Thinner film-containing columns are durable at high oven temperatures (>300 °C); thus, it allows to analyze less volatile compounds like pesticides, fatty acid methyl esters, and other semi-volatiles.

The temperature at which the column is operated is another important parameter that influences separation. Isothermal oven temperatures are applied when the analyte boiling points are uniform across most analytes in the sample. However, when the sample composition is complex, and the sample contains high and low boiling point constituents together, a gradient temperature profile during sample elution from the column is often applied.

Detector The detector on a GC allows for the discovery/recognition and allows for the quantification of compounds that were separated on the column. Detector can be also coupled with single/tandem mass spectroscopy and time of flight mass spectrometer.

Gas chromatography coupled with flame ionization detector (GC-FID) is used for targeted volatile compounds' quantitation. The FID uses a mixture of hydrogen and air/O_2 mixture to generate a flame, and the separated analytes from the column are combusted to produce gas-phase ions whose presence is detected using an electrode that converts into the electrical signal. Gas chromatography equipped with mass spectroscopy (GC-MS) allows both targeted and non-targeted analyses. Separated analytes pass through a plasma which causes a molecular degradation of the original analyte molecules into molecular fragments, which are subsequently presented to a mass selective detector that is used to identify compounds based on the molecular fingerprint using mass spectral databases. Gas chromatography-quadrupole-time of flight mass spectroscopy (GC-QTOF-MS) is a hybrid technology which provides high resolution mass spectra and accurate mass on targeted, non-targeted, and unknown analytes. After column separation, individual analytes are ionized in the ion source. Ions are focused into a beam and transmitted through the quadrupole and followed by TOF measuring the time taken for the ions travel through the flight tube. In comparison to GC-MS, GC-QTOF-MS has been an important tool for identification unknown analytes.

1.1.4 Compound Identification

GC-FID The identification of the target analytes is done by comparing the retention times of compound in the sample and known analytical standards. Sample and the analytical standard should be analyzed using the same analytical method.

GC-MS, GC-MS/MS, GC-QTOF-MS, GC-HRMS Analyte identification relies on the mass spectrum comparison with an external (NIST) or in-house mass spectra (MS) database/library, Kovats retention indices (RI)* (calculated from alkane series analysis with the same analytical method), and pure standards.

*Kovats retention indices (RIs) are calculated using the alkane series. Selected alkane series that fit the purpose of the research is analyzed with the sample method, and further RI indices are calculated with the instrument software. The unknown analyte's carbon number is considered to be between the lowest and the highest carbon number of the selected alkane series (C8-C40, C7-C40, C8-C20, etc.).

1.1.5 Compound Quantification

Targeted compounds are quantitated with a calibration curve plotted using the signal response of the standard material (peak area/ height) at various analyte concentrations. Calibration curves can be prepared in an external solvent, model sample (artificially prepared samples that consist of major sample ingredients), or de-aromatized samples.

A calibration curve prepared in a solvent is called external standard quantitation/calibration method. As the solvent does not represent the original sample matrix, the prepared curve is free of matrix effect. Hence, the results calculated with an external calibration curve may yield to lower than actual concentration of compounds due to the standards not having background/matrix interference. The external calibration curve is used for isolated and purified compounds that are further dissolved in a particular solvent and/or quantification of the matrix effect together with matrix-matched calibration that is explained below.

If a sample is comprised of a composite material, i.e., food, the sample will contain the targeted compounds but also other constituents, which can be detected in the volatile analysis method. This phenomenon is called a "matrix effect" that affects the targeted compound(s) resolution either suppressing or enhancing the signal response. In order to overcome background effect, a matrix-matched calibration curve is used. Separate or combined solutions of targeted standard materials are prepared in an external solvent at high concentration (stock solution). Then, these stock solutions are artificially added into de-aromatized or model samples at various concentrations to achieve calibration curve standard points. Alternatively, only the highest calibration point is prepared by transferring the stock standard solution into the de-aromatized sample which is then diluted to lower concentrations proportionally using the identical de-aromatized/model sample matrix. An example of dataset is shown in Table 2. Calibration standards of 2,3-butanediol are prepared at five different concentrations, and the signal responses at each concentration are recorded for calibration curve.

Using the given dataset, calibration curve is presented in Fig. 1.

Internal standards (ISTD) may be used to quantitate the relative target compound concentration, validate extraction efficiency for accuracy and precision, or control instrument response efficiency. Isotopically labelled analogues are generally used when the

Table 2
Example of calibration dataset

2,3-Butanediol	
Concentration (mg/L)	**Signal response (peak area)**
0.3	1.40E+04
1	3.66E+04
3.3	1.38E+05
10	3.86E+05
33.3	1.35E+06

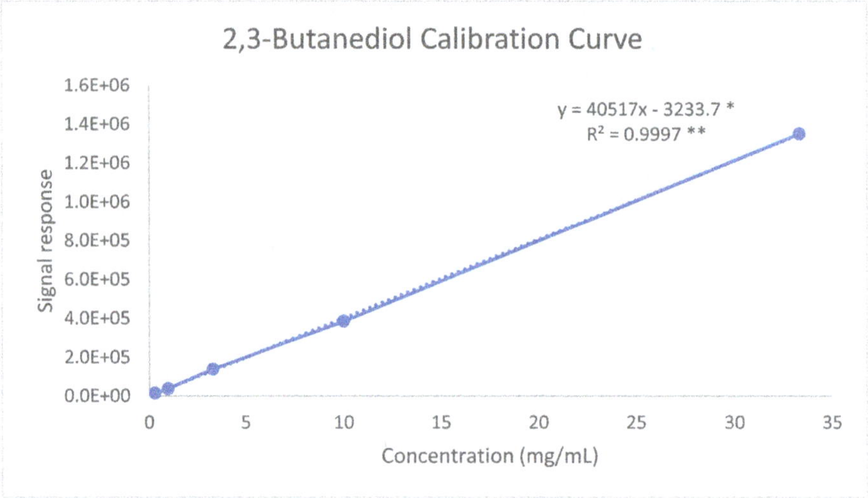

Fig. 1 Calibration curve plot of 2,3-butanediol between the range of 0.3–33.3 mg/mL. *Linear equation. **Linearity (regression co-efficient)

volatile analysis is performed in mass spectroscopy (GC-MS, GC/MS-MS, and GC-QTOF/MS). If the isotopically labelled standards are not available or analysis is performed with GC-FID, the selected internal standard should meet some criteria. An ideal internal standard must not be present in the sample. Additionally, it should be in pure form and preferably structurally similar to the target analytes, well-resolved (should not overlap with other compounds), compatible with used detector and elute after the target analytes.

Relative Compound Quantification: An internal standard, added into the sample at certain concentration, can provide relative analyte amount which can be calculated using the following equation:

$$\text{Relative Analyte Concentration} = \frac{\text{Analyte area}}{\text{IS peak area}} \times \text{IS concentration}$$

Validation of Extraction Method Internal standards are also used to validate the sample extraction efficiency for accuracy and precision of the analysis. For this purpose, a certain volume of the stock internal standard solution is pipetted into the sample prior to the extraction at a known concentration.

The concentration of transferred ISTD should be conformable with the targeted analyte concentrations. Furthermore, the volume of ISTD used must be taken into consideration, especially if the liquid extraction method is performed. When any solvent-free extraction methods are applied, the solvent peak will also appear in the mass spectra or chromatogram.

Once the extraction is performed, the ISTD recovery rate must be calculated. For this purpose, the multiple replicates of ISTD are prepared in the reference solvent at the same concentration as samples and analyzed with the identical method (no extraction is applied). After signal responses are collected, obtained IS signal responses are compared. Extraction process is considered as successful if the IS recovery is >90%.

Instrument/Process Control Internal standard can also be used for instrument and process control. This method is specifically used when the external interferences are strong enough to affect the results. Non-automated headspace sampling and manual injections may be given as an example. When fully automated systems are used, the use of ISTD is not necessary for instrument control.

If the ISTD is used in the quantitative analysis, ISTD should also be added into the calibration standard series at the same concentration added to sample. Therefore, calibration curve must be prepared from response factors at each standard concentration level. Response factor is calculated as follows:

$$\text{Response factor (RF)} = \frac{\text{Analyte peak area}}{\text{IS peak area}}$$

An example of dataset (Table 3) using the internal standard to create the calibration curve (Fig. 2) is shown before.

Table 3
Calibration standard dataset using the ISTD

Concentration (mg/mL)	Analyte signal response	ISTD-signal response	Response factor (RF)
0.03	4.32E+04	1.52E+06	2.84E-02
0.1	1.38E+05	1.67E+06	8.26E-02
0.3	4.52E+05	1.75E+06	2.58E-01
1	1.41E+06	1.73E+06	8.15E-01
3.3	4.75E+06	1.69E+06	2.81E+00
10	1.41E+07	1.48E+06	9.53E+00
33.3	4.48E+07	1.62E+06	2.77E+01

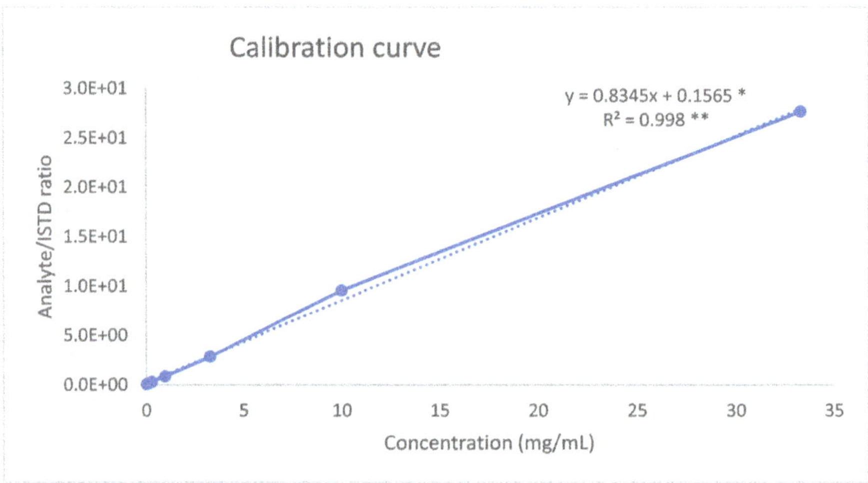

Fig. 2 Calibration curve plot of 2,3-butanediol between the range of 0.03–33.3 mg/mL. *Linear equation. **linearity (regression co-efficient)

2 Materials

This section aims to illustrate previously mentioned extraction and analytical methods. Therefore, a series of examples using different volatile extraction and analysis techniques at varied samples are described. All equipment and consumables mentioned in this chapter can be replaced with substitutes from other manufacturers.

2.1 Reagents

All mentioned solvents below used at extraction and gas chromatography analysis must be chromatography grade purity (like *n*-hexane, for gas chromatography MS Suprasolv®, 1007951000).

Solid-state chemicals must be dried overnight above 200 °C to remove the air-borne contamination as these chemicals have also strong odor and are moisture absorbents. Other liquids like water or buffers should be degassed prior to use.

Citrate buffer (pH 5.0) [20] can be prepared as follows: Prepare a solution containing 20.1 g/L citric acid and 8.0 g/L sodium hydroxide solution in distilled water. Adjust the pH with dilute hydrochloric acid. Buffer should be autoclaved or filtered degassed in order to remove dissolved volatiles before use.

2.2 Instrumentation

2.2.1 Analysis of Semi-volatile Phenolic Compounds in Spirits

Instrumental and Analytical Conditions Semi-volatile phenolic compounds in spirits can be analyzed by gas chromatography coupled with mass spectrometer (GC/MS). Injection port temperature is set at 240 °C. Ultra-pure helium is used as carrier gas at the 1.4 mL/min flow rate. DB-Wax capillary column (60 m × 0.32 mm, 0.5 μm film thickness) is used to separate the analytes. The column temperature program is set at 40 °C and held for a minute, follow by temperature increase to 250 °C with a 5 °C/min ramp rate when the temperature reaches to 250 °C and hold isothermally for 11 minutes. The interface temperature is set at 250 °C. The mass spectrum of the analytes is collected in the electron impact (EI) mode using the 35–400 amu mass range with 2 scans/s.

2.2.2 Analysis of Free Terpenoids in Fermented Alcoholic Beverages with SPE Extraction and GC-MS

Instrumental and Analytical Conditions Volatile terpene compounds in wine can be analyzed using GC-single quadrupole-MS equipped with an autosampler. Injection port is set at 250 °C, and chromatographic separation can be performed using Zebron ZB-FFAP capillary column (60 m × 0.32 mm, 0.5 μm film thickness) with a temperature program. Helium is used as a carrier gas at 1.2 mL/min constant flow. Initial column temperature is set at 40 ° C and held at this temperature for 5 min, then ramped up to 120 °C at 8 °C/min rate, and remained isothermally at 120 °C for 4. Then temperature is raised to 170 °C with a 5 °C/min ramp rate and held for 4 minutes. At final, column oven temperature is increased to 240 °C at 5 °C/min and held for 3 min. A post run can be included at 240 °C for 5 min with an increased helium flow rate (4 mL/min) to clean the column.

The interface (MS transfer line) temperature is set at 250 °C. Data is acquired in the single-ion monitoring (SIM) mode with a solvent delay set at 12.5 min. MS source and quadrupole temperatures are maintained at 230 and 150 °C, respectively. The mass spectrum of the analytes is collected in electron ionization (EI) mode (70 eV) between 50–250 amu mass range.

2.2.3 Determination of Targeted Volatile Content in Spice Samples

Instrumental and Analytical Conditions Targeted volatile content analysis can be performed using gas chromatography equipped with flame ionization detector. Injection port is set at 220 °C, and helium is used as a carrier gas at 1 mL/min flow rate. Chromatographic separation is performed using HP-5-MS capillary column (30 m × 0.25 mm i.d, 0.25 μm film thickness). Column oven temperature program is set as follows: maintain initial temperature at 50 °C for 3 min, and then gradually increase to 180 °C using 3 °C/min ramp rate. Ultimate column temperature is increased to 250 °C with 15 °C/min ramp rate and held at 250 °C for 5 min. Detector temperature is set at 290 °C.

2.2.4 Characterization of Volatiles from Different Samples (Unknown Analytes)

Method 1: Fruit Wines

Instrumental and Analytical Conditions Untargeted analysis can be performed on an accurate mass GC-QTOF-MS instrument that operated in electron impact ionization (EI) mode at 70 eV. Helium is used as carrier gas at constant flow rate of 1.2 mL/min. Chromatographic separation is carried out with HP-5-MS (30 m × 250 μm, 0.25 μm film thickness) column with a gradient temperature program set as follows: column temperature is held at 40 °C for 5 min and increased to 180 °C at 3 °C/min ramp rate and held at 180 °C for a minute; then temperature is raised to 300 °C at 30 °C/min ramp rate, and column is maintained at 300 °C for 2 min. The interface and ion source temperatures are set to 300 °C and 250 °C, respectively. The scan was performed at 35 to 350 m/z mass range.

Method 2: Vinegar

Instrumental and Analytical Conditions Gas chromatography coupled with a mass spectrometer is also used for untargeted analysis. The instrument is operated in electron impact ionization (EI+) mode at 70 eV. Helium is used as a carrier gas at 1.1 mL/min flow rate. Chromatographic separation is carried out with DB-Wax capillary column (60 m × 0.25 mm, 0.25 μm film thickness) with a gradient temperature program as follows: column oven is held at 30 °C for 10 min, ramped to 100 °C at 5 °C/min rate, then increased to 210 °C at 3 °C/min ramp rate, and held at this temperature for 40 min. Mass scan was performed in the 30–450 amu range.

3 Methods

3.1 Analysis and Quantitation of Semi-volatiles with GC-MS [21]

- *Method 1*: Analysis of semi-volatile phenolic compounds in spirits.
  - *Sample*: Spirit.
  - *Target analytes*: Phenol compounds.
  - *Extraction method*: Solid phase microextraction.

- *Injection type*: SPME injection.
- *Sample preparation*: Transfer an aliquot of 3 mL spirit into the 10 mL GC vials, and dilute the ethanol content to 20% with 3 mL ultra-pure water. Internal standard (IS; 2,3,4-trimethylphenol) should be added into the sample at this stage (50 μL). Stock IS solution at 100 mg/L is prepared in 70% ethanol in water.
- *Standard preparation*: Targeted individual analyte standard solutions should be prepared in 70% ethanol in ultra-pure (degassed) water (v/v) mix. Standards should be diluted in 70% ethanol in water mixture to create a six-point calibration curve.
- In order to maintain the same alcohol content as the samples, transfer 2 mL of prepared standards at relevant concentration into the GC vials, and dilute the ethanol content to 20% with 4 mL ultra-pure water. Internal standard should be added to the calibration standards at same volume with sample (50 μL). Total liquid volume must be 6 mL in 10 mL GC vials.
- *Calibration*: Calibration curves must be created by plotting the linearly increasing signal response (peak area) at relevant concentrations. Calibration curve must be plotted with signal responses from at least five different standard concentration points, and regression coefficient (R^2) must be greater than 0.99.
- *Sampling and injection*: Sampling and injection may vary based on the existing infrastructure. The following methodology describes the manual sample extraction and injection to the injection port.
- Prepared samples and calibration standards should be pre-incubated at 30 °C for 5 min before inserting the SPME fiber into the vial. After the tube content reaches the temperature, SPME fiber inserted into GC vial and analytes are extracted for 15 min at 30 °C. After extraction, SPME fiber is injected into the injection port.

• ***Alternative method***: Analysis of terpenoids in wines with SPE extraction and GC-MS [22].

- *Target analytes*: Terpenes in wine.
- *Extraction method*: Solid-phase extraction (SPE) (LRC C18-OH SPE cartridges, 500 mg).
- *Injection type*: Liquid injection.
- *Sample preparation*: Internal standard (3-octanol and 2,6-dimethyl-6-dimethyl-6-hepten-2-ol) is also prepared in

model wine at 400 mg/L concentration, and 25 μL of IS is added to representative wine sample (50 mL).

- SPE cartridges are conditioned using 4 mL dichloromethane (DCM), 4 mL methanol (MeOH), and 4 mL of model wine (*see following standard preparation), respectively. Following SPE conditioning, 50 mL wine sample is loaded into the cartridge in small amounts. After sample transfer, cartridge is washed with 6 mL water and dried under vacuum (−60 KPa) for 25 min. The sample is eluted with 2 mL of DCM. The DCM fraction should be dried using anhydrous sodium sulfate, and prior to GC analysis, sodium sulfate must be separated with centrifuge.
- *Standard preparation*: A model wine consisting of 12% ethanol and 5 g/L tartaric acid at pH 3.5 is prepared. Stock internal and calibration standards are prepared in the model wine and can further be diluted to working concentrations. Calibration standards at seven concentration levels are prepared using the stock standard mixture by progressive dilution between the 0.88–500 μg/L concentration range.
- *Calibration*: The peak area of each analyte is normalized by its corresponding internal standard. Calibration curve is also plotted by the ratio of standard analyte peak area to IS peak area at each concentration.
- *Sampling and injection*: Dried DCM fractions are collected in a 2 mL GC vial with a seal cap. The vial is inserted into autosampler rack, and 1 μL of sample is injected into the injection port set at 250 °C.

3.2 Determination of Targeted Volatile Content in Spice Samples

- ***Extraction method***: Simultaneous distillation–solvent extraction (in low-density solvent configuration) [23].
  - *Injection type*: Liquid injection.
  - *Sample preparation*: Representative 5 g of spice sample is weighed into the Likens–Nickerson apparatus sample container, and 50 mL of deionized water is added. Transfer an aliquot of 5 mL diethyl ether as an extraction solvent. Simultaneous extraction and distillation procedure is carried out for 2 h. The condenser can be cooled with water/glycol mixture for steam condensation. The final volume of collected diethyl ether should be 5 mL.
  - *Standard preparation*: The serial dilutions of targeted standards are prepared in diethyl ether.
  - *Calibration*: Calibration curve is created with plotting the signal response against each concentration level.

  - *Sampling and injection*: Extracts are transferred into 2 mL crimped-cap GC vials and injected into the injection port at 1 μL in the splitless mode.

- ***Alternative method***: For heat-labile volatile compounds.
  - *Sample preparation*: Fresh mushrooms representing 50 g dry mass were blended in 250 mL of distilled water. This mixture was poured into in a 2 L round-bottom flask, and the extraction carried out, under vacuum at −40 kPa, by simultaneous extraction and distillation in a modified Likens–Nickerson apparatus with 10 mL of hexane as the extracting solvent. The simultaneous extraction and distillation were allowed to continue for 2 h after which the system was cooled, and the total extracted volume reduced to 2 mL with a rotary evaporator at 30 °C.

3.3 Characterization of the Volatile from Different Samples (Unknown Analysis)

- ***Method 1***: Characterization of blueberry wines' cultivar differences with GC-QTOF/MS [11].
  - *Sample*: Fruit wines.
  - *Extraction method*: Headspace SPME.
  - *Injection type*: SPME injection.
  - *Sample preparation*: Citrate buffer (pH 5.0) is prepared (*see* Subheading 2.1.) degassed with vacuum filtration prior to analysis. Transfer an aliquot of 9 mL citrate buffer, and weigh 3 g of NaCl into the 20 mL GC vials. Pipette 1 mL of representative wine sample into the GC vial, and cap tightly with magnetic screw cap equipped with silicon/PTFE septa. Then vial is placed into a thermostatic water bath at 50 °C for 15 min until vial temperature reaches the equilibrium. Insert the SPME fiber (DVB/CAR/PDMS, 50/30 μm) into the vial, and extract the analytes for 45 min with constant stirring at 500 rpm. After extraction is completed, fiber is inserted into the injection port; set at 250 °C to desorb the analytes. Injection is performed in 1:10 split mode without solvent delay.
  - *Compound identification*: Background subtraction was performed using dedicated instrument software. The GCs equipped with TOF are able to produce accurate mass of the unknown compound. Hence, each data file from individual samples is converted to the sure mass files, and tentative annotation of the compound peaks without standard was made by using spectral comparison (mass difference of less than 5 ppm theoretical value, at least one indicative fragment and isotopic pattern), literature information about chromatographic properties, and mass spectra records of in-house fragrance and flavor compounds database and

external databases such as the National Institute of Standards and Technology (NIST) and by comparing Kovats retention indices (RI) in NIST database.

- ***Alternative method***: Stirring plate-headspace SPME extraction [9].
  - *Sample*: Vinegar.
  - *Injection type*: SPME injection.
  - Weigh 6.14 g of NaCl into a 50 mL GC vial prior to sample addition. Separately, internal standard (4-methyl-2-pentanol) is prepared at 2.27 g/L concentration in 8% acetic acid in water (model sample). Before capping, magnetic stirrer is placed in the GC vial and capped with screw cap equipped with PTFE silicone septa. Vial is placed onto a magnetic stirrer with heating function set at 70 °C for 5 min. Then, SPME fiber (CAR-PDMS, 75 μm) is inserted into the vial, and headspace analytes are extracted for 60 min. During extraction, the sample is stirred at constant speed. In the end of extraction, the SPME is removed from sample, and fiber is injected into injection port for analyte desorption in splitless mode for 2 min.
  - *Compound identification*: The peak identification is carried out by comparing the mass spectra with library (NIST) and retention indices of the standards (if available) or by literature.

4 Notes

1. Phenolic compounds are semi-volatile analytes; hence polyacrylate (PA) SPME fiber (85 μm) is used.
2. Salt addition to the vials allows to increase volatile analyte concentrations via lowering the partitioning coefficient (pK). Salt can be added up to 30% of the sample weight.
3. Buffer pH must be compatible with the sample pH for optimal analyte extraction.
4. Used liner in the injection port should be compatible with headspace-SPME analysis. Glass wool should not be used to avoid fiber damages.
5. Column names may vary from brand to brand (HP, DB, etc.). Therefore, second half of the name (Wax, 5-MS, etc.) should be taken into consideration to decide which column can work well with the analytes.
6. Longer capillary column provides better chromatographic separation.

7. Calibration curves must be generate the data obtained from identical extraction method with sample.
8. Calibration curves are recommended to plot at least five different data points, and regression coefficient (R^2) must be ≥ 0.99.

References

1. Hazelwood LA, Daran JM, Van Maris AJA, Pronk JT, Dickinson JR (2008) The Ehrlich pathway for fusel alcohol production: a century of research on Saccharomyces cerevisiae metabolism. Appl Environ Microbiol 74:2259–2266
2. Starowicz M (2021) Analysis of volatiles in food products. vol 8, Separations. MDPI
3. Willaert R, Verachtert H, Van Den Bremt K, Delvaux F, Derdelinckx G (2005) Bioflavouring of foods and beverages
4. Poisson L, Schieberle P (2008) Characterization of the most odor-active compounds in an American Bourbon whisky by application of the aroma extract dilution analysis. J Agric Food Chem 56(14):5813–5819
5. Maga JA, Sizer CE (1973) Pyrazines in foods. C R C Crit Rev Food Technol 4(1):39–115
6. Maga JA, Katz I (1979) Furans in foods. C R C Crit Rev Food Sci Nutr 11(4):355–400
7. Maga JA (1981) Agricultural and food chemistry pyrroles in foods. 29:691
8. Alvarez P, Malcrops P, Sa Almeida A, Ferreira A, Meyer AM, Dufour JP (1994) Analysis of free fatty acids, fusel alcohols, and esters in beer: an alternative to CS 2 extraction. J Am Soc Brew Chem 52
9. Natera R, Castro Mejıas R, de Valme M, Moreno G, Rowe FG et al (2002) Headspace solid-phase microextraction analysis of aroma compounds in vinegar Validation study [Internet]. J Chromatogr A. Available from: www.elsevier.com/locate/chroma
10. Daute M, Jack F, Harrison B, Walker G (2021) Experimental whisky fermentations: influence of wort pretreatments. Food Secur 10(11)
11. Yuan F, Cheng K, Gao J, Pan S (2018) Characterization of cultivar differences of blueberry wines using GC-QTOF-MS and metabolic profiling methods. Molecules 23(9)
12. Braggins TJ, Grimm CC, Visser FR (1999) Analysis of food volatiles using SPME. In: Pawliszyn J (ed) Applications of solid phase microextraction, 1st edn, pp 407–422
13. Rocha S, Ramalheira V, Barros A, Delgadillo I, Coimbra MA (2001) Headspace solid phase microextraction (SPME) analysis of flavor compounds in wines. Effect of the matrix volatile composition in the relative response factors in a wine model. J Agric Food Chem 49(11): 5142–5151
14. Alves AAR, Barros EBP, Rezende CM (2014) Method development and optimization of liquid–liquid extraction for the quantitative analysis of volatile compounds from Brazilian grape juices. In: Flavour science. Elsevier, pp 417–421
15. Chaintreau A (2001) Simultaneous distillation-extraction: from birth to maturity – review. Flavour Fragr J 16:136–148
16. Likens ST, Nickerson GB (1964) Detection of certain hop oil constituents in brewing products. Proc Annu Meet Am Soc Brew Chem 22(1):5–13
17. Mayr CM, Geue JP, Holt HE, Pearson WP, Jeffery DW, Francis IL (2014) Characterization of the key aroma compounds in Shiraz wine by quantitation, aroma reconstitution, and omission studies. J Agric Food Chem 62(20):4528–4536
18. Lahne J (2010) Aroma characterization of American Rye Whiskey by chemical and sensory analysis [master of science]. University of Illinois, Urbana/Illinois
19. Kerscher R, Grosch W (1997) Comparative evaluation of potent odorants of boiled beef by aroma extract dilution and concentration analysis. vol 204, Z Lebensm Unters Forsch A. Springer
20. European Pharmacopoeia Buffer Solutions (2005) European Pharmacopoeia 50. European Pharmacopoeia, pp 430–435
21. Yang T (2014) The impact of whisky blend matrices on the sensory perception of peaty flavours
22. Williams C, Buica A (2020) Comparison of an offline SPE–GC–MS and online HS–SPME–GC–MS method for the analysis of volatile terpenoids in wine. Molecules 25(3)
23. Kanakis CD, Daferera DJ, Tarantilis PA, Polissiou MG (2004) Qualitative determination of volatile compounds and quantitative evaluation of safranal and 4-hydroxy-2,6,6-trimethyl-cyclohexene-1-carboxaldehyde (HTCC) in Greek saffron. J Agric Food Chem 52(14): 4515–4521

Index

Lilian Mariutti and Frank Vriesekoop (eds.), *Food Chemistry: Basic Protocols*, Methods and Protocols in Food Science, https://doi.org/10.1007/978-1-0716-4678-6,

GPSR Compliance
The European Union's (EU) General Product Safety Regulation (GPSR) is a set of rules that requires consumer products to be safe and our obligations to ensure this.

If you have any concerns about our products, you can contact us on

ProductSafety@springernature.com

In case Publisher is established outside the EU, the EU authorized representative is:

Springer Nature Customer Service Center GmbH
Europaplatz 3
69115 Heidelberg, Germany

www.ingramcontent.com/pod-product-compliance
Lightning Source LLC
Chambersburg PA
CBHW081225101125
35216CB00006B/635
* 9 7 8 1 0 7 1 6 4 6 7 7 9 *